TECHNOLOGY AND THE FUTURE

SEVENTH EDITION

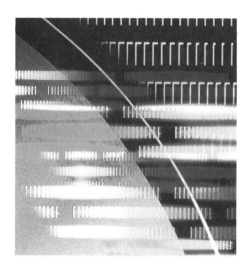

ALBERT H. TEICH

EDITOR

American Association for
the Advancement of Science

St. Martin's Press
New York

To Jill, Sammi, Mitch, and Ken

Sponsoring editor: Beth Gillett
Development editor: Meg Spilleth
Managing editor: Patricia Mansfield Phelan
Project editor: Nathan Saunders
Production supervisor: Melissa Kaprelian
Art director: Lucy Krikorian
Cover design: Michael Jung
Cover photo: Steven Hunt/The Image Bank

Library of Congress Catalog Card Number: 95-73201

Manufactured in the United States of America.

1 0 9 8 7
f e d c b a

For information, write:
St. Martin's Press, Inc.
175 Fifth Avenue
New York, NY 10010

ISBN: 0-312-11612-8

Preface

Each decade in our technological age seems to have its technological icon—not necessarily the technology that is promoted as having the greatest impact on people's lives, nor that which may ultimately have the most historical significance, but the technology that seems most to capture the public imagination and the spirit of the time. In the 1950s, it was certainly atomic energy, with the awesome destructive power of the Hiroshima and Nagasaki bombs still fresh in the minds of many and the prospect of "Atoms for Peace" and electricity "too cheap to meter" tantalizingly just around the corner. In the 1960s, it was unquestionably space, as we raced the Soviet Union to the moon, watched one televised space spectacular after another, gained our first close-up views of the nearby planets, and began to reap the benefits of communications, weather, and remote-sensing satellites.

By the 1970s, we had become disillusioned with many of these technologies, and our symbols took on a negative cast—the anti-nuclear movement and the opposition to atomic power; the worries over resource shortages, the "limits to growth," and the energy crisis; the growing concerns over environmental pollution; and the somewhat overblown fears (at least in retrospect) of dangerous mutant organisms escaping from gene-splicing laboratories. While these anxieties did not disappear in the 1980s, they were at least partially overshadowed by the appearance of a new technological icon—the computer, which had evolved from a room-sized behemoth, found only in large organizations, to a ubiquitous "user-friendly" desktop (and laptop) appliance.

And what of the 1990s? There seems little doubt that the technology of this decade is the next stage in computer evolution: the Internet, and especially the Internet's "World Wide Web," the system that has transformed this once obscure network used mainly by scientists and academics into the darling of Wall Street and the popular press.

The incredibly rapid growth and increasing pervasiveness of the Internet brings to mind a piece of science fiction, a short story that I read many years ago in the days when UNIVACs and enormous IBM mainframes represented the popular image of computers. In the story, a group of scientists decides to wire together all of the world's giant computers in order to seek the answers to humanity's ultimate questions. Having finally reached their goal after long and difficult labors, the scientists throw a switch activating their gargantuan electronic brain. They then approach the machine and ask it their first question: "Is there a God?" Suddenly, a bolt of lightning comes down from the sky, striking the device and fusing the switch in the "on" position. "There is now!" comes a booming voice in reply.

While such an exaggerated view of the power of networked computers may now seem charmingly quaint, it is not that far beyond some of the wilder claims one hears for the future of the global information superhighway, of which the Internet is widely regarded as the prototype. And if those claims are exaggerated, they at least reflect the extent to which this technology has caught hold of the popular and commercial imagination.

The Internet is essentially a worldwide collection of networks through which millions of computers and the people who operate them can communicate with one another. Its most widely used application is electronic mail (e-mail), which for many people is increasingly taking the place of conventional mail ("snail mail"), fax, and even the telephone. What has propelled the Internet into popular culture, however, is the World Wide Web.

Fundamentally, the Web is an information service that operates on the Internet, allowing information providers to place text, pictures, movies, and sound on "Web sites" or "home pages" on their computers. Anyone with a computer, an Internet connection, and a "Web browser" (a specialized software program such as Netscape Navigator) can gain access to it. Two features of the Web make it particularly engaging. First, it operates through hypertext—that is, most sites contain information allowing users to jump to other sites and thereby "surf" from one location to another, often with unexpected results. Second, using a browser requires relatively little expert knowledge, making such surfing fairly effortless and accessible to a wide audience.

As the Web emerged into public view in 1993 and 1994 and millions of people started to explore it, information providers, more and more of them profit-oriented, began to make available on the Web an astonishingly wide range of material. This stimulated more interest, and a positive feedback loop was established, fueling additional growth. Thus, by mid-1996, the number of sites accessible through the Web had reached an estimated 200,000 and was growing at a rate of more than 20 percent a month. In the space of a few minutes, a Web surfer can "visit" sites that offer, for example, the full text of current congressional legislation; live radio broadcasts of professional basketball games; a model railroad setup in Germany whose trains can be controlled remotely by the "visitor"; home pages for hundreds of elementary and high schools and for thousands of individuals; a wide range of pornographic materials; and the pope's most recent Christmas message (in Italian, English, French, Spanish, German, and Portuguese).

In some ways this remarkable development was not entirely unforeseen. As far back as the 1960s (and perhaps earlier), futurists were forecasting the development of data banks that would allow people to access information from their homes or offices. The first edition of *Technology and the Future* (1972) included a study by two futurists predicting that, within a decade or so, the nation would see the "establishment of a central data storage facility (or several regional facilities) with wide public access (perhaps in the home) for general or specialized information retrieval, primarily in the areas of library, medical, and legal data." If this forecast seems to have more in common with my "deus ex machina" science fiction tale than with the decentralization, variety, and even anarchy of today's World Wide Web, it is perhaps worth remembering the following: the prediction was made long before the invention of PCs and modems, in the days when most computers were used for accounting or scientific calculations, most data were entered on punched cards, and the concepts of computer games and computer graphics were just glimmers in the eyes of a few computer scientists and engineers.

The Web is not only an icon for today's technology (as well as an icon, literally, on millions of computer screens); it is an apt metaphor for the pervasiveness and interconnectedness of technology and human life. We exist in a web of technology—a set of tools, machines, and human infrastructure based to a large extent on the products of scientific research and development. As a result, we now have the means both to control our destinies to a greater extent than ever before and to destroy ourselves and most of the life on our planet. Understanding our relationship with technology is essential if humans are to make responsible use of the power they have created. Helping to broaden this understanding has been the goal of *Technology and the Future* since I first conceived the book as a young academic in 1971.

My aim over the years has been to present a balanced set of readings on technology and society—to give students from both technical and nontechnical backgrounds an opportunity to explore the nuances and subtleties of the many differing views on this subject. At the same time, I have sought to relate these views to policy perspectives, suggesting avenues of public action that might influence the future in positive ways.

Through twenty-five years and six revisions, *Technology and the Future* has changed a great deal. This seventh edition includes just four readings that appeared in the original version of the book and only a handful of others from early editions. Ten readings are new to this edition; a number of these are brand-new, published within the past year or two. The selections are—by design—a mixed bag. Not all students and not all instructors will like all of the readings. Most readers will probably love some, hate others, find some fascinating, others tedious. I myself do not like or agree with all of the selections. I do, however, feel they are all important and worth reading—most because of what they say, others because of what they stand for or who their authors are. If readers are troubled or offended by a selection, I hope they will recognize that its inclusion in the book reflects not an endorsement of its point of view, but rather my belief that it is something to which students of technology and society should be exposed.

The structure of the book is the same as it has been in recent editions. There are four major sections. Part I raises the big questions: Is technology good, bad, or neutral? Is it synonymous with progress? How is it influencing culture, social relations, the distribution of political and economic power? Part II looks at forecasts of technology and of social and environmental trends. It considers how well we can predict the course of technologies and their impacts on society and what the consequences of such predictions and assessments are. In Part III, the authors challenge the status quo. Their essays discuss developing alternatives to contemporary mainstream technology or view mainstream technology from unorthodox perspectives. Finally, in Part IV, the authors address questions related to individual technologies or areas of technology rather than broad conceptual issues. Recognizing that not all instructors or students will choose to move through the book from front to back, I have provided, in addition to the standard table of contents, a topical table of contents, which groups the selections somewhat differently and which may be useful to those who wish to pursue specific topics, such as risk, energy and environment, or women and technology.

Each of the four parts of the book opens with a brief introduction discussing the theme that ties the chapters in that part together. Each chapter, in turn, is introduced by a headnote that puts it in context and provides some background about its author. These headnotes are particularly important for the older selections, some of which may easily be misunderstood if taken out of their historical context.

Technology and the Future has been a part of my life throughout nearly all of my professional career. It is gratifying, therefore, to have watched the growing interest in the study of science, technology, and society in American colleges and universities over the past twenty-five years and to feel that the book may have made a modest contribution to this important intellectual development.

I am grateful to many people who have contributed to the success of the book. My deepest appreciation goes to the authors and publishers of the essays included for allowing me to reprint their work. In many cases, the selection that appears in this volume represents only a brief introduction to a rich body of thought and writing. I hope that the exposure to these authors gained here stimulates readers to seek out some of their other writings.

I want to express my appreciation to users of the book for their interest and helpful feedback. In particular, I offer my thanks to: Stephen Frantzich, United States Naval Academy; Thomas Misa, Illinois Institute of Technology; Matthew Novak, California Polytechnic State University; Robert Rydell, Montana State University; and Aaron Segal, University of Texas at El Paso. The following individuals also provided suggestions through responses to questionnaires: Sister Caroline Marie Sloan, Mount Mary College; Dalton Smart, Millersville University; Javier Ibanez-Noe, Marquette University; Terry Richardson, Northern State University; Daniel W. Hackmann, Kirkwood College; Nancy Rose MacKenzie, Mankato State University; George Gmelch, Union College; Catherine Hobbs, University of Oklahoma; Thomas A. Easton, Thomas College; Daniel W. Pound, University of Alabama at Tuscaloosa; Thomas Ilgen, Pitzer College; Reza Rezazadeh, University of Wisconsin; M. Comninou, University of Michigan; and Margaret Lang Ott, formerly of the University of Alabama at Huntsville.

A special note of thanks goes to the staff of the college division of St. Martin's Press, for their vision in publishing the first edition in 1972 and for their continuing interest and support. My editor for several years was Don Reisman. I look forward to working with his successor, Beth Gillett, and I am grateful for the many ideas and suggestions of Meg Spilleth, who has worked with me closely in preparing this edition. Finally, I must acknowledge the roles of my wife, Jill Pace, my sons Mitch and Ken, and my daughter Samantha, for the ideas and the love they share and for the meaning they give to my life.

A SPECIAL NOTE

The emergence of the World Wide Web provides an opportunity for me to add another dimension to this edition of *Technology and the Future:* I have set up a Web site for the book. The site includes basic information about the book (this preface, the table of contents and topical table of contents, and ordering instructions), images of the covers of all seven editions, and a personal homepage with some information about me. Also included are hypertext links to sources of additional information about the authors represented in the book and to other sources of information on the study of technology and society, as well as an e-mail link for readers to provide me with feedback on the book and suggestions for future editions. Eventually, I hope to create an interactive forum on the site in which to discuss the issues raised by the book. The URL (Web address) for the site is http://www.intr.net/ateich. If, for any reason, this address should change, it should still be possible to locate the site through links on various related sites; by using one of the increasingly sophisticated Internet search engines, such as Yahoo, Lycos, InfoSeek, or AltaVista; or by contacting either the College Division of St. Martin's Press or me (preferably by e-mail at ateich@aaas.org).

ALBERT H. TEICH

Contents

Topical Contents

TECHNOLOGY AND SOCIAL PROBLEMS

WOMEN AND TECHNOLOGY

Part I
THINKING ABOUT
TECHNOLOGY

Technology is more than just machines. It is a pervasive, complex system whose cultural, social, political, and intellectual elements are manifest in virtually every aspect of our lives. Small wonder, then, that it has attracted the attention of such a large and diverse group of writers and commentators. A small sampling of the range of writings on the social dimensions of technology is contained in this first section of *Technology and the Future*. All of the writers represented here are attempting to understand—from one perspective or another—these social dimensions of technology. Little else ties them together. Their points of view are vastly different. Their common goal is not so much to prescribe particular courses of action as it is to explore the conceptual, metaphysical issues underlying technology–society interactions.

In the opening selection, historian Leo Marx explores the development of the American notion of progress and looks at its connections with technological advance. Futurist Alvin Toffler writes breathlessly about the transformation of power and wealth in society at the edge of the twenty-first century. Neil Postman, humanist and media critic, following in the tradition of Paul Goodman (see his essay in Part III) and French scholar Jacques Ellul, sees modern society as a "technopoly" in which technology is the supreme force and supreme authority—a situation he finds extremely troublesome. A very different way of looking at technology's role in society is presented by Thomas Hughes in "The Technological Torrent." Taking a broad view of the surge of invention and technological enthusiasm that simultaneously shaped and was shaped by American culture in the century from 1870 to 1970, Hughes's essay provides a useful and balanced introduction to the notion of "social construction" of technology.

Following Hughes, in a thirty-year-old essay that captures some of the heights of this technological enthusiasm, physicist Alvin Weinberg suggests that we can find shortcuts to the solution of social problems by transforming them into technological ones—since technological problems are much easier to solve. This selection, which appears more than a little dated in the mid-1990s, is important for the perspective it represents, as are the two pieces that follow it. In them, Emmanuel Mesthene and John Mc-Dermott engage in a classic debate (first published in 1969) on the politics of technology. Mesthene (an IBM-funded "establishment" academic) coolly describes its moral neutrality, while McDermott (a radical antiwar activist during the Vietnam conflict) bitterly criticizes what he sees as the right-wing bias of contemporary technology. Completing Part I, Samuel Florman, a practicing engineer as well as a humanist, proposes an alternative approach—a "tragic" view that recognizes the role of technology in human life, including its limits.

The reader looking for unequivocal answers to the problems posed by technology will not find them here. On the whole, the readings in this section, like those in the remainder of the book, raise many more questions than they answer.

1. Does Improved Technology Mean Progress?

LEO MARX

The concepts of technology *and* progress *have been firmly linked in the minds of most Americans for the past 150 years. Only in the past three decades, however, has the question that Leo Marx asks in his essay "Does Improved Technology Mean Progress?" begun to receive serious attention in our culture. This question is the perfect starting point for* Technology and the Future. *Deceptive in its simplicity, it underlies most of what follows in this book.*

Leo Marx is William R. Kenan Professor of American Cultural History Emeritus at MIT. He is the author of The Machine in the Garden: Technology and the Pastoral Ideal in America *(1964) and is coeditor, with Merritt Roe Smith, of* Does Technology Drive History? *(Cambridge, MA: MIT Press, 1995). He holds a Ph.D. in history of American civilization from Harvard and has taught at that institution and at the University of Minnesota and Amherst College. He has twice been a Guggenheim Fellow and was a Rockefeller Humanities Fellow in 1983–84. Marx was born in New York City in 1919.*

In this reading (first published in Technology Review *in 1987), he examines how the concept of progress has itself evolved since the early days of the Republic and what that evolution means for understanding the technological choices that confront us today. Improved technology could mean progress, Marx concludes, but "only if we are willing and able to answer the next question: progress toward what?"*

Does improved technology mean progress? If some variant of this question had been addressed to a reliable sample of Americans at any time since the early nineteenth century, the answer of a majority almost

Source: *Technology Review* (January 1987), pp. 33–41, 71. Copyright ©1987. Reprinted with permission from *Technology Review*.

certainly would have been an unequivocal "yes." The idea that technological improvements are a primary basis for—and an accurate gauge
of—progress has long been a fundamental belief in the United States.
In the last half-century, however, that belief has lost some of its
credibility. A growing minority of Americans has adopted a skeptical,
even negative, view of technological innovation as an index of social
progress.

The extent of this change in American attitudes was brought home to
me when I spent October 1984 in China. At that time the announced
goal of the People's Republic was to carry out (in the popular slogan)
"Four Modernizations"—agriculture, science and technology, industry,
and the military. What particularly struck our group of Americans was
the seemingly unbounded, largely uncritical ardor with which the Chinese were conducting their love affair with Western-style modernization—individualistic, entrepreneurial, or "capitalist," as well as scientific and technological. Like early nineteenth-century visitors to the
United States, we were witnessing a society in a veritable transport of
improvement: long pent-up, innovative energies were being released,
everyone seemed to be in motion, everything was eligible for change. It
was assumed that any such change almost certainly would be for the
better.

Most of the Chinese we came to know best—teachers and students
of American studies—explicitly associated the kind of progress represented by the four modernizations with the United States. This respect
for American wealth and power was flattering but disconcerting, for we
often found ourselves reminding the Chinese of serious shortcomings,
even some terrible dangers, inherent in the Western mode of industrial
development. Like the Americans whom European travelers met 150
years ago, many of the Chinese seemed to be extravagantly, almost
blindly, credulous and optimistic.

Our reaction revealed, among other things, a change in our own
culture and, in some cases, in our own personal attitudes. We came
face to face with the gulf that separates the outlook of many contemporary Americans from the old national faith in the advance of technology as the basis of social progress.

The standard explanation for this change includes that familiar lit-
any of death and destruction that distinguishes the recent history of the
West: two barbaric world wars, the Nazi holocaust, the Stalinist terror,
and the nuclear arms race. It is striking to note how many of the fearful
events of our time involve the destructive use or misuse, the unforeseen

consequences, or the disastrous malfunction of modern technologies: Hiroshima and the nuclear threat; the damage inflicted upon the environment by advanced industrial societies; and spectacular accidents like Three Mile Island.

Conspicuous disasters have helped to undermine the public's faith in progress, but there also has been a longer-term change in our thinking. It is less obvious, less dramatic and tangible than the record of catastrophe that distinguishes our twentieth-century history, but I believe it is more fundamental. Our very conception—our chief criterion—of progress has undergone a subtle but decisive change since the founding of the Republic, and that change is at once a cause and a reflection of our current disenchantment with technology. To chart this change in attitude, we need to go back at least as far as the first Industrial Revolution.

THE ENLIGHTENMENT BELIEF IN PROGRESS

The development of radically improved machinery (based on mechanized motive power) used in the new factory system of the late eighteenth century coincided with the formulation and diffusion of the modern Enlightenment idea of history as a record of progress. This conception became the fulcrum of the dominant American worldview. It assumes that history, or at least modern history, is driven by the steady, cumulative, and inevitable expansion of human knowledge of and power over nature. The new scientific knowledge and technological power was expected to make possible a comprehensive improvement in all the conditions of life—social, political, moral, and intellectual as well as material.

The modern idea of progress, as developed by its radical French, English, and American adherents, emerged in an era of political revolution. It was a revolutionary doctrine, bonded to the radical struggle for freedom from feudal forms of domination. To ardent republicans like the French philosopher Condorcet, the English chemist Priestley, and Benjamin Franklin, a necessary criterion of progress was the achievement of political and social liberation. They regarded the new sciences and technologies not as ends in themselves, but as instruments for carrying out a comprehensive transformation of society. The new knowledge and power would provide the basis for alternatives to the deeply entrenched authoritarian, hierarchical institutions of l'ancien régime: monarchical, aristocratic, and ecclesiastical. Thus in 1813

Thomas Jefferson wrote to John Adams describing the combined effect of the new science and the American Revolution on the minds of Europeans:

> Science had liberated the ideas of those who read and reflect, and the American example had kindled feelings of right in the people. An insurrection has consequently begun, of science, talents, and courage, against rank and birth, which have fallen into contempt. . . . Science is progressive.

Admittedly, the idea of history as endless progress did encourage extravagantly optimistic expectations, and in its most extreme form, it fostered some wildly improbable dreams of the "perfectability of Man" and of humanity's absolute mastery of nature. Yet the political beliefs of the radical republicans of the eighteenth century, such as the principle of making the authority of government dependent upon the consent of the governed, often had the effect of limiting those aspirations to omnipotence.

The constraining effect of such ultimate, long-term political goals makes itself felt, for example, in Jefferson's initial reaction to the prospect of introducing the new manufacturing system to America. "Let our work-shops remain in Europe," he wrote in 1785.

Although a committed believer in the benefits of science and technology, Jefferson rejected the idea of developing an American factory system on the ground that the emergence of an urban proletariat, which he then regarded as an inescapable consequence of the European factory system, would be too high a price to pay for any potential improvement in the American material standard of living. He regarded the existence of manufacturing cities and an industrial working class as incompatible with republican government and the happiness of the people. He argued that it was preferable, even if more costly in strictly economic terms, to ship raw materials to Europe and import manufactured goods. "The loss by the transportation of commodities across the Atlantic will be made up in happiness and permanence of government." In weighing political, moral, and aesthetic costs against economic benefits, he anticipated the viewpoint of the environmentalists and others of our time for whom the test of a technological innovation is its effect on the overall quality of life.

Another instance of the constraining effect of republican political ideals is Benjamin Franklin's refusal to exploit his inventions for private profit. Thus Franklin's reaction when the governor of Pennsylvania urged him to accept a patent for his successful design of the "Franklin stove":

> Governor Thomas was so pleased with the construction of this stove as described in . . . [the pamphlet] that . . . he offered to give me a patent for the sole vending of them for a term of years; but I declined it from a principle which has ever weighed with me on such occasions, namely; viz., *that as we enjoy great advantages from the inventions of others, we should be glad of an opportunity to serve others by any invention of ours, and this we should do freely and generously* [emphasis in original].

What makes the example of Franklin particularly interesting is the fact that he later came to be regarded as the archetypal self-made American and the embodiment of the Protestant work ethic. When Max Weber sought out of all the world *the* exemplar of that mentality for his seminal study, *The Protestant Ethic and the Spirit of Capitalism*, whom did he choose but our own Ben? But Franklin's was a principled and limited self-interest. In his *Autobiography*, he told the story of his rise in the world not to exemplify a merely personal success, but rather to illustrate the achievements of a "rising people." He belonged to that heroic revolutionary phase in the history of the bourgeoisie when that class saw itself as the vanguard of humanity and its principles as universal. He thought of his inventions as designed not for his private benefit but for the benefit of all.

THE TECHNOCRATIC CONCEPT OF PROGRESS

With the further development of industrial capitalism, a quite different conception of technological progress gradually came to the fore in the United States. Americans celebrated the advance of science and technology with increasing fervor, but they began to detach the idea from the goal of social and political liberation. Many regarded the eventual attainment of that goal as having been assured by the victorious American Revolution and the founding of the Republic.

The difference between this later view of progress and that of Jefferson's and Franklin's generation can be heard in the rhetoric of Daniel Webster. He and Edward Everett were perhaps the leading public communicators of this new version of the progressive ideology. When Webster decided to become a senator from Massachusetts instead of New

Hampshire, the change was widely interpreted to mean that he had become the quasi-official spokesman for the new industrial manufacturing interests. Thus Webster, who was generally considered the nation's foremost orator, was an obvious choice as the speaker at the dedication of new railroads. Here is a characteristic peroration of one such performance in 1847:

It is an extraordinary era in which we live. It is altogether new. The world has seen nothing like it before. I will not pretend, no one can pretend, to discern the end; but everybody knows that the age is remarkable for scientific research into the heavens, the earth, and what is beneath the earth; and perhaps more remarkable still for the application of this scientific research to the pursuits of life. . . . We see the ocean navigated and the solid land traversed by steam power, and intelligence communicated by electricity. Truly this is almost a miraculous era. What is before us no one can say, what is upon us no one can hardly realize. The progress of the age has almost outstripped human belief; the future is known only to Omniscience.

By the 1840s, as Webster's rhetoric suggests, the idea of progress was already being dissociated from the Enlightenment vision of political liberation. He invests the railroad with a quasi-religious inevitability that lends force to the characterization of his language as the rhetoric of the technological sublime. Elsewhere in the speech, to be sure, Webster makes the obligatory bow to the democratic influence of technological change, but it is clear that he is casting the new machine power as the prime exemplar of the overall progress of the age, quite apart from its political significance. Speaking for the business and industrial elite, Webster and Everett thus depict technological innovation as a sufficient cause, *in itself*, for the fact that history assumes the character of continuous, cumulative progress.

At the same time, discarding the radical political ideals of the Enlightenment allowed the idea of technological progress to blend with other grandiose national aspirations. Webster's version of the "rhetoric of the technological sublime" is of a piece with the soaring imperial ambitions embodied in the slogan "Manifest Destiny," and by such tacit military figurations of American development as the popular notion of the "conquest of nature" (including Native Americans) by the increasingly technologized forces of advancing European-American "civilization." These future-oriented themes easily harmonized with the belief in the coming of the millennium that characterized evangelical Protestantism, the most popular American religion at the time. Webster indicates as much when, at the end of his tribute to the new railroad, he glibly brings in "Omniscience" as the ultimate locus of the meaning of progress.

The difference between the earlier Enlightenment conception of progress and that exemplified by Webster is largely attributable to the difference between the groups they represented. Franklin, Jefferson, and the heroic generation of founding revolutionists constituted a distinct, rather unusual social class in that for a short time the same men possessed authority and power in most of its important forms: economic, social, political, and intellectual. The industrial capitalists for whom Daniel Webster spoke were men of a very different stripe. They derived their status from a different kind of wealth and power, and their conception of progress, like their economic and social aspirations, was correspondingly different. The new technology and the immense profits it generated belonged to them, and since they had every reason to assume that they would retain their property and power, they had a vested interest in technological innovation. It is not surprising, under the circumstances, that as industrialization proceeded these men became true believers in technological improvement as the primary basis for—as virtually tantamount to—universal progress.

This dissociation of technological and material advancement from the larger political vision of progress was an intermediate stage in the eventual impoverishment of that radical eighteenth-century worldview. This subtle change prepared the way for the emergence, later in the century, of a thoroughly technocratic idea of progress. It was "technocratic" in that it valued improvements in power, efficiency, rationality as ends in themselves. Among those who bore witness to the widespread diffusion of this concept at the turn of the century were Henry Adams and Thorstein Veblen, who were critical of it, and Andrew Carnegie, Thomas Edison, and Frederick Winslow Taylor and his followers, who lent expression to it. Taylor's theory of scientific management embodies the quintessence of the technocratic mentality, "the idea," as historian Hugh Aitken describes it, "that human activity could be measured, analyzed, and controlled by techniques analogous to those that had proved so successful when applied to physical objects."

The technocratic idea of progress is a belief in the sufficiency of scientific and technological innovation as the basis for general progress. It says that if we can ensure the advance of science-based technologies, the rest will take care of itself. (The "rest" refers to nothing less than a corresponding degree of improvement in the social, political, and cultural conditions of life.) Turning the Jeffersonian ideal on its head, this view makes instrumental values fundamental to social progress, and relegates what formerly were considered primary, goal-setting values (justice, freedom, harmony, beauty, or self-fulfillment) to a secondary status.

In this century, the technocratic view of progress was enshrined in Fordism and an obsessive interest in economies of scale, standardization of process and product, and control of the workplace. This shift to mass production was accompanied by the more or less official commitment of the U.S. government to the growth of the nation's wealth, productivity, and global power, and to the most rapid possible rate of technological innovation as the essential criterion of social progress.

But the old republican vision of progress—the vision of advancing knowledge empowering humankind to establish a less hierarchical, more just and peaceful society—did not disappear. If it no longer inspired Webster and his associates, it lived on in the minds of many farmers, artisans, factory workers, shopkeepers, and small-business owners, as well as in the beliefs of the professionals, artists, intellectuals, and other members of the lower middle and middle classes. During the late nineteenth century, a number of disaffected intellectuals sought new forms for the old progressive faith. They translated it into such political idioms as utopian socialism, the single-tax movement, the populist revolt, Progressivism in cities, and Marxism and its native variants.

THE ROOTS OF OUR ADVERSARY CULTURE

Let me turn to a set of these late-eighteenth-century ideas that was to become the basis for a powerful critique of the culture of advanced industrial society. Usually described as the viewpoint of the "counter-Enlightenment" or the "romantic reaction," these ideas have formed the basis for a surprisingly long-lived adversarial culture.

According to conventional wisdom, this critical view originated in the intellectual backlash from the triumph of the natural sciences we associate with the great discoveries of Galileo, Kepler, Harvey, and Newton. Put differently, this tendency was a reaction against the extravagant claims of the universal, not to say exclusive, truth of "the Mechanical Philosophy." That term derived from the ubiquity of the machine metaphor in the work of Newton and other natural scientists ("celestial mechanics") and many of their philosophic allies, notably Descartes, all of whom tended to conceive of nature itself as a "great engine" and its subordinate parts (including the human body) as lesser machines.

By the late eighteenth century, a powerful set of critical, antimechanistic ideas was being developed by Kant, Fichte, and other German idealists, and by great English poets like Coleridge and Wordsworth. But in their time the image of the machine also was being invested with greater tangibility and social import. The Industrial Revolution was gaining momentum, and as power machinery was more widely diffused in Great Britain, Western Europe, and North America, the machine acquired much greater resonance: it came to represent both the new technologies based on mechanized motive power and the mechanistic mindset of scientific rationalism. Thus the Scottish philosopher and historian Thomas Carlyle, who had been deeply influenced by the new German philosophy, announced in his seminal 1829 essay, "Signs of the Times," that the right name for the dawning era was the "Age of Machinery." It was to be the Age of Machinery, he warned, in every "inward" and "outward" sense of the word, meaning that it would be dominated by mechanical (utilitarian) thinking as well as by actual machines.

In his criticism of this new era, Carlyle took the view that neither kind of "machinery" was inherently dangerous. In his opinion, indeed, they represented *potential* progress as long as neither was allowed to become the exclusive or predominant mode in its respective realm.

In the United States a small, gifted, if disaffected minority of writers, artists, and intellectuals adopted this ideology. Their version of Carlyle's critical viewpoint was labeled "romantic" in reference to its European strains, or "transcendentalist" in its native use. In the work of writers like Emerson and Thoreau, Hawthorne and Melville, we encounter critical responses to the onset of industrialism that cannot be written off as mere nostalgia or primitivism. These writers did not hold up an idealized wilderness, a pre-industrial Eden, as preferable to the world they saw in the making. Nor did they dismiss the worth of material improvement as such. But they did regard the dominant view, often represented (as in Webster's speech) by the appearance of the new machine power in the American landscape, as dangerously shallow, materialistic, and one-sided. Fear of "mechanism," in the several senses of that word—especially the domination of the individual by impersonal systems—colored all of their thought. In their work, the image of the machine-in-the-landscape, far from being an occasion for exultation, often seems to arouse anxiety, dislocation, and foreboding. Henry Thoreau's detailed, carefully composed account of the intrusion of the railroad into the Concord woods is a good example; it bears out his delineation of the new inventions as "improved means to unimproved ends."

This critical view of the relationship between technological means and social ends did not merely appear in random images, phrases, and narrative episodes. Indeed, the whole of *Walden* may be read as a sustained attack on a culture that had allowed itself to become confused about the relationship of ends and means. Thoreau's countrymen are depicted as becoming "the tools of their tools." Much the same argument underlies Hawthorne's satire, "The Celestial Railroad," a modern replay of *Pilgrim's Progress* in which the hero, Christian, realizes too late that his comfortable railroad journey to salvation is taking him to hell, not heaven. Melville incorporates a similar insight into his characterization of Captain Ahab, who is the embodiment of the Faustian aspiration toward domination and total control given credence by the sudden emergence of exciting new technological capacities. Ahab exults in his power over the crew, and he explicitly identifies it with the power exhibited by the new railroad spanning the North American continent. In reflective moments, however, he also acknowledges the self-destructive nature of his own behavior: "Now in his heart, Ahab had some glimpse of this, namely, all my means are sane, my motive and my object mad."

Of course there was nothing new about the moral posture adopted by these American writers. Indeed, their attitude toward the exuberant national celebration of the railroad and other inventions is no doubt traceable to traditional moral and religious objections to such an exaggeration of human powers. In this view, the worshipful attitude of Americans toward these new instruments of power had to be recognized for what it was: idolatry like that attacked by Old Testament prophets in a disguised, new-fashioned form. This moral critique of the debased, technocratic version of the progressive worldview has slowly gained adherents since the mid-nineteenth century, and by now it is one of the chief ideological supports of an adversary culture in the United States.

The ideas of writers like Hawthorne, Melville, and Thoreau were usually dismissed as excessively idealistic, nostalgic, or sentimental, hence impractical and unreliable. They were particularly vulnerable to that charge at a time when the rapid improvement in the material conditions of American life lent a compelling power to the idea that the meaning of history is universal progress. Only in the late twentieth century, with the growth of skepticism about scientific and technological progress, and with the emergence of a vigorous adversary culture in the 1960s, has the standpoint of that earlier eccentric minority been accorded a certain intellectual respect. To be sure, it is still chiefly the viewpoint of a relatively small minority, but there have been times, like the Vietnam upheaval of the 1960s, when that minority has won the temporary support of, or formed a tacit coalition with, a remarkably large number of other disaffected Americans. Much the same antitechnocratic viewpoint has made itself felt in various dissident movements and intellectual tendencies since the 1960s: the antinuclear movements (against both nuclear power and nuclear weaponry); some branches of the environmental and feminist movements; the "small is beautiful" and "stable-state" economic theories, as well as the quest for "soft energy paths" and "alternative (or appropriate) technologies."

TECHNOCRATIC VERSUS SOCIAL PROGRESS

Perhaps this historical summary will help explain the ambivalence toward the ideal of progress expressed by many Americans nowadays. Compared with prevailing attitudes in the U.S. in the 1840s, when the American situation was more like that of China today, the current mood in this country would have to be described as mildly disillusioned.

To appreciate the reasons for that disillusionment, let me repeat the distinction between the two views of progress on which this analysis rests. The initial Enlightenment belief in progress perceived science and technology to be in the service of liberation from political oppression. Over time that conception was transformed, or partly supplanted, by the now familiar view that innovations in science-based technologies are in themselves a sufficient and reliable basis for progress. The distinction, then, turns on the apparent loss of interest in, or unwillingness to name, the social ends for which the scientific and technological instruments of power are to be used. What we seem to have instead of a guiding political goal is a minimalist definition of civic obligation.

The distinction between two versions of the belief in progress helps sort out reactions to the many troubling issues raised by the diffusion of high technology. When, for example, the introduction of some new labor-saving technology is proposed, it is useful to ask what the purpose of this new technology is. Only by questioning the assumption that innovation represents progress can we begin to judge its worth. The aim may well be to reduce labor costs, yet in our society the personal costs to the displaced workers are likely to be ignored.

The same essential defect of the technocratic mindset also becomes evident when the president of the United States calls upon those who devise nuclear weapons to provide an elaborate new system of weaponry, the Strategic Defense Initiative, as the only reliable means of avoiding nuclear war. Not only does he invite us to put all our hope in a "techno-logical fix,"[1] but he rejects the ordinary but indispensable method of international negotiation and compromise. Here again, technology is thought to obviate the need for political ideas and practices.

One final word. I perhaps need to clarify the claim that it is the modern, technocratic worldview of Webster's intellectual heirs, not the Enlightenment view descended from the Jeffersonians, that encour-ages the more dangerous contemporary fantasies of domination and total control. The political and social aspirations of the generation of Benjamin Franklin and Thomas Jefferson *provided tacit limits to, as well as ends for, the progressive vision of the future.* But the technocratic version so popular today entails a belief in the worth of scientific and technological innovations as ends in themselves.

All of which is to say that we urgently need a set of political, social, and cultural goals comparable to those formulated at the beginning of the industrial era if we are to accurately assess the worth of new tech-nologies. Only such goals can provide the criteria required to make rational and humane choices among alternative technologies and, more important, among alternative long-term policies.

Does improved technology mean progress? Yes, it certainly *could* mean just that. But only if we are willing and able to answer the next question: progress toward what? What is it that we want our new technologies to accomplish? What do we want beyond such immediate, limited goals as achieving efficiencies, decreasing financial costs, and eliminating the troubling human element from our workplaces? In the absence of an-swers to these questions, technological improvements may very well turn out to be incompatible with genuine, that is to say *social,* progress.

---

[1] See Alvin M. Weinberg, "Can Technology Replace Social Engineering?" (Chapter 5), for a discussion of the concept of the "technological fix."—Ed.

2. The Powershift Era

ALVIN TOFFLER

For the past twenty-five years, Alvin Toffler's writings have brought the study of the future into the mass market. Toffler, who at one time was a journalist and associate editor of Fortune magazine, gained worldwide attention in 1970 with his book Future Shock. He followed up that success in 1980 with the best-selling The Third Wave. Powershift, which is subtitled Knowledge, Wealth, and Violence at the Edge of the 21st Century, is his book for the 1990s. Some see Toffler as a visionary whose writings can help foretell the future and whose interpretations of current trends can help make that future a reality. Others see his writings as superficial "pop sociology" that offers little of substance. These critics accuse him of substituting glib clichés and jargon— "premature arrival of the future," "massive adaptational breakdown," "social future assemblies," and so on—for serious thought, and they take him to task for his shallow (or nonexistent) reading of history.

While Toffler's star may have dimmed somewhat since the days when brightly colored paperback copies of Future Shock could be found on supermarket book racks, his reputation (and book sales) certainly received a boost in 1995 from his friendship with House Speaker Newt Gingrich—a futurism buff on whose thinking Toffler is reported to have a substantial influence. The following selection, "The Powershift Era," is the opening of Powershift. It is fairly representative of Toffler's writing, suggesting the range of his thought, the ease with which he integrates disparate ideas, and either his global vision or his superficiality, depending on your point of view. Toffler works closely with his wife, Heidi Toffler, who has coauthored a number of his works. Apart from his writing, he serves as a lecturer and consultant to governments and corporations.

[*Powershift* is] about power at the edge of the 21st century. It deals with violence, wealth, and knowledge and the roles they play in our lives. It is about the new paths to power opened by a world in upheaval.

Despite the bad odor that clings to the very notion of power because of the misuses to which it has been put, power in itself is neither good nor bad. It is an inescapable aspect of every human relationship, and it influences everything from our sexual relations to the jobs we hold, the cars we drive, the television we watch, the hopes we pursue. To a greater degree than most imagine, we are the products of power.

Yet of all the aspects of our lives, power remains one of the least understood and most important—especially for our generation.

For this is the dawn of the Powershift Era. We live at a moment when the entire structure of power that held the world together is now disintegrating. A radically different structure of power is taking form. And this is happening at every level of human society.

In the office, in the supermarket, at the bank, in the executive suite, in our churches, hospitals, schools, and homes, old patterns of power are fracturing along strange new lines. Campuses are stirring from Berkeley to Rome and Taipei, preparing to explode. Ethnic and racial clashes are multiplying.

In the business world we see giant corporations taken apart and put back together, their CEOs often dumped, along with thousands of their employees. A "golden parachute" or goodbye package of money and benefits may soften the shock of landing for a top manager, but gone are the appurtenances of power: the corporate jet, the limousine, the conferences at glamorous golf resorts, and above all, the secret thrill that many feel in the sheer exercise of power.

Power isn't just shifting at the pinnacle of corporate life. The office manager and the supervisor on the plant floor are both discovering that workers no longer take orders blindly, as many once did. They ask questions and demand answers. Military officers are learning the same thing about their troops. Police chiefs about their cops. Teachers, increasingly, about their students.

This crackup of old-style authority and power in business and daily life is accelerating at the very moment when global power structures are disintegrating as well.

Ever since the end of World War II, two superpowers have straddled the earth like colossi. Each had its allies, satellites, and cheering section. Each balanced the other, missile for missile, tank for tank, spy for spy. Today, of course, that balancing act is over.

As a result, "black holes" are already opening up in the world system: great sucking power vacuums, in Eastern Europe for example, that could sweep nations and peoples into strange new—or, for that matter, ancient—alliances and collisions. Power is shifting at so astonishing a rate that world leaders are being swept along by events, rather than imposing order on them.

There is strong reason to believe that the forces now shaking power at every level of the human system will become more intense and pervasive in the years immediately ahead.

Out of this massive restructuring of power relationships, like the shifting and grinding of tectonic plates in advance of an earthquake, will come one of the rarest events in human history: a revolution in the very nature of power.

A "powershift" does not merely transfer power. It transforms it.

THE END OF EMPIRE

The entire world watched awestruck as a half-century-old empire based on Soviet power in Eastern Europe suddenly came unglued in 1989. Desperate for the Western technology needed to energize its rust-belt economy, the Soviet Union itself plunged into a period of near-chaotic change.

Slower and less dramatically, the world's other superpower also went into relative decline. So much has been written about America's loss of global power that it bears no repetition here. Even more striking, however, have been the many shifts of power away from its once-dominant domestic institutions.

Twenty years ago General Motors was regarded as the world's premier manufacturing company, a gleaming model for managers in countries around the world and a political powerhouse in Washington. Today, says a high GM official, "We are running for our lives." We may well see, in the years ahead, the actual breakup of GM.[1]

Twenty years ago IBM had only the feeblest competition and the United States probably had more computers than the rest of the world combined. Today computer power has spread rapidly around the world, the U.S. share has sagged, and IBM faces stiff competition from companies like NEC, Hitachi, and Fujitsu in Japan; Groupe Bull in France; ICL in Britain; and many others. Industry analysts speculate about the post-IBM era.[2]

Nor is all this a result of foreign competition. Twenty years ago three television networks, ABC, CBS, and NBC, dominated the American airwaves. They faced no foreign competition at all. Yet today they are shrinking so fast, their very survival is in doubt.

Twenty years ago, to choose a different kind of example, medical doctors in the United States were white-coated gods. Patients typically accepted their word as law. Physicians virtually controlled the entire American health system. Their political clout was enormous.

Today, by contrast, American doctors are under siege. Patients talk back. They sue for malpractice. Nurses demand responsibility and re-spect. Pharmaceutical companies are less deferential. And it is insur-ance companies, "managed care groups," and government, not doctors, who now control the American health system.

Across the board, then, some of the most powerful institutions and professions inside the most powerful of nations saw their dominance decline in the same twenty-year period that saw America's external power, relative to other nations, sink.

Lest these immense shake-ups in the distribution of power seem a disease of the aging superpowers, a look elsewhere proves otherwise.

While U.S. economic power faded, Japan's skyrocketed. But success, too, can trigger significant power shifts. Just as in the United States, Japan's most powerful Second Wave or rust-belt industries declined in importance as new Third Wave industries rose. Even as Japan's eco-nomic heft increased, however, the three institutions perhaps most responsible for its growth saw their own power plummet. The first was the governing Liberal-Democratic Party. The second was the Ministry of International Trade and Industry (MITI), arguably the brain behind the Japanese economic miracle. The third was Keidanren, Japan's most politically potent business federation.

Today the LDP is in retreat, its elderly male leaders embarrassed by financial and sexual scandals. It is faced, for the first time, by outraged and increasingly active women voters, by consumers, taxpayers, and farmers who formerly supported it. To retain the power it has held since 1955, it will be compelled to shift its base from rural to urban voters, and deal with a far more heterogeneous population than ever before. For Japan, like all the high-tech nations, is becoming a de-massified society, with many more actors arriving on the political scene. Whether the LDP can make this long-term switch is at issue. What is not at issue is that significant power has shifted away from the LDP.

As for MITI, even now many American academics and politicians urge the United States to adopt MITI-style planning as a model.[3] Yet today, MITI itself is in trouble. Japan's biggest corporations once danced attendance on its bureaucrats and, willingly or not, usually followed its "guidelines." Today MITI is a fast-fading power as the corporations themselves have grown strong enough to thumb their noses at it.[4] Japan remains economically powerful in the outside world but politically weak at home. Immense economic weight pivots around a shaky political base.

Even more pronounced has been the decline in the strength of Keidanren, still dominated by the hierarchs of the fast-fading smoke-stack industries.

Even those dreadnoughts of Japanese fiscal power, the Bank of Japan and the Ministry of Finance, whose controls guided Japan through the high-growth period, the oil shock, the stock market crash, and the yen rise, now find themselves impotent against the turbulent market forces destabilizing the economy.

Still more striking shifts of power are changing the face of Western Europe. Thus power has shifted away from London, Paris, and Rome as the German economy has outstripped all the rest. Today, as East and West Germany progressively fuse their economies, all Europè once more fears German domination of the continent.

To protect themselves, France and other West European nations, with the exception of Britain, are hastily trying to integrate the European community politically as well as economically. But the more successful they become, the more of their national power is transfused into the veins of the Brussels-based European Community, which has progressively stripped away bigger and bigger chunks of their sovereignty.

The nations of Western Europe thus are caught between Bonn or Berlin on the one side and Brussels on the other. Here, too, power is shifting rapidly away from its established centers.

The list of such global and domestic power shifts could be extended indefinitely. They represent a remarkable series of changes for so brief a peacetime period. Of course, some power shifting is normal at any time.

Yet only rarely does an entire globe-girdling *system* of power fly apart in this fashion. It is an even rarer moment in history when all the rules of the power game change at once, and the very nature of power is revolutionized.

Yet that is exactly what is happening today. Power, which to a large extent defines us as individuals and as nations, is itself being redefined.

GOD-IN-A-WHITE-COAT[5]

A clue to this redefinition emerges when we look more closely at the above list of apparently unrelated changes. For we discover that they are not as random as they seem. Whether it is Japan's meteoric rise, GM's embarrassing decline, or the American doctor's fall from grace, a single common thread unites them.

Take the punctured power of the god-in-a-white-coat.

Throughout the heyday of doctor-dominance in America, physicians kept a tight choke-hold on medical knowledge. Prescriptions were written in Latin, providing the profession with a semi-secret code, as it were, which kept most patients in ignorance. Medical journals and texts were restricted to professional readers. Medical conferences were closed to the laity. Doctors controlled medical-school curricula and enrollments.

Contrast this with the situation today, when patients have astonishing access to medical knowledge. With a personal computer and a modem, anyone from home can access data bases like Index Medicus, and obtain scientific papers on everything from Addison's disease to zygomycosis, and, in fact, collect more information about a specific ailment or treatment than the ordinary doctor has time to read.

Copies of the 2,354-page book known as the PDR or *Physicians' Desk Reference* are also readily available to anyone. Once a week on the Lifetime cable network, any televiewer can watch twelve uninterrupted hours of highly technical television programming designed specifically to educate doctors. Many of these programs carry a disclaimer to the effect that "some of this material may not be suited to a general audience." But that is for the viewer to decide.

The rest of the week, hardly a single newscast is aired in America without a medical story or segment. A video version of material from the *Journal of the American Medical Association* is now broadcast by three hundred stations on Thursday nights. The press reports on medical malpractice cases. Inexpensive paperbacks tell ordinary readers what drug side effects to watch for, what drugs not to mix, how to raise or lower cholesterol levels through diet. In addition, major medical breakthroughs, even if first published in medical journals, are reported on the evening television news almost before the M.D. has even taken his subscription copy of the journal out of the in-box.

In short, the knowledge monopoly of the medical profession has been thoroughly smashed. And the doctor is no longer a god.

This case of the dethroned doctor is, however, only one small example of a more general process changing the entire relationship of knowledge to power in the high-tech nations.

In many other fields, too, closely held specialists' knowledge is slipping out of control and reaching ordinary citizens. Similarly, inside major corporations, employees are winning access to knowledge once monopolized by management. And as knowledge is redistributed, so, too, is the power based on it.

BOMBARDED BY THE FUTURE

There is, however, a much larger sense in which changes in knowledge are causing or contributing to enormous power shifts. The most important economic development of our lifetime has been the rise of a new system for creating wealth, based no longer on muscle but on mind. Labor in the advanced economy no longer consists of working on "things," writes historian Mark Poster of the University of California (Irvine), but of "men and women acting on other men and women, or . . . people acting on information and information acting on people."[6]

The substitution of information or knowledge for brute labor, in fact, lies behind the troubles of General Motors and the rise of Japan as well. For while GM still thought the earth was flat, Japan was exploring its edges and discovering otherwise.

As early as 1970, when American business leaders still thought their smokestack world secure, Japan's business leaders, and even the general public, were being bombarded by books, newspaper articles, and television programs heralding the arrival of the "information age" and focusing on the 21st century. While the end-of-industrialism concept was dismissed with a shrug in the United States, it was welcomed and embraced by Japanese decision-makers in business, politics, and the media. Knowledge, they concluded, was the key to economic growth in the 21st century.

It was hardly surprising, therefore, that even though the United States started computerizing earlier, Japan moved more quickly to substitute the knowledge-based technologies of the Third Wave for the brute muscle technologies of the Second Wave past.

Robots proliferated. Sophisticated manufacturing methods, heavily dependent on computers and information, began turning out products whose quality could not be easily matched in world markets. Moreover, recognizing that its old smokestack technologies were ultimately doomed, Japan took steps to facilitate the transition to the new and to buffer itself against the dislocations entailed in such a strategy. The contrast with General Motors—and American policy in general—could not have been sharper.

If we also look closely at many of the other power shifts cited above, it will become apparent that in these cases, too, the changed role of knowledge—the rise of the new wealth-creation system—either caused or contributed to major shifts of power.

The spread of this new knowledge economy is, in fact, the explosive new force that has hurled the advanced economies into bitter global competition, confronted the socialist nations with their hopeless obsolescence, forced many "developing nations" to scrap their traditional economic strategies, and is now profoundly dislocating power relationships in both personal and public spheres.

In a prescient remark, Winston Churchill once said that "empires of the future are empires of the mind." Today that observation has come true. What has not yet been appreciated is the degree to which raw, elemental power—at the level of private life as well as at the level of empire—will be transformed in the decades ahead as a result of the new role of "mind."

THE MAKING OF A SHABBY GENTILITY

A revolutionary new system for creating wealth cannot spread without triggering personal, political, and international conflict. Change the way wealth is made and you immediately collide with all the entrenched interests whose power arose from the prior wealth-system. Bitter conflicts erupt as each side fights for control of the future.

It is this conflict, spreading around the world today, that helps explain the present power shake-up. To anticipate what might lie ahead for us, therefore, it is helpful to glance briefly backward at the last such global conflict.

Three hundred years ago the industrial revolution also brought a new system of wealth creation into being. Smokestacks speared the skies where fields once were cultivated. Factories proliferated. These "dark Satanic mills" brought with them a totally new way of life—and a new system of power.

Peasants freed from near-servitude on the land turned into urban workers subordinated to private or public employers. With this change came changes in power relations in the home as well. Agrarian families, several generations under a single roof, all ruled by a bearded patriarch, gave way to stripped-down nuclear families from which the elderly were soon extruded or reduced in prestige and influence. The family itself, as an institution, lost much of its social power as many of its functions were transferred to other institutions—education to the school, for example.

Sooner or later, too, wherever steam engines and smokestacks multiplied, vast political changes followed. Monarchies collapsed or shriveled into tourist attractions. New political forms were introduced.

If they were clever and farsighted enough, rural landowners, once dominant in their regions, moved into the cities to ride the wave of industrial expansion, their sons becoming stockbrokers or captains of industry. Most of the landed gentry who clung to their rural way of life wound up as shabby gentility, their mansions eventually turned into museums or into money-raising lion parks.

Against their fading power, however, new elites arose: corporate chieftains, bureaucrats, media moguls. Mass production, mass distribution, mass education, and mass communication were accompanied by mass democracy, or dictatorships claiming to be democratic.

These internal changes were matched by gigantic shifts in global power, too, as the industrialized nations colonized, conquered, or dominated much of the rest of the world, creating a hierarchy of world power that still exists in some regions.

In short, the appearance of a new system for creating wealth undermined every pillar of the old power system, ultimately transforming family life, business, politics, the nation-state, and the structure of the global power itself.

Those who fought for control of the future made use of violence, wealth, and knowledge. Today a similar, though far more accelerated, upheaval has started. The changes we have recently seen in business, the economy, politics, and at the global level are only the first skirmishes of far bigger power struggles to come. For we stand at the edge of the deepest powershift in human history.

NOTES

1. "GM Is Tougher Than You Think," by Anne B. Fisher, *Fortune,* November 10, 1986.
2. *Datamation,* June 15, 1988.
3. "Gephardt Plans to Call for Japan-Style Trade Agency," *Los Angeles Times,* October 4, 1989.
4. On MITI, see following from *Japan Economic Journal:* "MITI Fights to Hold Influence as Japanese Firms Go Global," April 1, 1989; "Icy Welcome for MITI's Retail Law Change," October 21, 1989; "Japan Carmakers Eye Growth Despite MITI Warning," October 21, 1989; "Trade Policy Flip-Flop Puts MITI on Defensive," January 20, 1990.
5. Medical material based in part on interviews with staff of The Wilkerson Group, medical management consulting organization, New York; also Wendy Borow, Director of American Medical Association Division of Television, Radio and Film; and Barry Cohn, television news producer, AMA, Chicago.
6. Mark Poster, *Foucault, Marxism and History* (Oxford: Polity Press, 1984), p. 53.

3. Technopoly: The Broken Defenses

NEIL POSTMAN

Neil Postman is a technological critic in the tradition of Jacques Ellul, Lewis Mumford, and Paul Goodman. His concern, like those of his predecessors, is with the destructive moral and cultural consequences of technology in modern society. His book Technopoly, *subtitled* The Surrender of Culture to Technology, *is an attempt (in his own words) "to describe when, how, and why technology became a particularly dangerous enemy." Postman recognizes the benefits of technology, acknowledging that "it makes life easier, cleaner, and longer." But he suggests that these benefits are seductive, because they lead us to overlook the negative side of technology—the fact that it "creates a culture without a moral foundation" and undermines the very things that make life worth living. The chapter included here describes how, in Postman's view, bureaucracies have emerged to deal with the information glut in society and how they undermine traditional social institutions.*

Postman is chair of the Department of Communications Arts and Sciences at New York University, where he received the Distinguished Professor Award in 1989. He has published numerous books, among the best known of which are Teaching as a Subversive Activity, The Disappearance of Childhood, *and* Amusing Ourselves to Death.

Technopoly is a state of culture. It is also a state of mind. It consists in the deification of technology, which means that the culture seeks its authorization in technology, finds its satisfactions in technology, and takes its orders from technology. This requires the development of a new kind of social order, and of necessity leads to the rapid dissolution of much that is associated with traditional beliefs. Those who feel most comfortable in Technopoly are those who are convinced that technical progress is humanity's supreme achievement and the instrument by which our most profound dilemmas may be solved. They also believe that information is an unmixed blessing, which through its continued and uncontrolled production and dissemination offers increased freedom, creativity, and peace of mind. The fact that information does none of these things—but quite the opposite—seems to change few opinions, for such unwavering beliefs are an inevitable product of the structure of Technopoly. In particular, Technopoly flourishes when the defenses against information break down.

The relationship between information and the mechanisms for its control is fairly simple to describe: Technology increases the available supply of information. As the supply is increased, control mechanisms are strained. Additional control mechanisms are needed to cope with new information. When additional control mechanisms are themselves technical, they in turn further increase the supply of information. When the supply of information is no longer controllable, a general breakdown in psychic tranquillity and social purpose occurs. Without defenses, people have no way of finding meaning in their experiences, lose their capacity to remember, and have difficulty imagining reasonable futures.

One way of defining Technopoly, then, is to say it is what happens to society when the defenses against information glut have broken down. It is what happens when institutional life becomes inadequate to cope with too much information. It is what happens when a culture, overcome by information generated by technology, tries to employ technology itself as a means of providing clear direction and humane purpose. The effort is mostly doomed to failure. Though it is sometimes possible to use a disease as a cure for itself, this occurs only when we are fully aware of the processes by which disease is normally held in check. My purpose here is to describe the defenses that in principle are available and to suggest how they have become dysfunctional.

The dangers of information on the loose may be understood by the analogy I suggested [elsewhere] with an individual's biological immune system, which serves as a defense against the uncontrolled growth of cells. Cellular growth is, of course, a normal process without which organic life cannot survive. But without a well-functioning immune system, an organism cannot manage cellular growth. It becomes disordered and destroys the delicate interconnectedness of essential organs.

An immune system, in short, destroys unwanted cells. All societies have institutions and techniques that function as does a biological immune system. Their purpose is to maintain a balance between the old and the new, between novelty and tradition, between meaning and conceptual disorder, and they do so by "destroying" unwanted information.

I must emphasize that social institutions of all kinds function as control mechanisms. This is important to say, because most writers on the subject of social institutions (especially sociologists) do not grasp the idea that any decline in the force of institutions makes people vulnerable to information chaos.[1] To say that life is destabilized by weakened institutions is merely to say that information loses its use and therefore becomes a source of confusion rather than coherence.

Social institutions sometimes do their work simply by denying people access to information, but principally by directing how much weight and, therefore, value one must give to information. Social institutions are concerned with the *meaning* of information and can be quite rigorous in enforcing standards of admission. Take as a simple example a court of law. Almost all rules for the presentation of evidence and for the conduct of those who participate in a trial are designed to limit the amount of information that is allowed entry into the system. In our system, a judge disallows "hearsay" or personal opinion as evidence except under strictly controlled circumstances, spectators are forbidden to express their feelings, a defendant's previous convictions may not be mentioned, juries are not allowed to hear arguments over the admissibility of evidence—these are instances of information control. The rules on which such control is based derive from a theory of justice that defines what information may be considered relevant and, especially, what information must be considered irrelevant. The theory may be deemed flawed in some respects— lawyers, for example, may disagree over the rules governing the flow of information—but no one disputes that information must be regulated in some manner. In even the simplest law case, thousands of events may have had a bearing on the dispute, and it is well understood that, if they were all permitted entry, there could be no theory of due process, trials would have no end, law itself would be reduced to meaninglessness. In short, the rule of law is concerned with the "destruction" of information.

It is worth mentioning here that, although legal theory has been taxed to the limit by new information from diverse sources—biology, psychology, and sociology among them—the rules governing relevance have remained fairly stable. This may account for Americans' overuse of the courts as a means of finding coherence and stability. As other institutions become unusable as mechanisms for the control of wanton information, the courts stand as a final arbiter of truth. For how long, no one knows.

I have referred [elsewhere] to the school as a mechanism for information control. What its standards are can usually be found in a curriculum or, with even more clarity, in a course catalogue. A college catalogue lists courses, subjects, and fields of study that, taken together, amount to a certified statement of what a serious student ought to think about. More to the point, in what is omitted from a catalogue, we may learn what a serious student ought *not* to think about. A college catalogue, in other words, is a formal description of an information management program; it defines and categorizes knowledge, and in so doing systematically excludes, demeans, labels as trivial—in a word, disregards certain kinds of information. That is why it "makes sense" (or, more accurately, used to make sense). By what it includes/ excludes it reflects a theory of the purpose and meaning of education. In the university where I teach, you will not find courses in astrology or dianetics or creationism. There is, of course, much available information about these subjects, but the theory of education that sustains the university does not allow such information entry into the formal structure of its courses. Professors and students are denied the opportunity to focus their attention on it, and are encouraged to proceed as if it did not exist. In this way, the university gives expression to its idea of what constitutes legitimate knowledge. At the present time, some accept this idea and some do not, and the resulting controversy weakens the university's function as an information control center.

The clearest symptom of the breakdown of the curriculum is found in the concept of "cultural literacy," which has been put forward as an organizing principle and has attracted the serious attention of many educators.[2] If one is culturally literate, the idea goes, one should master a certain list of thousands of names, places, dates, and aphorisms; these are supposed to make up the content of the literate American's mind. But . . . cultural literacy is not an organizing principle at all; it represents, in fact, a case of calling the disease the cure. The point to be stressed here is that any educational institution, if it is to function well in the management of information, must have a theory about its purpose and meaning, must have the means to give clear expression to its theory, and must do so, to a large extent, by excluding information.

As another example, consider the family. As it developed in Europe in the late eighteenth century, its theory included the premise that individuals need emotional protection from a cold and competitive society. The family became, as Christopher Lasch calls it, a haven in a heartless world.[3] Its program included (I quote Lasch here) preserving "separatist religious traditions, alien languages and dialects, local lore and other traditions." To do this, the family was required to take charge of the socialization of children; the family became a structure, albeit an informal one, for the management of information. It controlled what "secrets" of adult life would be allowed entry and what "secrets" would not. There may be readers who can remember when in the presence of children adults avoided using certain words and did not discuss certain topics whose details and ramifications were considered unsuitable for children to know. A family that does not or cannot control the information environment of its children is barely a family at all, and may lay claim to the name only by virtue of the fact that its members share biological information through DNA. In fact, in many societies a family was just that—a group connected by genetic information, itself controlled through the careful planning of marriages. In the West, the family as an institution for the management of nonbiological information began with the ascendance of print. As books on every conceivable subject became available, parents were forced into the roles of guardians, protectors, nurturers, and arbiters of taste and rectitude. Their function was to define what it means to be a child by excluding from the family's domain information that would undermine its purpose. That the family can no longer do this is, I believe, obvious to everyone.

Courts of law, the school, and the family are only three of several control institutions that serve as part of a culture's information immune system. The political party is another. As a young man growing up in a Democratic household, I was provided with clear instructions on what value to assign to political events and commentary. The instructions did not require explicit statement. They followed logically from theory, which was, as I remember it, as follows: Because people need protection, they must align themselves with a political organization. The Democratic Party was entitled to our loyalty because it represented the social and economic interests of the working class, of which our family, relatives, and neighbors were members (except for one uncle who, though a truck driver, consistently voted Republican and was therefore thought to be either stupid or crazy). The Republican Party represented the interests of the rich, who, by definition, had no concern for us.

The theory gave clarity to our perceptions and a standard by which to judge the significance of information. The general principle was that information provided by Democrats was always to be taken seriously and, in all probability, was both true and useful (except if it came from Southern Democrats, who were helpful in electing presidents but were otherwise never to be taken seriously because of their special theory of race). Information provided by Republicans was rubbish and was useful only to the extent that it confirmed how self-serving Republicans were.

I am not prepared to argue here that the theory was correct, but to the accusation that it was an oversimplification I would reply that all theories are oversimplifications, or at least lead to oversimplification. The rule of law is an oversimplification. A curriculum is an oversimplification. So is a family's conception of a child. That is the function of theories—to oversimplify, and thus to assist believers in organizing, weighting, and excluding information. Therein lies the power of theories. Their weakness is that precisely because they oversimplify, they are vulnerable to attack by new information. When there is too much information to sustain *any* theory, information becomes essentially meaningless.

The most imposing institutions for the control of information are religion and the state. They do their work in a somewhat more abstract way than do courts, schools, families, or political parties. They manage information through the creation of myths and stories that express theories about fundamental questions: why are we here, where have we come from, and where are we headed? I have [elsewhere] alluded to the comprehensive theological narrative of the medieval European world and how its great explanatory power contributed to a sense of well-being and coherence. Perhaps I have not stressed enough the extent to which the Bible also served as an information control mechanism, especially in the moral domain. The Bible gives manifold instructions on what one must do and must not do, as well as guidance on what language to avoid (on pain of committing blasphemy), what ideas to avoid (on pain of committing heresy), what symbols to avoid (on pain of committing idolatry). Necessarily but perhaps unfortunately, the Bible also explained how the world came into being in such literal detail that it could not accommodate new information produced by the telescope and subsequent technologies. The trials of Galileo and, three hundred years later, of Scopes were therefore about the admissibility of certain kinds of information. Both Cardinal Bellarmine and William Jennings Bryan were fighting to maintain the authority of the Bible to control information about the profane world as well as the sacred. In their defeat, more was lost than the Bible's claim to explain the origins and structure of nature. The Bible's authority in defining and categorizing moral behavior was also weakened.

Nonetheless, Scripture has at its core such a powerful mythology that even the residue of that mythology is still sufficient to serve as an exacting control mechanism for some people. It provides, first of all, a theory about the meaning of life and therefore rules on how one is to conduct oneself. With apologies to Rabbi Hillel, who expressed it more profoundly and in the time it takes to stand on one leg, the theory is as follows: There is one God, who created the universe and all that is in it. Although humans can never fully understand God, He has revealed Himself and His will to us throughout history, particularly through His commandments and the testament of the prophets as recorded in the Bible. The greatest of these commandments tells us that humans are to love God and express their love for Him through love, mercy, and justice to our fellow humans. At the end of time, all nations and humans will appear before God to be judged, and those who have followed His commandments will find favor in His sight. Those who have denied God and the commandments will perish utterly in the darkness that lies outside the presence of God's light.

To borrow from Hillel: That is the theory. All the rest is commentary.

Those who believe in this theory—particularly those who accept the Bible as the literal word of God—are free to dismiss other theories about the origin and meaning of life and to give minimal weight to the facts on which other theories are based. Moreover, in observing God's laws, and the detailed requirements of their enactment, believers receive guidance about what books they should not read, about what plays and films they should not see, about what music they should not hear, about what subjects their children should not study, and so on. For strict fundamentalists of the Bible, the theory and what follows from it seal them off from unwanted information, and in that way their actions are invested with meaning, clarity, and, they believe, moral authority.

Those who reject the Bible's theory and who believe, let us say, in the theory of Science are also protected from unwanted information. Their theory, for example, instructs them to disregard information about astrology, dianetics, and creationism, which they usually label as medieval

superstition or subjective opinion. Their theory fails to give any guidance about moral information and, by definition, gives little weight to
information that falls outside the constraints of science. Undeniably,
fewer and fewer people are bound in any serious way to Biblical or other
religious traditions as a source of compelling attention and authority, the
result of which is that they make no moral decisions, only practical ones.
This is still another way of defining Technopoly. The term is aptly used
for a culture whose available theories do not offer guidance about what is
acceptable information in the moral domain.

I trust the reader does not conclude that I am making an argument
for fundamentalism of any kind. One can hardly approve, for example,
of a Muslim fundamentalism that decrees a death sentence to someone
who writes what are construed as blasphemous words, or a Christian
fundamentalism that once did the same or could lead to the same. I
must hasten to acknowledge, in this context, that it is entirely possible
to live as a Muslim, a Christian, or a Jew with a modified and temper-
ate view of religious theory. Here, I am merely making the point that
religious tradition serves as a mechanism for the regulation and valua-
tion of information. When religion loses much or all of its binding
power—if it is reduced to mere rhetorical ash—then confusion inevitably follows about what to attend to and how to assign it significance.

Indeed, as I write, another great world narrative, Marxism, is in the
process of decomposing. No doubt there are fundamentalist Marxists
who will not let go of Marx's theory, and will continue to be guided by
its prescriptions and constraints. The theory, after all, is sufficiently
powerful to have engaged the imagination and devotion of more than a
billion people. Like the Bible, the theory includes a transcendent idea,
as do all great world narratives. With apologies to a century and a half
of philosophical and sociological disputation, the idea is as follows: All
forms of institutional misery and oppression are a result of class conflict, since the consciousness of all people is formed by their material
situation. God has no interest in this, because there is no God. But
there *is* a plan, which is both knowable and beneficent. The plan
unfolds in the movement of history itself, which shows unmistakably
that the working class, in the end, must triumph. When it does, with
or without the help of revolutionary movements, class itself will have
disappeared. All will share equally in the bounties of nature and creative production, and no one will exploit the labors of another.

It is generally believed that this theory has fallen into disrepute
among believers because information made available by television,

films, telephone, fax machines, and other technologies has revealed that the working classes of capitalist nations are sharing quite nicely in the bounties of nature while at the same time enjoying a considerable measure of personal freedom. Their situation is so vastly superior to those of nations enacting Marxist theory that millions of people have concluded, seemingly all at once, that history may have no opinion whatever on the fate of the working class or, if it has, that it is moving toward a final chapter quite different in its point from what Marx prophesied.

All of this is said provisionally. History takes a long time, and there may yet be developments that will provide Marx's vision with fresh sources of verisimilitude. Meanwhile, the following points need to be made: Believers in the Marxist story were given quite clear guidelines on how they were to weight information and therefore to understand events. To the extent that they now reject the theory, they are threatened with conceptual confusion, which means they no longer know who to believe or what to believe. In the West, and especially in the United States, there is much rejoicing over this situation, and assurances are given that Marxism can be replaced by what is called "liberal democracy." But this must be stated more as a question than an answer, for it is no longer entirely clear what sort of story liberal democracy tells.

A clear and scholarly celebration of liberal democracy's triumph is found in Francis Fukuyama's essay "The End of History?" Using a somewhat peculiar definition of history, Fukuyama concludes that there will be no more ideological conflicts, all the competitors to modern liberalism having been defeated. In support of his conclusion, Fukuyama cites Hegel as having come to a similar position in the early nineteenth century, when the principles of liberty and equality, as expressed in the American and French revolutions, emerged triumphant. With the contemporary decline of fascism and communism, no threat now remains. But Fukuyama pays insufficient attention to the changes in meaning of liberal democracy over two centuries. Its meaning in a technocracy is quite different from its meaning in Technopoly; indeed, in Technopoly it comes much closer to what Walter Benjamin called "commodity capitalism." In the case of the United States, the great eighteenth-century revolution was not indifferent to commodity capitalism but was nonetheless infused with profound moral content. The United States was not merely an experiment in a new form of governance; it was the fulfillment of God's plan. True, Adams, Jefferson, and Paine rejected the supernatural elements in the Bible, but they never doubted that their experiment had the imprimatur of Providence. People were to be free but for a purpose. Their God-given rights implied obligations and responsibilities, not only to God but to other nations, to which the new republic would be a guide and a showcase of what is possible when reason and spirituality commingle.

It is an open question whether or not "liberal democracy" in its present form can provide a thought-world of sufficient moral substance to sustain meaningful lives. This is precisely the question that Václav Havel, then newly elected as president of Czechoslovakia, posed in an address to the U.S. Congress. "We still don't know how to put morality ahead of politics, science, and economics," he said. "We are still incapable of understanding that the only genuine backbone of our actions—if they are to be moral—is responsibility. Responsibility to something higher than my family, my country, my firm, my success." What Havel is saying is that it is not enough for his nation to liberate itself from one flawed theory; it is necessary to find another, and he worries that Technopoly provides no answer. To say it in still another way: Francis Fukuyama is wrong. There *is* another ideological conflict to be fought—between "liberal democracy" as conceived in the eighteenth century, with all its transcendent moral underpinnings, and Technopoly, a twentieth-century thought-world that functions not only without a transcendent narrative to provide moral underpinnings but also without strong social institutions to control the flood of information produced by technology.

Because that flood has laid waste the theories on which schools, families, political parties, religion, nationhood itself are based, American Technopoly must rely, to an obsessive extent, on technical methods to control the flow of information. Three such means merit special attention. They are interrelated but for purposes of clarity may be described separately.

The first is bureaucracy, which James Beniger in *The Control Revolution* ranks as "foremost among all technological solutions to the crisis of control."[4] Bureaucracy is not, of course, a creation of Technopoly. Its history goes back five thousand years, although the word itself did not appear in English until the nineteenth century. It is not unlikely that the ancient Egyptians found bureaucracy an irritation, but it is certain that, beginning in the nineteenth century, as bureaucracies became more important, the complaints against them became more insistent. John Stuart Mill referred to them as "administrative tyranny." Carlyle called them "the Continental nuisance." In a chilling paragraph, Tocqueville warned about them taking hold in the United States:

> I have previously made the distinction between two types of cen-
> tralization, calling one governmental and the other administrative.
> Only the first exists in America, the second being almost unknown.
> If the directing power in American society had both these means of
> government at its disposal and combined the right to command with
> the faculty and habit to perform everything itself, if having estab-
> lished the general principles of the government, it entered into the
> details of their application, and having regulated the great interests
> of the country, it came down to consider even individual interest,
> then freedom would soon be banished from the New World.[5]

Writing in our own time, C. S. Lewis believed bureaucracy to be the technical embodiment of the Devil himself:

> I live in the Managerial Age, in a world of "Admin." The greatest
> evil is not now done in those sordid "dens of crime" that Dickens
> loved to paint. It is not done even in concentration camps and
> labour camps. In those we see its final result. But it is conceived and
> ordered (moved, seconded, carried, and minuted) in clean, car-
> peted, warmed, and well-lighted offices, by quiet men with white
> collars and cut fingernails and smooth-shaven cheeks who do not
> need to raise their voices. Hence, naturally enough, my symbol for
> Hell is something like the bureaucracy of a police state or the office
> of a thoroughly nasty business concern.[6]

Putting these attacks aside for the moment, we may say that in principle a bureaucracy is simply a coordinated series of techniques for reducing the amount of information that requires processing. Beniger notes, for example, that the invention of the standardized form—a staple of bureaucracy—allows for the "destruction" of every nuance and detail of a situation. By requiring us to check boxes and fill in blanks, the standardized form admits only a limited range of formal, objective, and impersonal information, which in some cases is precisely what is needed to solve a particular problem. Bureaucracy is, as Max Weber described it, an attempt to rationalize the flow of information, to make its use efficient to the highest degree by eliminating informa-tion that diverts attention from the problem at hand. Beniger offers as a prime example of such bureaucratic rationalization the decision in 1884 to organize time, on a worldwide basis, into twenty-four time zones. Prior to this decision, towns only a mile or two apart could and did differ on what time of day it was, which made the operation of railroads and other businesses unnecessarily complex. By simply ignoring the fact that solar time differs at each node of a transportation system, bureaucracy eliminated a problem of information chaos, much to the satisfaction of most people. But not of everyone. It must be noted that the idea of "God's own time" (a phrase used by the novelist Marie Corelli in the early twentieth century to oppose the introduction of Summer Time) had to be considered irrelevant. This is important to say, because, in attempting to make the most rational use of information, bureaucracy ignores all information and ideas that do not contribute to efficiency. The idea of God's time made no such contribution.

Bureaucracy is not in principle a social institution; nor are all institutions that reduce information by excluding some kinds or sources necessarily bureaucracies. Schools may exclude dianetics and astrology; courts exclude hearsay evidence. They do so for substantive reasons having to do with the theories on which these institutions are based. But bureaucracy has no intellectual, political, or moral theory—except for its implicit assumption that efficiency is the principal aim of all social institutions and that other goals are essentially less worthy, if not irrelevant. That is why John Stuart Mill thought bureaucracy a "tyranny" and C. S. Lewis identified it with Hell.

The transformation of bureaucracy from a set of techniques designed to serve social institutions to an autonomous meta-institution that largely serves itself came as a result of several developments in the mid- and late-nineteenth century: rapid industrial growth, improvements in transportation and communication, the extension of government into ever-larger realms of public and business affairs, the increasing centralization of governmental structures. To these were added, in the twentieth century, the information explosion and what we might call the "bureaucracy effect": as techniques for managing information became more necessary, extensive, and complex, the number of people and structures required to manage those techniques grew, and so did the amount of information *generated* by bureaucratic techniques. This created the need for bureaucracies to manage and coordinate bureaucracies, then for additional structures and techniques to manage the bureaucracies that coordinated bureaucracies, and so on—until bureaucracy became, to borrow . . . Karl Kraus's comment on psychoanalysis, the disease for which it purported to be the cure. Along the way, it ceased to be merely a servant of social institutions and became their master. Bureaucracy now not only solves problems but creates them. More important, it defines what our problems are—and they are always, in the bureaucratic view, problems of efficiency. As Lewis suggests, this makes bureaucracies exceedingly dangerous, because, though they were originally designed to process only technical information, they now are commonly employed to address problems of a moral, social, and political nature. The bureaucracy of the nineteenth century was largely concerned with making transportation, industry, and the distribution of goods more efficient. Technopoly's bureaucracy has broken loose from such restrictions and now claims sovereignty over all of society's affairs.

The peril we face in trusting social, moral, and political affairs to bureaucracy may be highlighted by reminding ourselves what a bureaucrat does. As the word's history suggests, a bureaucrat is little else than a glorified counter. The French word *bureau* first meant a cloth for covering a reckoning table, then the table itself, then the room in which the table was kept, and finally the office and staff that ran the entire counting room or house. The word "bureaucrat" has come to mean a person who by training, commitment, and even temperament is indifferent to both the content and the totality of a human problem. The bureaucrat considers the implications of a decision only to the extent that the decision will affect the efficient operations of the bureaucracy, and takes no responsibility for its human consequences. Thus, Adolf Eichmann becomes the basic model and metaphor for a bureaucrat in the age of Technopoly.[7] When faced with the charge of crimes against humanity, he argued that he had no part in the formulation of Nazi political or sociological theory; he dealt only with the technical problems of moving vast numbers of people from one place to another. Why they were being moved and, especially, what would happen to them when they arrived at their destination were not relevant to his job. Although the jobs of bureaucrats in today's Technopoly have results far less horrific, Eichmann's answer is probably given five thousand times a day in America alone: I have no responsibility for the human consequences of my decisions. I am only responsible for the efficiency of my part of the bureaucracy, which must be maintained at all costs.

Eichmann, it must also be noted, was an expert. And expertise is a second important technical means by which Technopoly strives furiously to control information. There have, of course, always been experts, even in tool-using cultures. The pyramids, Roman roads, the

Strasbourg Cathedral, could hardly have been built without experts. But the expert in Technopoly has two characteristics that distinguish him or her from experts of the past. First, Technopoly's experts tend to be ignorant about any matter not directly related to their specialized area. The average psychotherapist, for example, barely has even superficial knowledge of literature, philosophy, social history, art, religion, and biology, and is not expected to have such knowledge. Second, like bureaucracy itself (with which an expert may or may not be connected), Technopoly's experts claim dominion not only over technical matters but also over social, psychological, and moral affairs. In the United States, we have experts in how to raise children, how to educate them, how to be lovable, how to make love, how to influence people, how to make friends. There is no aspect of human relations that has not been technicalized and therefore relegated to the control of experts.

These special characteristics of the expert arose as a result of three factors. First, the growth of bureaucracies, which, in effect, produced the world's first entirely mechanistic specialists and thereby gave credence and prestige to the specialist-as-ignoramus. Second, the weakening of traditional social institutions, which led ordinary people to lose confidence in the value of tradition. Third, and underlying everything else, the torrent of information which made it impossible for anyone to possess more than a tiny fraction of the sum total of human knowledge. As a college undergraduate, I was told by an enthusiastic professor of German literature that Goethe was the last person who knew everything. I assume she meant, by this astounding remark, less to deify Goethe than to suggest that by the year of his death, 1832, it was no longer possible for even the most brilliant mind to comprehend, let alone integrate, what was known.

The role of the expert is to concentrate on one field of knowledge, sift through all that is available, eliminate that which has no bearing on a problem, and use what is left to assist in solving a problem. This process works fairly well in situations where only a technical solution is required and there is no conflict with human purposes—for example, in space rocketry or the construction of a sewer system. It works less well in situations where technical requirements may conflict with human purposes, as in medicine or architecture. And it is disastrous when applied to situations that cannot be solved by technical means and where efficiency is usually irrelevant, such as in education, law, family life, and problems of personal maladjustment. I assume I do not need to convince the reader that there are no experts—there can be no experts—in child-rearing and lovemaking and friend-making. All of this is a figment of the Technopolist's imagination, made plausible by the use of technical machinery, without which the expert would be totally disarmed and exposed as an intruder and an ignoramus.

Technical machinery is essential to both the bureaucrat and the expert, and may be regarded as a third mechanism of information control. I do not have in mind such "hard" technologies as the computer—which must, in any case, be treated separately, since it embodies all that Technopoly stands for. I have in mind "softer" technologies such as IQ tests, SATs, standardized forms, taxonomies, and opinion polls. . . . I mention them here because their role in reducing the types and quantity of information admitted to a system often goes unnoticed, and therefore their role in redefining traditional concepts also goes unnoticed. *There is, for example, no test that can measure a person's intelligence.* Intelligence is a general term used to denote one's capacity to solve real-life problems in a variety of novel contexts. It is acknowledged by everyone except experts that each person varies greatly in such capacities, from consistently effective to consistently ineffective, depending on the kinds of problems requiring solution. If, however, we are made to believe that a test can reveal precisely the quantity of intelligence a person has, then, for all institutional purposes, a score on a test becomes his or her intelligence. The test transforms an abstract and multifaceted meaning into a technical and exact term that leaves out everything of importance. One might even say that an intelligence test is a tale told by an expert, signifying nothing. Nonetheless, the expert relies on our believing in the reality of technical machinery, which means we will reify the answers generated by the machinery. We come to believe that our score *is* our intelligence, or our capacity for creativity or love or pain. We come to believe that the results of opinion polls *are* what people believe, as if our beliefs can be encapsulated in such sentences as "I approve" and "I disapprove."

When Catholic priests use wine, wafers, and incantations to embody spiritual ideas, they acknowledge the mystery and the metaphor being used. But experts of Technopoly acknowledge no such overtones or nuances when they use forms, standardized tests, polls, and other machinery to give technical reality to ideas about intelligence, creativity, sensitivity, emotional imbalance, social deviance, or political opinion.

They would have us believe that technology can plainly reveal the true nature of some human condition or belief because the score, statistic, or taxonomy has given it technical form.

There is no denying that the technicalization of terms and problems is a serious form of information control. Institutions can make decisions on the basis of scores and statistics, and there certainly may be occasions where there is no reasonable alternative. But unless such decisions are made with profound skepticism—that is, acknowledged as being made for administrative convenience—they are delusionary. In Technopoly, the delusion is sanctified by our granting inordinate prestige to experts who are armed with sophisticated technical machinery. Shaw once remarked that all professions are conspiracies against the laity. I would go further: in Technopoly, all experts are invested with the charisma of priestliness. Some of our priest-experts are called psychiatrists, some psychologists, some sociologists, some statisticians. The god they serve does not speak of righteousness or goodness or mercy or grace. Their god speaks of efficiency, precision, objectivity. And that is why such concepts as sin and evil disappear in Technopoly. They come from a moral universe that is irrelevant to the theology of expertise. And so the priests of Technopoly call sin "social deviance," which is a statistical concept, and they call evil "psychopathology," which is a medical concept. Sin and evil disappear because they cannot be measured and objectified, and therefore cannot be dealt with by experts.

As the power of traditional social institutions to organize perceptions and judgment declines, bureaucracies, expertise, and technical machinery become the principal means by which Technopoly hopes to control information and thereby provide itself with intelligibility and order.

NOTES

1. An emphatic exception among those sociologists who have written on this subject is Arnold Gehlen. See his *Man in the Age of Technology* (New York: Columbia University Press, 1980).
2. Though this term is by no means original with E. D. Hirsch, Jr., its current popularity is attributable to Hirsch's book, *Cultural Literacy: What Every American Needs to Know* (Boston: Houghton Mifflin Co., 1987).
3. This poignant phrase is also the title of one of Lasch's most important books.
4. James Beniger, *The Control Revolution: Technological and Economic Origins of the Information Society* (Cambridge, Mass., and London: Harvard University

Press, 1986), p. 13. Beniger's book is the best source for an understanding of the technical means of eliminating—i.e., controlling—information.

5. A. de Tocqueville, *Democracy in America* (New York: Anchor Books [Doubleday & Co., Inc.], 1969), p. 262.

6. C. S. Lewis, *The Screwtape Letters* (New York: Macmillan, 1943), p. x.

7. See H. Arendt, *Eichmann in Jerusalem: A Report on the Banality of Evil* (New York: Penguin Books, 1977).

4.   The Technological Torrent

THOMAS P. HUGHES

For some years, researchers in the sociology of science have pursued the notion of "social construction" of science, in which it is argued that scientific knowledge does not exist in some absolute sense, but is created within a community of scientists. According to this approach one can understand the development of that knowledge by examining the structure of the community and the relations among the scientists. More recently, scholars of technology and society have adapted the approach to the study of technological development, and the idea of "social construction of technology" has received increasing attention. Social constructivists contend that the nature of society and of the individuals who develop technology shapes the characteristics of technological systems just as the technologies they create influence the shape of society.

This is a provocative idea and one that offers the prospect of transcending simplistic notions of technology as an autonomous force running amok in society. Unfortunately for students of technology and society, much of the literature on social construction of technology has been written by sociologists for other sociologists; it is rather arcane and jargon-laden, and is accessible mainly by those with a deep and abiding interest in the subject. A striking exception to this is Thomas Hughes's superb history of U.S. technology, American Genesis. In this book, for which the following selection serves as the introduction, Hughes examines how inventors, entrepreneurs, and technological enthusiasts shaped the character of American society during the period from 1870 to 1970. This is not a tale of technology imposed on a society from outside, but of dynamic interaction among individuals, artifacts, and social institutions. Hughes's argument, to use his own words, is that "inventors, industrial *scientists, engineers, and system builders have been the makers of modern America. The values of order, system, and control that they embedded in machines, devices, processes, and systems have become the values of modern technological culture."*

Thomas P. Hughes is Mellon Professor of History and Sociology of Science at the University of Pennsylvania and holds the Torsten Althin Chair at the Royal Institute of Technology in Stockholm, Sweden. His best-known work has been on the development of large-scale electric power systems, and he has twice received the Dexter Prize for the outstanding book in the history of technology.

This [chapter] is about an era of technological enthusiasm in the United States, an era now passing into history. Literary critic and historian Perry Miller provides a marvelous image of Americans exhilarated by the thrill of the technological transformation. They "flung themselves into the technological torrent, how they shouted with glee in the midst of the cataract, and cried to each other as they went headlong down the chute that here was their destiny. . . . "[1] By 1900 they had reached the promised land of the technological world, the world as artifact. In so doing they had acquired traits that have become characteristically American. A nation of machine makers and system builders, they became imbued with a drive for order, system, and control.

Most Americans, however, still see themselves primarily as a democratic people dedicated to the doctrine of free enterprise. They celebrate the founding fathers and argue that the business of America is business. They celebrate technological achievements, too, but they see these as fruits of free enterprise and democratic politics. They commonly assume that Americans are primarily dedicated to money making and business dealing. Americans rarely think of themselves principally as builders, a people whose most notable and character-forming achievement for almost three centuries has been to transform a wilderness into a building site. A major reason that a nation of builders does not know itself is that most of the history it reads and hears instructs otherwise.

Perceptive foreigners are not so prone to sentimentalize America's founding fathers, frontiersmen, and business moguls. Other peoples have looked to the United States as the land of Thomas Edison, Henry Ford, the Tennessee Valley Authority, and the Manhattan Project. Foreigners have made the second discovery of America, not nature's nation but technology's nation. Foreigners have come to Philadelphia to see Independence Hall, but those who wish to understand the foundations of U.S. power have asked to see Pittsburgh when it was the steel capital of the world, Detroit when most automobiles were made there, the Tennessee Valley Authority when engineering was transforming a poverty-stricken valley into a thriving one, and New York City because its skyscrapers symbolized the technological power of the nation. The Manhattan Project, which produced the atom bomb, reinforced the belief throughout the world that America was the technological giant. Until the space-shuttle disasters and an embarrassing series of launching failures, the National Aeronautics and Space Administration symbolized America's technological creativity.

Americans rightly admire the founding fathers, who displayed extraordinary inventiveness as they conceived the Declaration of Independence and framed the Constitution, but Americans have embodied comparable, if not greater, inventiveness in the material constitution, the technological systems of the nation. Perhaps the myth that they are essentially a political and business people may be emended, if they reflect more on the technological enthusiasm and activity they have displayed throughout their history, but most obviously during the century from about 1870 to 1970. The enthusiasm reached its height during the middle decades of the period, then subsided, especially after World War II. [The book from which this chapter is drawn], despite its emphasis on invention, development, and technological-system building, is not a history of technology, a work of specialization outside the mainstream of American history. To the contrary, it is mainstream American history, an exploration of the American nation involved in its most characteristic activity. Historians looking back a century from now on the sweep of American history may well decide that the century of technological enthusiasm was the most characteristic and impressively achieving century in the nation's history, an era comparable to the Renaissance in Italian history, the era of Louis XIV in France, or the Victorian period in British history. During the century after 1870, Americans created the modern technological nation; this was the American genesis.[2]

In popular accounts of technology, inventions of the late nineteenth century, such as the incandescent light, the radio, the airplane, and the gasoline-driven automobile, occupy center stage, but these inventions were embedded within technological systems. Such systems involve far more than the so-called hardware, devices, machines and processes, and the transportation, communication, and information networks that interconnect them. Such systems consist also of people and organizations. An electric light-and-power system, for instance, may involve generators, motors, transmission lines, utility companies, manufacturing enterprises, and banks. Even a regulatory body may be co-opted into the system. During the era of technological enthusiasm, the characteristic endeavor was inventing, developing, and organizing large technological systems—production, communication, and military.

The development of massive systems for producing and using automobiles and for generating and utilizing electric power, the making of telephone and wireless networks, and the organization of complex systems for making war reveal the creative drive of inventors, engineers, industrial scientists, managers, and entrepreneurs possessed of the system builder's instincts and mentality. The remarkably prolific inventors of the late nineteenth century, such as Edison, persuaded us that we were involved in a second creation of the world. The system builders, like Ford, led us to believe that we could rationally organize the second creation to serve our ends. Only after World War II did a handful of philosophers and publicists whom we now associate with a counterculture raise doubts about the rationality and controllability of a nation organized into massive military, production, and communication systems. Their doubts increased as the nation's technological preeminence waned.

If the nation, then, has been essentially a technological one characterized by a creative spirit manifesting itself in the building of a human-made world patterned by machines, megamachines, and systems, Americans need to fathom the depths of the technological society, to identify currents running more deeply than those conventionally associated with politics and economics. Indeed, many of the forces that Americans need to understand and control in order to shape their destiny, insofar as that is possible, are now not primarily natural or political but technological. We celebrate Charles Darwin for discerning patterns in the natural world; we do not yet sufficiently appreciate the importance of finding patterns in the human-made, or technological, world.[3] The purpose of the understanding is not simply to comprehend the impressively ordered, systematized, and controlled, but to exercise the civic responsibility of shaping those forces that in turn shape our lives so intimately, deeply, and lastingly.

A history stressing the technology of an era of technological enthusiasm should be no more celebratory than a history stressing the politics and business of a gilded age. The tendency of popular histories and of museum exhibits of technology uncritically to unfold a story of problem-free achievement unfortunately leaves readers and viewers naive about the nature of technological change. When more histories of technology that take the critical stance of the best histories of politics are written, Americans will realize that not only their remarkable achievements but many of their deep and persistent problems arise, in the name of order, system, and control, from the mechanization and systematization of life and from the sacrifice of the organic and the spontaneous.

This history, then, argues that inventors, industrial scientists, engineers, and system builders have been the makers of modern America. The values of order, system, and control that they embedded in machines, devices, processes, and systems have become the values of modern technological culture. These values are embedded in the artifacts, or hardware. Modern inventors, engineers, industrial scientists, and system builders, those who flourished in the century of technological enthusiasm, concerned themselves with the production of goods and services and with preparations for, and the waging of, war. Their influence, however, did not end with these activities. Their numerous and enthusiastic supporters from many levels of society believed their methods and values applicable and beneficial when applied to such other realms of social activity as politics, business, architecture, and art.

This history, however, does not argue technological determinism. The creators of modern technology and the makers of the modern world expressed long-held human values and aspirations. Although the inventors, engineers, industrial scientists, and system builders created order, control, and system, in so doing they responded to a fundamental human longing for a world in which these characteristics prevail. They became the instruments of all those, including themselves, who were uneasy in a seemingly chaotic and purposeless world and who searched for compensatory order. In this sense, technology was, and is, socially constructed. As historian and social critic Lewis Mumford so eloquently insisted decades ago, technology is both a shaper of, and is shaped by, values.[4] It is value-laden.

Despite the drive for order, system, and control among the practitioners and enthusiasts of technology, the history of technology, like the history of politics, is complex and contradictory. Framers of constitutions have also tried to establish timeless, all-embracing systems of checks and balances. Neither they nor the designers of machines, devices, and processes have found the one best solution that pleases everyone and resists change. Contrary to popular myth, technology does not result from a series of searches for the "one best solution" to a problem. [American Genesis] does not present technology as engineers are taught even today to think about it: as an absolutely one-best-way solution to problems. Instead, it presents practitioners of technology confronting insolvable issues, making mistakes, and causing controversies and failures. It shows the practitioners creating new problems as they solve old ones. [It] intends to present the history of modern technology and society in all its vital, messy complexity.

Technology in the age of technological enthusiasm meant then, as now, different things to different people. The efforts of textbook writers notwithstanding, technology can be defined no more easily than politics. Rarely do we ask for a definition of politics. To ask for *the* definition of technology is to be equally innocent of complex reality. For many people, technology is goods and services to be consumed by the affluent, to be longed for by the poor. Others, such as inventors and engineers, see technology as the creation of the means of production for these goods and services. Further up the ladder of power and control, the great system builders, people like Ford, find consumingly interesting the organizing of the material world into great systems of production. Still others analyzing modern technology find rational method, efficiency, order, control, and system to be its essence. Taking into consideration the infinite aspects of technology, the best that I can do is to fall back on a general definition that covers much of the activity described in this book. Technology is the *effort* to organize the world for problem solving so that goods and services can be invented, developed, produced, and used.[5] The reader, however, can accept instead of a definition the historian's traditional approach of naming a subject and defining it by examples of his or her choice.

[American Genesis] centers less on ideas and more on people, especially American inventors, engineers, system builders, architects, artists, and social critics. The organizations and movements of a modern culture, the institutional frameworks and symbolic structures in which inventors, system builders, and others acted are not, however, neglected. Among the organizations considered are the inventor's workshop, the industrial research laboratory, the business corporation, the government agency, and the military-industrial complex. Among the movements included are the international style in architecture; the

Futurists, Constructivists, Dadaists, and Precisionists in art; scientific management and progressivism in production and politics; and the conservationist and counterculture advocates among the social critics. Throughout, references to modern culture refer to the devices, machines, processes, values, organizations, symbols, and forms expressing the order, system, and control of modern technology, and to the thought and behavior mediated by these and their expression.[6]

There is a pattern to [American Genesis] analogous to that of the growth of the large technological systems about which it is written. The early chapters treat the invention of systems; the middle section deals with the spread of large systems; and the final chapters recount the emergence of a technological culture, of mammoth government systems, and counterculture reaction to systems. The remarkable achievements of independent inventors and industrial research opened and greatly shaped the age of technological enthusiasm. Philosopher Alfred North Whitehead believed that the invention of a method of invention was the greatest invention of the era.[7] Men and women assumed, as never before, that they had the power to create a world of their own design. Independent inventors experienced their heyday during a gilded era after the United States had emerged from the Civil War, and they forged a massive productive enterprise that ended up dominated by giant corporations. The historians Charles and Mary Beard called this the era of "the Second American Revolution," referring to the momentous technological, economic, political, and social changes.[8] Mumford saw it as the beginning of the modern, or neo-technic, era in the history of technology and society.[9] The inventions of the independents provided the foundations for the rise of the industrial giants, especially the newly burgeoning electrical industry. Edison, the Wizard of Menlo Park, became the heroic figure of the era, but there were other independent inventors, such as Elmer Sperry, who were impressively creative and more professional. The inventors continued to flourish as their country competed successfully with the great European powers for industrial supremacy. As World War I approached, the inventors became involved in inventing for the military. The military establishment funded their inventive activity and used their creations to develop new weapons, strategy, and tactics.

By the beginning of World War I, American inventors had helped to establish the United States as the most inventive of all nations. Only Germany, recently united, seemed a competitor for the title. Inspired by German achievements, leading American corporations such as General Electric, Du Pont, General Motors, and Bell Telephone also established industrial research laboratories. Industrial scientists widely criticized the haphazard methods of the independent inventors and claimed that the mantle of creativity had fallen onto their own shoulders. Yet there flowed from the industrial laboratories inventions with a conservative cast—improvements rather than dramatic innovations. During World War I in the United States, the scientists, especially those with graduate training in physics, effectively challenged the role of the independent inventors as the source of improvements in military systems. The war-waging nations, dependent on their inventors and scientists, innovated and counterinnovated with the submarine, airplane, tank, and poison gas much as large corporations in peacetime contended for market advantages with innovations. Technology was capable of creating not only a new life-supporting world, but also a deadly environment.

The inventions and discoveries of the inventors and the industrial scientists became part of large systems of production that expanded impressively during the interwar years. These systems were the work of the system builders, whose creative drive surpassed in scope and magnitude that of the inventors. Designing a machine or a power-and-light system that functioned in an orderly, controllable, and predictable way delighted Edison the inventor; designing a technological system made up of machines, chemical and metallurgical processes, mines, manufacturing plants, railway lines, and sales organizations to function rationally and efficiently exhilarated Ford the system builder. The achievements of the system builders help us understand why their contemporaries believed not only that they could create a new world, but that they also knew how to order and control it. Frederick W. Taylor, father of scientific management, became famous, or notorious, throughout the industrial world for his techniques of order and control.

American technology, especially its systems of production, fascinated European industrial managers, bureaucrats, social scientists, and social critics. Fordism and Taylorism for them symbolized the essence of the modern American achievement. Fordism and Taylorism spread throughout Europe, much as Japanese managerial techniques would into the United States after World War II. Lenin and other leaders of the Soviet Union displayed even greater enthusiasm for Fordism and Taylorism than the Americans had. When the Soviet Union embarked on a Five-Year Plan that specified mammoth regional systems of technology based on hydroelectric power and prodigiously rich stores of

Siberian natural resources, it turned to American consulting engineers and industrial corporations for advice and equipment. The Soviets constructed entire industrial systems modeled on the steel works in Gary, Indiana, and hydroelectric projects on the Mississippi. In Weimar Germany after World War I, many persons believed that Taylor and Ford had the answer not only to production problems but to labor and social unrest as well. They labeled Ford's ideas white socialism, believing this to be an answer to Marxism. Many Europeans, especially Weimar Germans, decided that democracy, American technology, and a new European and modern culture could restore war-devastated Europe and create a good society. In the Soviet Union, Lenin predicted that Soviet politics, Prussian railway management, American technology, and the organizational forms of the trust-building entrepreneurs would bring the new socialist society.[10]

Modern technology was made in America. Even the Germans who developed it so well acknowledged the United States as the prime source. During the interwar years, the industrial world recognized the United States as the preeminent technological nation, and the era of technological enthusiasm reached its apogee. Modern technological culture, however, was defined in Europe. The Europeans held up a mirror in which the Americans could see themselves as the raw materials of modernity which the Europeans wanted to fashion into modern culture. European engineers, industrialists, artists, and architects came to America to admire its "plumbing and its bridges"[11] and made, as we have observed, the second discovery of America—the great systems of production.

From the turn of the century on, avant-garde European architects and industrial designers searched for ways to combine American modes of mass production and the principles of quality design. In so doing they were inventing the forms and symbols for a modern technological culture. In the 1920s, at the Bauhaus in Dessau, Walter Gropius and his architect and artist associates brought the movement to a climax by contributing greatly to the establishment of the modern or international style of architecture and design. This style expressed in construction methods and in formal design the principles of modern American technology. The dire housing shortage following World War I spurred Gropius and other avant-garde architects to apply the mass-production methods attributed to Ford and the scientific management methods of Taylor. A description of the construction of great housing settlements in Dessau and Berlin makes this clear. In France, Le Corbusier fervently and eloquently articulated the technological age. In his journal *L'Esprit nouveau*, published in the 1920s, he sought to define verbally and visually the modern in art, architecture, and interior and industrial design. He believed that American engineers had found the heart of modern design by joining, in their bridges, ocean steamers, grain silos, and automobiles, a mathematical exactness with rational methods of production and rational design. The architects who adopted the engineers' techniques and infused them with the aesthetics of the artist were, he was convinced, creating the modern style. Le Corbusier was more enamored of order and system than the engineers themselves.

Painters, too, became self-consciously modern. The Italian Futurists, around the turn of the century, saw modern technology as a way of destroying traditional culture in Italy. Social and artistic radicals, they found Italy backward and oppressive. Modern motor cars, not Renaissance museums, held the key to the future for Italians. The Futurists celebrated the dramatic and dynamic artifacts of modern technology—"adventurous steamers that sniff the horizon . . . deep-chested locomotives whose wheels paw the tracks . . . the sleek flight of planes. . . ."[12] After the Russian Revolution of 1917, the Soviet artists of the Constructivist movement, several of whom were graduate engineers, also envisioned art as a means of radically transforming culture, of bringing into being the new Soviet society. Vladimir Tatlin conceived of "machine art" and El Lissitzky of new elements of style from which a modern art and architecture could be created that would influence the character of the new man in the modern social system. In Germany after the war, the artifacts and order of the technological world fascinated the artists of the Neue Sachlichkeit school. Their visual vocabulary included "order," "clarity," and "harmony." They thought these to be the principles of technological rationality and the governing principles of the human-made world.

In 1915 Marcel Duchamp and Francis Picabia came to New York and emboldened a few American artists to look to technological America rather than Europe for the subject matter, forms, and symbols of the modern. The American Precisionists Charles Sheeler and Charles Demuth, and the Russian American Louis Lozowick painted technological landscapes and objects inspired by the development of modern systems of production. Their work was exemplified by Sheeler's series of paintings and photographs of Ford's River Rouge plant.

Leading American architects did not adopt a formal vocabulary characterized by a technological or machine aesthetic until the 1930s, when

Gropius, Ludwig Mies van der Rohe, and other avant-garde architects, emigrating from Nazi Germany to the United States, brought with them the international style. The paradox remains that, although modern technology originated in America, modern painting and architecture inspired by it germinated and took root first in Europe.

The Great Depression and the violence and destruction made possible by modern technology during World War II dampened technological enthusiasm, but technological systems entered a new stage in the United States when the government became involved in their cultivation. Franklin Delano Roosevelt inaugurated the Tennessee Valley Authority, a government-funded, -designed, -constructed, and -operated project that systematically developed the resources of an extensive river valley. Once again the United States provided the world a model of modern technology. During World War II, the United States poured unprecedented resources into the Manhattan Project, a technological system of unprecedented size. When President Dwight Eisenhower later warned his nation about the increasing momentum of the military-industrial complex, he referred to the rise of great systems of armament production modeled on the Manhattan Project. The Strategic Defense Initiative, or Starwars, exemplifies the most recent military-industrial (and university) complex.

The dropping of the bombs on Hiroshima and Nagasaki starkly revealed for many the threat of uncontrolled, destructive, technological creativity and the massive size of technological projects and systems in which the government was involved. Subsequent and largely unsuccessful efforts to bring about control of the nuclear arsenal heightened these anxieties. Rachel Carson in *Silent Spring* and others who followed her lead stimulated an increased concern about environmental costs of large-scale production technology. The wasting of Vietnam by military technology brought the growing reaction to a head. A counterculture erupted. Reflective radicals of the 1960s, both in America and abroad, attacked modern technology and the order, system, and control associated with it. The counterculture called for the organic instead of the mechanical; small and beautiful technology, not centralized systems; spontaneity instead of order; and compassion, not efficiency. Paul Goodman, Herbert Marcuse, and other intellectual leaders of the counterculture unerringly aimed their attacks at technological rationality and system. Mumford, whose critical concern about technology and society antedated that of the counterculture, also wrote of megamachines. Jacques Ellul also criticized the technological systems that he and Mumford feared were determining the course of history.

Time has dampened the bitterness and vision of the counterculture. Today technological enthusiasm, although much muted as compared with the 1920s, survives among engineers, managers, system builders, and others with vested interests in technological systems. The systems spawned by that enthusiasm, however, have acquired a momentum—almost a life—of their own. They involve the surviving technological enthusiasts, persons whose income derives from the systems, large corporations, government agencies, and politicians beholden to those with vested interests in the systems. A multitude of persons persuaded that armaments and the producers of them are critical for the nation's defense and survival adds to the momentum of military-industrial systems. The age of technological enthusiasm has passed, but it has left behind a burden of history. Those who know the history and the burden may be able to rid themselves of it or turn it to their ends.[13]

NOTES

1. Perry Miller, "The Responsibility of Mind in a Civilization of Machines," *The American Scholar,* XXXI (Winter 1961–1962), 51–69.
2. I am indebted to Elaine Scarry of the University of Pennsylvania for the notion of the invention of the nation's "material constitution" and to Jaroslav Pelikan of Yale University for reminding me of the Genesis story as "Urmythus," of God as the first to practice technology (Genesis III: 21), and of Cain bearing the curse of work and creation (Genesis III: 23).
3. Thomas P. Hughes, "The Order of the Technological World," *History of Technology,* V (1980), 1–16.
4. Casey Blake, "Lewis Mumford: Values over Technique," *Democracy* (Spring 1983), pp. 125–37.
5. Martin Heidegger, *The Question Concerning Technology and Other Essays,* trans. W. Lovitt (New York: Harper & Row, 1977), p. 19. In *American Genesis* I have concentrated on the means of production, especially the mechanical and electrical. Also of surpassing importance during the age of technological enthusiasm were the great works of civil engineering. See, for instance, David McCullough, *The Great Bridge* (New York: Simon & Schuster, 1972), and McCullough, *The Path Between the Seas: The Creation of the Panama Canal, 1870–1914* (New York: Simon & Schuster, 1977).
6. Sidney W. Mintz, "Culture: An Anthropological View," *The Yale Review,* 71 (1982), 499–512.
7. Alfred North Whitehead, *Science and the Modern World* (London: Free Association Books, 1985; first ed. 1926), p. 120.

8. Charles A. Beard and Mary R. Beard, *The Rise of American Civilization* (New York: Macmillan, 1930), pp. 52–121.
9. Lewis Mumford, *Technics and Civilization* (New York: Harcourt, Brace, 1934), pp. 215–21. More recently Louis Galambos has described "The Emerging Organizational Synthesis in Modern American History" (*Business History Review*, XLIV [1970], 279–90), and discerned three major characteristics of modern America as "Technology, Political Economy, and Professionalization: Central Themes of the Organizational Synthesis" (*Business History Review*, LVII [1983], 471–93).
10. "Soviet power + Prussian railroad administration + American technology and monopolistic industrial organization . . . = Socialism," *Leninskij Sbornik*, XXXVI: 37, cited in Eckhart Gillen, "Die Sachlichkeit der Revolutionäre," in *Wem gehört die Welt: Kunst und Gesellschaft in der Weimarer Republik* (Berlin: Neue Gesellschaft für Bildende Kunst, 1977), p. 214.
11. Marcel Duchamp quoted in Stanislaus von Moos, "Die Zweite Entdeckung Amerikas," afterword to Sigfried Giedion, *Die Herrschaft der Mechanisierung* (Frankfurt am Main: Europäische Verlagsanstalt, 1982), p. 807.
12. F. T. Marinetti, "The Founding and Manifesto of Futurism 1909," in *Futurist Manifestos*, ed. and intro. by Umbro Apollonio (London: Thames and Hudson, 1973), p. 22.
13. C. Vann Woodward, *The Burden of Southern History* (Baton Rouge: Louisiana State University Press, 1968), pp. 187–211.

5. Can Technology Replace Social Engineering?

ALVIN M. WEINBERG

Since Alvin Weinberg's essay "Can Technology Replace Social Engineering?" was first published in the mid-1960s, it has become something of a classic in the literature of technology and society. Indeed, the term technological fix, *introduced here, has become part of the lexicon of the field. Weinberg, one of the pioneers of large-scale atomic energy R&D and an inveterate technological optimist, argues that technology is capable of finding shortcuts (technological fixes) to the solution of social problems. For example, faced with a shortage of fresh water, he suggests, society can try either social engineering—altering life-styles and the ways people use water—or a technological fix, such as the provision of additional fresh water through nuclear-powered desalting of sea water. Although some aspects of this selection may seem almost quaint and out of tune with contemporary views of technology and politics, the questions it raises are as relevant today as they were thirty years ago.*

Alvin M. Weinberg is a physicist who joined the World War II Manhattan Project early in his career. He went to Oak Ridge National Laboratory in 1945 and served as its director from 1955 through 1973. He is currently on the staff of Oak Ridge Associated Universities. Weinberg was a member of the President's Science Advisory Committee in 1960–1962 and is the recipient of many awards, including the President's Medal of Science and the Enrico Fermi Award. He was born in Chicago in 1915 and holds A.B., A.M., and Ph.D. degrees from the University of Chicago.

Source: *University of Chicago Magazine,* LIX (October 1966), pp. 6–10. Reprinted by permission of the author.

During World War II, and immediately afterward, our federal government mobilized its scientific and technical resources, such as the Oak Ridge National Laboratory, around great technological problems. Nuclear reactors, nuclear weapons, radar, and space are some of the miraculous new technologies that have been created by this mobilization of federal effort. In the past few years there has been a major change in focus of much of our federal research. Instead of being preoccupied with technology, our government is now mobilizing around problems that are largely social. We are beginning to ask what can we do about world population, about the deterioration of our environment, about our educational system, our decaying cities, race relations, poverty. Recent administrations have dedicated the power of a scientifically oriented federal apparatus to finding solutions for these complex social problems.

Social problems are much more complex than are technological problems. It is much harder to identify a social problem than a technological problem: how do we know when our cities need renewing, or when our population is too big, or when our modes of transportation have broken down? The problems are, in a way, harder to identify just because their solutions are never clear-cut: how do we know when our cities are renewed, or our air clean enough, or our transportation convenient enough? By contrast, the availability of a crisp and beautiful technological *solution* often helps focus on the problem to which the new technology is the solution. I doubt that we would have been nearly as concerned with an eventual shortage of energy as we now are if we had not had a neat solution—nuclear energy—available to eliminate the shortage.

There is a more basic sense in which social problems are much more difficult than are technological problems. A social problem exists because many people behave, individually, in a socially unacceptable way. To solve a social problem one must induce social change—one must persuade many people to behave differently than they have behaved in the past. One must persuade many people to have fewer babies, or to drive more carefully, or to refrain from disliking blacks. By contrast, resolution of a technological problem involves many fewer individual decisions. Once President Roosevelt decided to go after atomic energy, it was by comparison a relatively simple task to mobilize the Manhattan Project.

The resolution of social problems by the traditional methods—by motivating or forcing people to behave more rationally—is a frustrating business. People don't behave rationally; it is a long, hard business to persuade individuals to forgo immediate personal gain or pleasure (as seen by the individual) in favor of longer term social gain. And indeed, the aim of social engineering is to invent the social devices—usually legal, but also moral and educational and organizational—that will change each person's motivation and redirect his activities along ways that are more acceptable to the society.

The technologist is appalled by the difficulties faced by the social engineer; to engineer even a small social change by inducing individuals to behave differently is always hard even when the change is rather neutral or even beneficial. For example, some rice eaters in India are reported to prefer starvation to eating wheat which we send to them. How much harder it is to change motivations where the individual is insecure and feels threatened if he acts differently, as illustrated by the poor white's reluctance to accept the black as an equal. By contrast, technological engineering is simple: the rocket, the reactor, and the desalination plants are devices that are expensive to develop, to be sure, but their feasibility is relatively easy to assess, and their success relatively easy to achieve once one understands the scientific principles that underlie them. It is, therefore, tempting to raise the following question: In view of the simplicity of technological engineering, and the complexity of social engineering, to what extent can social problems be circumvented by reducing them to technological problems? Can we identify Quick Technological Fixes for profound and almost infinitely complicated social problems, "fixes" that are within the grasp of modern technology, and which would either eliminate the original social problem without requiring a change in the individual's social attitudes, or would so alter the problem as to make its resolution more feasible? To paraphrase Ralph Nader, to what extent can technological *remedies* be found for social problems without first having to remove the *causes* of the problem? It is in this sense that I ask, "Can technology replace social engineering?"

THE MAJOR TECHNOLOGICAL FIXES OF THE PAST

To better explain what I have in mind, I shall describe how two of our profoundest social problems—poverty and war—have in some limited degree been solved by the Technological Fix, rather than by the methods of social engineering. Let me begin with poverty.

The traditional Marxian view of poverty regarded our economic ills as being primarily a question of maldistribution of goods. The Marxist recipe for elimination of poverty, therefore, was to eliminate profit, in the erroneous belief that it was the loss of this relatively small increment from the worker's paycheck that kept him poverty-stricken. The Marxist dogma is typical of the approach of the social engineer: one tries to convince or coerce many people to forgo their short-term profits in what is presumed to be the long-term interest of the society as a whole.

The Marxian view seems archaic in this age of mass production and automation not only to us, but apparently to many Eastern bloc economists. For the brilliant advances in the technology of energy, of mass production, and of automation have created the affluent society. Technology has expanded our productive capacity so greatly that even though our distribution is still inefficient, and unfair by Marxian precepts, there is more than enough to go around. Technology has provided a "fix"—greatly expanded production of goods—which enables our capitalistic society to achieve many of the aims of the Marxist social engineer without going through the social revolution Marx viewed as inevitable. Technology has converted the seemingly intractable social problem of *widespread* poverty into a relatively tractable one.

My second example is war. The traditional Christian position views war as primarily a moral issue: if men become good, and model themselves after the Prince of Peace, they will live in peace. This doctrine is so deeply ingrained in the spirit of all civilized men that I suppose it is a blasphemy to point out that it has never worked very well—that men have not been good, and that they are not paragons of virtue or even of reasonableness.

Though I realize it is terribly presumptuous to claim, I believe that Edward Teller may have supplied the nearest thing to a Quick Technological Fix to the problem of war. The hydrogen bomb greatly increases the provocation that would precipitate large-scale war—and not because men's motivations have been changed, not because men have become more tolerant and understanding, but rather because the appeal to the primitive instinct of self-preservation has been intensified far beyond anything we could have imagined before the H-bomb was invented. To point out these things today, with the United States involved in a shooting war, may sound hollow and unconvincing; yet the desperate and partial peace we have now is much better than a full-fledged exchange of thermonuclear weapons. One cannot deny that the Soviet leaders now recognize the force of H-bombs, and that this has surely contributed to the less militant attitude of the USSR. One can only hope that the Chinese leadership, as it acquires familiarity with H-bombs, will also become less militant. If I were to be asked who has given the world a more effective means of achieving peace, our great religious leaders who urge men to love their neighbors and, thus, avoid fights, or our weapons technologists who simply present men with no rational alternative to peace, I would vote for the weapons technologist. That the peace we get is at best terribly fragile, I cannot deny; yet, as I shall explain, I think technology can help stabilize our imperfect and precarious peace.

THE TECHNOLOGICAL FIXES OF THE FUTURE

Are there other Technological Fixes on the horizon, other technologies that can reduce immensely complicated social questions to a matter of "engineering"? Are there new technologies that offer society ways of circumventing social problems and at the same time do *not* require individuals to renounce short-term advantage for long-term gain?

Probably the most important new Technological Fix is the Intra-Uterine Device for birth control. Before the IUD was invented, birth control demanded very strong motivation of countless individuals. Even with the pill, the individual's motivation had to be sustained day in and day out; should it flag even temporarily, the strong motivation of the previous month might go for naught. But the IUD, being a one-shot method, greatly reduces the individual motivation required to induce a social change. To be sure, the mother must be sufficiently motivated to accept the IUD in the first place, but, as experience in India already seems to show, it is much easier to persuade the Indian mother to accept the IUD once, than it is to persuade her to take a pill every day. The IUD does not completely replace social engineering by technology; and indeed, in some Spanish American cultures where the husband's manliness is measured by the number of children he has, the IUD attacks only part of the problem. Yet, in many other situations, as in India, the IUD so reduces the social component of the problem as to make an impossibly difficult social problem much less hopeless.

Let me turn now to problems which from the beginning have had both technical and social components—broadly, those concerned with conservation of our resources: our environment, our water, and our raw materials for production of the means of subsistence. The social issue here arises because many people by their individual acts cause shortages and, thus, create economic, and ultimately social, imbalance. For example, people use water wastefully, or they insist on moving to California because of its climate, and so we have water shortages; or too many people drive cars in Los Angeles with its curious meteorology, and so Los Angeles suffocates from smog.

The water resources problem is a particularly good example of a complicated problem with strong social and technological connotations. Our management of water resources in the past has been based largely on the ancient Roman device, the aqueduct: every water shortage was to be relieved by stealing water from someone else who at the moment didn't need the water or was too poor or too weak to prevent the steal. Southern California would steal from Northern California, New York City from upstate New York, the farmer who could afford a cloud-seeder from the farmer who could not afford a cloud-seeder. The social engineer insists that such shortsighted expedients have got us into serious trouble; we have no water resources policy, we waste water disgracefully, and, perhaps, in denying the ethic of thriftiness in using water, we have generally undermined our moral fiber. The social engineer, therefore, views such technological shenanigans as being shortsighted, if not downright immoral. Instead, he says, we should persuade or force people to use less water, or to stay in the cold Middle West where water is plentiful instead of migrating to California where water is scarce.

The water technologist, on the other hand, views the social engineer's approach as rather impractical. To persuade people to use less water, to get along with expensive water, is difficult, time-consuming, and uncertain in the extreme. Moreover, say the technologists, what right does the water resources expert have to insist that people use water less wastefully? Green lawns and clean cars and swimming pools are part of the good life, American style, . . . and what right do we have to deny this luxury if there is some alternative to cutting down the water we use?

Here we have a sharp confrontation of the two ways of dealing with a complex social issue: the social engineering way which asks people to behave more "reasonably," and the technologists' way which tries to avoid changing people's habits or motivation. Even though I am a technologist, I have sympathy for the social engineer. I think we must use our water as efficiently as possible, that we ought to improve people's attitudes toward the use of water, and that everything that can be done to rationalize our water policy will be welcome. Yet as a technologist, I believe I see ways of providing more water more cheaply than the social engineers may concede is possible.

I refer to the possibility of nuclear desalination. The social engineer dismisses the technologist's simpleminded idea of solving a water shortage by transporting more water primarily because, in so doing, the water user steals water from someone else—possibly foreclosing the possibility of ultimately utilizing land now only sparsely settled. But surely water drawn from the sea deprives no one of his share of water. The whole issue is then a technological one; can fresh water be drawn from the sea cheaply enough to have a major impact on our chronically water-short areas like Southern California, Arizona, and the Eastern seaboard?

I believe the answer is yes, though much hard technical work remains to be done. A large program to develop cheap methods of nuclear desalting has been undertaken by the United States, and I have little doubt that within the next ten to twenty years we shall see huge dual-purpose desalting plants springing up on many parched seacoasts of the world.* At first these plants will produce water at municipal prices. But I believe, on the basis of research now in progress at ORNL [Oak Ridge National Laboratory] and elsewhere, water from the sea at a cost acceptable for agriculture—less than ten cents per 1,000 gallons—is eventually in the cards. In short, for areas close to the seacoasts, technology can provide water without requiring a great and difficult-to-accomplish change in people's attitudes toward the utilization of water.

The Technological Fix for water is based on the availability of extremely cheap energy from very large nuclear reactors. What other social consequences can one foresee flowing from really cheap energy eventually available to every country regardless of its endowment of conventional resources? Though we now see only vaguely the outlines of the possibilities, it does seem likely that from very cheap nuclear energy we shall get hydrogen by electrolysis of water, and, thence, the all important ammonia fertilizer necessary to help feed the hungry of the world; we shall reduce metals without requiring coking coal; we shall even power automobiles with electricity, via fuel cells or storage

*Here, as elsewhere, the reader should bear in mind that the essay dates from the mid-1960s.—Ed.

batteries, thus reducing our world's dependence on crude oil, as well as eliminating our air pollution insofar as it is caused by automobile exhaust or by the burning of fossil fuels. In short, the widespread availability of very cheap energy everywhere in the world ought to lead to an energy autarky in every country of the world; and eventually to an autarky in the many staples of life that should flow from really cheap energy.

WILL TECHNOLOGY REPLACE SOCIAL ENGINEERING?

I hope these examples suggest how social problems can be circumvented or at least reduced to less formidable proportions by the application of the Technological Fix. The examples I have given do not strike me as being fanciful, nor are they at all exhaustive. I have not touched, for example, upon the extent to which really cheap computers and improved technology of communication can help improve elementary teaching without having first to improve our elementary teachers. Nor have I mentioned Ralph Nader's brilliant observation that a safer car, and even its development and adoption by the auto company, is a quicker and probably surer way to reduce traffic deaths than is a campaign to teach people to drive more carefully. Nor have I invoked some really fanciful Technological Fixes: like providing air conditioners and free electricity to operate them for every black family in Watts on the assumption (suggested by Huntington) that race rioting is correlated with hot, humid weather; or the ultimate Technological Fix, Aldous Huxley's soma pills that eliminate human unhappiness without improving human relations in the usual sense.

My examples illustrate both the strength and the weakness of the Technological Fix for social problems. The Technological Fix accepts man's intrinsic shortcomings and circumvents them or capitalizes on them for socially useful ends. The Fix is, therefore, eminently practical and, in the short term, relatively effective. One does not wait around trying to change people's minds: if people want more water, one gets them more water rather than requiring them to reduce their use of water; if people insist on driving autos while they are drunk, one provides safer autos that prevent injuries even after a severe accident.

But the technological solutions to social problems tend to be incomplete and metastable, to replace one social problem with another.

Perhaps the best example of this instability is the peace imposed upon us by the H-bomb. Evidently the pax hydrogenica is metastable in two senses: in the short term, because the aggressor still enjoys such an advantage; in the long term, because the discrepancy between have and have-not nations must eventually be resolved if we are to have permanent peace. Yet, for these particular shortcomings, technology has something to offer. To the imbalance between offense and defense, technology says let us devise passive defense which redresses the balance. A world with H-bombs and adequate civil defense is less likely to lapse into thermonuclear war than a world with H-bombs alone, at least if one concedes that the danger of the thermonuclear war mainly lies in the acts of irresponsible leaders. Anything that deters the irresponsible leader is a force for peace: a technologically sound civil defense therefore would help stabilize the balance of terror.

To the discrepancy between haves and have-nots, technology offers the nuclear energy revolution, with its possibility of autarky for haves and have-nots alike. How this might work to stabilize our metastable thermonuclear peace is suggested by the possible political effect of the recently proposed Israeli desalting plant. The Arab states I should think would be much less set upon destroying the Jordan River Project if the Israelis had a desalination plant in reserve that would nullify the effect of such action. In this connection, I think countries like ours can contribute very much. Our country will soon have to decide whether to continue to spend 5.5×10^9 per year for space exploration after our lunar landing. Is it too outrageous to suggest that some of this money be devoted to building huge nuclear desalting complexes in the arid ocean rims of the troubled world? If the plants are powered with breeder reactors, the out-of-pocket costs, once the plants are built, should be low enough to make large-scale agriculture feasible in these areas. I estimate that for 4×10^9 per year we could build enough desalting capacity to feed more than ten million new mouths per year (provided we use agricultural methods that husband water), and we would, thereby, help stabilize the metastable, bomb-imposed balance of terror.

Yet, I am afraid we technologists shall not satisfy our social engineers, who tell us that our Technological Fixes do not get to the heart of the problem; they are at best temporary expedients; they create new problems as they solve old ones; to put a Technological Fix into effect requires a positive social action. Eventually, social engineering, like the Supreme Court decision on desegregation, must be invoked to solve social problems. And, of course, our social engineers are right.

Technology will never *replace* social engineering. But technology has provided and will continue to provide to the social engineer broader options, to make intractable social problems less intractable; perhaps, most of all, technology will buy time—that precious commodity that converts violent social revolution into acceptable social evolution.

Our country now recognizes and is mobilizing around the great social problems that corrupt and disfigure our human existence. It is natural that in this mobilization we should look first to the social engineer. But, unfortunately, the apparatus most readily available to the government, like the great federal laboratories, is technologically oriented, not socially oriented. I believe we have a great opportunity here; for, as I hope I have persuaded you, many of our seemingly social problems do admit of partial technological solutions. Our already deployed technological apparatus can contribute to the resolution of social questions. I plead, therefore, first for our government to deploy its laboratories, its hardware contractors, and its engineering universities around social problems. And I plead, secondly, for understanding and cooperation between technologist and social engineer. Even with all the help he can get from the technologist, the social engineer's problems are never really solved. It is only by cooperation between technologist and social engineer that we can hope to achieve what is the aim of all technologists and social engineers—a better society, and thereby, a better life, for all of us who are part of society.

6.   The Role of Technology in Society

EMMANUEL G. MESTHENE

Emmanuel Mesthene's essay, "The Role of Technology in Society," and the piece that follows it, "Technology: The Opiate of the Intellectuals" by John McDermott, constitute a classic debate over the role of technology in society. Both articles date from the late 1960s, when the war in Vietnam was in full swing and intellectual life in the United States was torn by bitter conflicts between the "establishment" and the "New Left."

Mesthene's perspective is that of the establishment. The selection originated as the overview section of the fourth annual report of the Harvard Program on Technology and Society, an interdisciplinary program of academic studies funded by a $5 million grant from IBM. Mesthene was the program's director, and this essay was his general statement of what the program had learned about the implications of technological change for society.

According to Mesthene, technology appears to induce social change in two ways: by creating new opportunities and by generating new problems for individuals and for societies. "It has both positive and negative effects, and it usually has the two at the same time and in virtue of each other." By enlarging the realm of goal choice, or by altering the relative costs associated with different values, technology can induce value change. In all areas, technology is seen to have two faces, one positive and one negative.

Emmanuel G. Mesthene directed the Harvard Program on Technology and Society from 1964 through 1974, following 11 years with the Rand Corporation. He joined Rutgers University in 1974, serving as the dean of Livingston College for several years, then as distinguished professor of

Source: *Technology and Culture* 10:4, (1969). Reprinted by permission of the author and The University of Chicago Press.

philosophy and professor of management. Mesthene died in 1990. Among his books are Technological Change: Its Impact on Man and Society *(1970) and* How Language Makes Us Know *(1964).*

SOCIAL CHANGE

Three Unhelpful Views about Technology

While a good deal of research is aimed at discerning the particular effects of technological change on industry, government, or education, systematic inquiry devoted to seeing these effects together and to assessing their implications for contemporary society as a whole is relatively recent and does not enjoy the strong methodology and richness of theory and data that mark more established fields of scholarship. It therefore often has to contend with facile or one-dimensional views about what technology means for society. Three such views, which are prevalent at the present time, may be mildly caricatured somewhat as follows.

The first holds that technology is an unalloyed blessing for man and society. Technology is seen as the motor of all progress, as holding the solution to most of our social problems, as helping to liberate the individual from the clutches of a complex and highly organized society, and as the source of permanent prosperity; in short, as the promise of utopia in our time. This view has its modern origins in the social philosophies of such 19th-century thinkers as Saint-Simon, Karl Marx, and Auguste Comte. It tends to be held by many scientists and engineers, by many military leaders and aerospace industrialists, by people who believe that man is fully in command of his tools and his destiny, and by many of the devotees of modern techniques of "scientific management."

A second view holds that technology is an unmitigated curse. Technology is said to rob people of their jobs, their privacy, their participation in democratic government, and even, in the end, their dignity as human beings. It is seen as autonomous and uncontrollable, as fostering materialistic values and as destructive of religion, as bringing about a technocratic society and bureaucratic state in which the individual is increasingly submerged, and as threatening, ultimately, to poison nature and blow up the world. This view is akin to historical "back-to-nature" attitudes toward the world and is propounded mainly by artists, literary commentators, popular social critics, and existentialist philosophers. It is becoming increasingly attractive to many of our youth, and it tends to be held, understandably enough, by segments of the population that have suffered dislocation as a result of technological change.

The third view is of a different sort. It argues that technology as such is not worthy of special notice, because it has been well recognized as a factor in social change at least since the Industrial Revolution, because it is unlikely that the social effects of computers will be nearly so traumatic as the introduction of the factory system in 18th-century England, because research has shown that technology has done little to accelerate the rate of economic productivity since the 1880s, because there has been no significant change in recent decades in the time periods between invention and widespread adoption of new technology, and because improved communications and higher levels of education make people much more adaptable than heretofore to new ideas and to new social reforms required by technology.

While this view is supported by a good deal of empirical evidence, however, it tends to ignore a number of social, cultural, psychological, and political effects of technological change that are less easy to identify with precision. It thus reflects the difficulty of coming to grips with a new or broadened subject matter by means of concepts and intellectual categories designed to deal with older and different subject matters. This view tends to be held by historians, for whom continuity is an indispensable methodological assumption, and by many economists, who find that their instruments measure some things quite well while those of the other social sciences do not yet measure much of anything.

Stripped of caricature, each of these views contains a measure of truth and reflects a real aspect of the relationship of technology and society. Yet they are oversimplifications that do not contribute much to understanding. One can find empirical evidence to support each of them without gaining much knowledge about the actual mechanism by which technology leads to social change or significant insight into its implications for the future. All three remain too uncritical or too partial to guide inquiry. Research and analysis lead to more differentiated conclusions and reveal more subtle relationships.

* * *

How Technological Change Impinges on Society

It is clearly possible to sketch a more adequate hypothesis about the interaction of technology and society than the partial views outlined above. Technological change would appear to induce or "motor" social change in two principal ways. New technology creates new opportunities for men and societies, and it also generates new problems for them. It has both positive and negative effects, and it usually has the two *at the same time and in virtue of each other.* Thus, industrial technology strengthens the economy, as our measures of growth and productivity show. . . . However, it also induces changes in the relative importance of individual supplying sectors in the economy as new techniques of production alter the amounts and kinds of materials, parts and components, energy, and service inputs used by each industry to produce its output. It thus tends to bring about dislocations of businesses and people as a result of changes in industrial patterns and in the structure of occupations.

The close relationship between technological and social change itself helps to explain why any given technological development is likely to have both positive and negative effects. The usual sequence is that (1) technological advance creates a new opportunity to achieve some desired goal; (2) this requires (except in trivial cases) alterations in social organization if advantage is to be taken of the new opportunity, (3) which means that the functions of existing social structures will be interfered with, (4) with the result that other goals which were served by the older structures are now only inadequately achieved.

As the Meyer-Kain[1] study has shown, for example, improved transportation technology and increased ownership of private automobiles have increased the mobility of businesses and individuals. This has led to altered patterns of industrial and residential location, so that older unified cities are being increasingly transformed into larger metropolitan complexes. The new opportunities for mobility are largely denied to the poor and black populations of the core cities, however, partly for economic reasons, and partly as a result of restrictions on choice of residence by blacks, thus leading to persistent black unemployment despite a generally high level of economic activity. Cities are thus increasingly unable to perform their traditional functions of providing employment opportunities for all segments of their populations and an integrated social environment that can temper ethnic and racial differences. The new urban complexes are neither fully viable economic units nor effective political organizations able to upgrade and integrate their core populations into new economic and social structures. The resulting instability is further aggravated by modern mass communications technology, which heightens the expectations of the poor and the fears of the well-to-do and adds frustration and bitterness to the urban crisis. . . .

In all such cases, technology creates a new opportunity and a new problem at the same time. That is why isolating the opportunity or the problem and construing it as the whole answer is ultimately obstructive of rather than helpful to understanding.

How Society Reacts to Technological Change

The heightened prominence of technology in our society makes the interrelated tasks of profiting from its opportunities and containing its dangers a major intellectual and political challenge of our time.

Failure of society to respond to the opportunities created by new technology means that much actual or potential technology lies fallow, that is, is not used at all or is not used to its full capacity. This can mean that potentially solvable problems are left unsolved and potentially achievable goals unachieved, because we waste our technological resources or use them inefficiently. A society has at least as much stake in the efficient utilization of technology as in that of its natural or human resources.

There are often good reasons, of course, for not developing or utilizing a particular technology. The mere fact that it can be developed is not sufficient reason for doing so. . . .

But there are also cases where technology lies fallow because existing social structures are inadequate to exploit the opportunities it offers. . . . Community institutions wither for want of interest and participation by residents. City agencies are unable to marshal the skills and take the systematic approach needed to deal with new and intensified problems of education, crime control, and public welfare. Business corporations, finally, which are organized around the expectation of private profit, are insufficiently motivated to bring new technology and management know-how to bear on urban projects where the benefits will be largely social. All these factors combine to dilute what may otherwise be a genuine desire to apply our best knowledge and adequate resources to the resolution of urban tensions and the eradication of poverty in the nation. . . .

Containing the Negative Effects of Technology

The kinds and magnitude of the negative effects of technology are no more independent of the institutional structures and cultural attitudes of society than is realization of the new opportunities that technology offers. In our society, there are individuals or individual firms always on the lookout for new technological opportunities, and large corporations hire scientists and engineers to invent such opportunities. In deciding whether to develop a new technology, individual entrepreneurs engage in calculations of expected benefits and expected costs to themselves, and proceed if the former are likely to exceed the latter. Their calculations do not take adequate account of the probable benefits and costs of the new developments to others than themselves or to society generally. These latter are what economists call external benefits and costs.

The external benefits potential in new technology will thus not be realized by the individual developer and will rather accrue to society as a result of deliberate social action, as has been argued above. Similarly with the external costs. In minimizing only expected costs to himself, the individual decision maker helps to contain only some of the potentially negative effects of the new technology. The external costs and therefore the negative effects on society at large are not of principal concern to him and, in our society, are not expected to be.

Most of the consequences of technology that are causing concern at the present time—pollution of the environment, potential damage to the ecology of the planet, occupational and social dislocations, threats to the privacy and political significance of the individual, social and psychological malaise—are negative externalities of this kind. They are with us in large measure because it has not been anybody's explicit business to foresee and anticipate them. They have fallen between the stools of innumerable individual decisions to develop individual technologies for individual purposes without explicit attention to what all these decisions add up to for society as a whole and for people as human beings. This freedom of individual decision making is a value that we have cherished and that is built into the institutional fabric of our society. The negative effects of technology that we deplore are a measure of what this traditional freedom is beginning to cost us. They are traceable, less to some mystical autonomy presumed to lie in technology, and much more to the autonomy that our economic and political institutions grant to individual decision making. . . .

Measures to control and mitigate the negative effects of technology, however, often appear to threaten freedoms that our traditions still take for granted as inalienable rights of men and good societies, however much they may have been tempered in practice by the social pressures of modern times; the freedom of the market, the freedom of private enterprise, the freedom of the scientist to follow truth wherever it may lead, and the freedom of the individual to pursue his fortune and decide his fate. There is thus set up a tension between the need to control technology and our wish to preserve our values, which leads some people to conclude that technology is inherently inimical to human values. The political effect of this tension takes the form of inability to adjust our decision-making structures to the realities of technology so as to take maximum advantage of the opportunities it offers and so that we can act to contain its potential ill effects before they become so pervasive and urgent as to seem uncontrollable.

To understand why such tensions are so prominent a social consequence of technological change, it becomes necessary to look explicitly at the effects of technology on social and individual values.

VALUES

* * *

Technology as a Cause of Value Change

Technology has a direct impact on values by virtue of its capacity for creating new opportunities. By making possible what was not possible before, it offers individuals and society new options to choose from. For example, space technology makes it possible for the first time to go to the moon or to communicate by satellite and thereby adds those two new options to the spectrum of choices available to society. By adding new options in this way, technology can lead to changes in values in the same way that the appearance of new dishes on the heretofore standard menu of one's favorite restaurant can lead to changes in one's tastes and choices of food. Specifically, technology can lead to value change either (1) by bringing some previously unattainable goal within the realm of choice or (2) by making some values easier to implement than heretofore, that is, by changing the costs associated with realizing them. . . .

One example related to the effect of technological change on values is implicit in our concept of democracy. The ideal we associate with the old New England town meeting is that each citizen should have a direct voice in political decisions. Since this has not been possible, we have elected representatives to serve our interests and vote our opinions. Sophisticated computer technology, however, now makes possible rapid and efficient collection and analysis of voter opinion and could eventually provide for "instant voting" by the whole electorate on any issue presented to it via television a few hours before. It thus raises the possibility of instituting a system of direct democracy and gives rise to tensions between those who would be violently opposed to such a prospect and those who are already advocating some system of participatory democracy.

This new technological possibility challenges us to clarify what we mean by democracy. Do we construe it as the will of an undifferentiated majority, as the resultant of transient coalitions of different interest groups representing different value commitments, as the considered judgment of the people's elected representatives, or as by and large the kind of government we actually have in the United States, minus the flaws in it that we would like to correct? By bringing us face to face with such questions, technology has the effect of calling society's bluff and thereby preparing the ground for changes in its values.

In the case where technological change alters the relative costs of implementing different values, it impinges on inherent contradictions in our value system. To pursue the same example, modern technology can enhance the values we associate with democracy. But it can also enhance another American value—that of "secular rationality," as sociologists call it—by facilitating the use of scientific and technical expertise in the process of political decision making. This can in turn further reduce citizen participation in the democratic process. Technology thus has the effect of facing us with contradictions in our value system and of calling for deliberate attention to their resolution.

* * *

ECONOMIC AND POLITICAL ORGANIZATION

The Enlarged Scope of Public Decision Making

When technology brings about social changes (as described in the first section of this essay) which impinge on our existing system of values (in ways reviewed in the second section), it poses for society a number of problems that are ultimately political in nature. The term "political" is used here in the broadest sense: it encompasses all of the decision-making structures and procedures that have to do with the allocation and distribution of wealth and power in society. The political organization of society thus includes not only the formal apparatus of the state but also industrial organizations and other private institutions that play a role in the decision-making process. It is particularly important to attend to the organization of the entire body politic when technological change leads to a blurring of once clear distinctions between the public and private sectors of society and to changes in the roles of its principal institutions.

It was suggested above that the political requirements of our modern technological society call for a relatively greater public commitment on the part of individuals than in previous times. The reason for this, stated most generally, is that technological change has the effect of enhancing the importance of public decision making in society, because technology is continually creating new possibilities for social action as well as new problems that have to be dealt with.

A society that undertakes to foster technology on a large scale, in fact, commits itself to social complexity and to facing and dealing with new problems as a normal feature of political life. Not much is yet known with any precision about the political imperatives inherent in technological change, but one may nevertheless speculate about the reasons why an increasingly technological society seems to be characterized by enlargement of the scope of public decision making.

For one thing, the development and application of technology seems to require large-scale, and hence increasingly complex, social concentrations, whether these be large cities, large corporations, big universities, or big government. In instances where technological advance appears to facilitate reduction of such first-order concentrations, it tends to instead enlarge the relevant *system* of social organization, that is, to lead to increased centralization. Thus, the physical dispersion made possible by transportation and communications technologies, as Meyer and Kain have shown, enlarges the urban complex that must be governed as a unit.

A second characteristic of advanced technology is that its effects cover large distances, in both the geographical and social senses of the term. Both its positive and negative features are more extensive. Horse-powered transportation technology was limited in its speed and capacity, but its nuisance value was also limited, in most cases to the owner and to the occupant of the next farm. The supersonic transport can carry hundreds across long distances in minutes, but its noise and vibration damage must also be suffered willy-nilly by everyone within the limits of a swath 3,000 miles long and several miles wide.

The concatenation of increased density (or enlarged system) and extended technological "distance" means that technological applications have increasingly wider ramifications and that increasingly large concentrations of people and organizations become dependent on technological systems. . . . The result is not only that more and more decisions must be social decisions taken in public ways, as already noted, but that, once made, decisions are likely to have a shorter useful life than heretofore. That is partly because technology is continually altering the spectrum of choices and problems that society faces, and partly because any decision taken is likely to generate a need to take ten more.

These speculations about the effects of technology on public decision making raise the problem of restructuring our decision-making mechanisms—including the system of market incentives—so that the increasing number and importance of social issues that confront us can be resolved equitably and effectively.

* * *

The Promise and Problems of Scientific Decision Making

There are two further consequences of the expanding role of public decision making. The first is that the latest information-handling devices and techniques tend to be utilized in the decision-making process. This is so (1) because public policy can be effective only to the degree that it is based on reliable knowledge about the actual state of the society, and thus requires a strong capability to collect, aggregate, and analyze detailed data about economic activities, social patterns, popular attitudes, and political trends, and (2) because it is recognized increasingly that decisions taken in one area impinge on and have consequences for other policy areas often thought of as unrelated, so that it becomes necessary to base decisions on a model of society that sees it as a system and that is capable of signaling as many as possible of the probable consequences of a contemplated action.

As Professor Alan F. Westin points out, reactions to the prospect of more decision making based on computerized data banks and scientific management techniques run the gamut of optimism to pessimism mentioned in the opening of this essay. Negative reactions take the form of rising political demands for greater popular participation in decision making, for more equality among different segments of the population, and for greater regard for the dignity of individuals. The increasing dependence of decision making on scientific and technological devices and techniques is seen as posing a threat to these goals, and pressures are generated in opposition to further "rationalization" of decision-making processes. These pressures have the paradoxical effect, however, not of deflecting the supporters of technological decision making from their course, but of spurring them on to renewed effort to save the society before it explodes under planlessness and inadequate administration.

The paradox goes further, and helps to explain much of the social discontent that we are witnessing at the present time. The greater complexity and the more extensive ramifications that technology brings about in society tend to make social processes increasingly circuitous and indirect. The effects of actions are widespread and difficult to keep track of, so that experts and sophisticated techniques are increasingly needed to detect and analyze social events and to formulate policies adequate to the complexity of social issues. The "logic" of modern decision making thus appears to require greater and greater dependence on the collection and analysis of data and on the use of technological devices and scientific techniques. Indeed, many observers would agree that there is an "increasing relegation of questions which used to be matters of political debate to professional cadres of technicians and experts which function almost independently of the democratic political process."[2] In recent times, that process has been most noticeable, perhaps, in the areas of economic policy and national security affairs.

This "logic" of modern decision making, however, runs counter to that element of traditional democratic theory that places high value on direct participation in the political processes and generates the kind of discontent referred to above. If it turns out on more careful examination that direct participation is becoming less relevant to a society in which the connections between causes and effects are long and often hidden—which is an increasingly "indirect" society, in other words—elaboration of a new democratic ethos and of new democratic processes more adequate to the realities of modern society will emerge as perhaps the major intellectual and political challenge of our time.

The Need for Institutional Innovation

The challenge is, indeed, already upon us, for the second consequence of the enlarged scope of public decision making is the need to develop new institutional forms and new mechanisms to replace established ones that can no longer deal effectively with the new kinds of problems with which we are increasingly faced. Much of the political ferment of the present time—over the problems of technology assessment, the introduction of statistical data banks, the extension to domestic problems of techniques of analysis developed for the military services, and the modification of the institutions of local government—is evidence of the need for new institutions. . . .

CONCLUSION

As we review what we are learning about the relationship of technological and social change, a number of conclusions begin to emerge. We find, on the one hand, that the creation of new physical possibilities and social options by technology tends toward and appears to require the emergence of new values, new forms of economic activity, and new political organizations. On the other hand, technological change also poses problems of social and psychological displacement.

The two phenomena are not unconnected, nor is the tension between them new: man's technical prowess always seems to run ahead of his ability to deal with and profit from it. In America, especially, we are becoming adept at extracting the new techniques, the physical power, and the economic productivity that are inherent in our knowledge and its associated technologies. Yet we have not fully accepted the fact that our progress in the technical realm does not leave our institutions, values, and political processes unaffected. Individuals will be fully integrated into society only when we can extract from our knowledge not only its technological potential but also its implications for a system of values and a social, economic, and political organization appropriate to a society in which technology is so prevalent. . . .

NOTES

1. Unless otherwise noted, studies referred to in this article are described in the Fourth Annual Report (1967–68) of the Harvard University Program on Technology and Society.
2. Harvey Brooks, "Scientific Concepts and Cultural Change," in G. Holton, ed., *Science and Culture* (Boston: Houghton Mifflin, 1965), p. 71.

7. Technology: The Opiate
of the Intellectuals

JOHN MCDERMOTT

Several months after the report containing Emmanuel Mesthene's article was published by Harvard, a sharply critical review-essay by John Mc-Dermott appeared in The New York Review of Books. *McDermott's piece, which follows here, is not a point-by-point analysis or rebuttal of the Mesthene work. Rather, it is McDermott's attempt to critique the entire point of view that he sees as epitomized by Mesthene—"a not new but . . . newly aggressive right-wing ideology in this country." McDermott focuses on a notion he calls* laissez innover, *which holds that technology is a self-correcting system. Mesthene, he claims, finds this principle acceptable because he defines technology abstractly. Mc-Dermott himself, however, rejects* laissez innover *because he claims to see specific characteristics in contemporary technology that contradict the abstraction.*

Concentrating on the application of technology to the war in Vietnam, McDermott examines its nature and concludes that "technology, in its concrete, empirical meaning, refers fundamentally to systems of rationalized control over large groups of men, events, and machines by small groups of technically skilled men operating through organized hierarchy." Using this definition, he proceeds to discuss the social effect of modern technology in America, concluding that the ideology of laissez innover *is attractive to those in power since they are in a position to reap technology's benefits while avoiding its costs.*

John McDermott has served on the faculty of the State University of New York at Old Westbury, in the Department of Labor Studies.

I

. . . If religion was formerly the opiate of the masses, then surely technology is the opiate of the educated public today, or at least of its favorite authors. No other single subject is so universally invested with high hopes for the improvement of mankind generally and of Americans in particular. . . .

These hopes for mankind's, or technology's, future, however, are not unalloyed. Technology's defenders, being otherwise reasonable men, are also aware that the world population explosion and the nuclear missile race are also the fruit of the enormous advances made in technology during the past half century or so. But here too a cursory reading of their literature would reveal widespread though qualified optimism that these scourges too will fall before technology's might. Thus population (and genetic) control and permanent peace are sometimes added to the already imposing roster of technology's promises. What are we to make of such extravagant optimism?

[In early 1968] Harvard University's Program on Technology and Society, ". . . an inquiry in depth into the effects of technological change on the economy, on public policies, and on the character of society, as well as into the reciprocal effects of social progress on the nature, dimension, and directions of scientific and technological development," issued its Fourth Annual Report to the accompaniment of full front-page coverage in *The New York Times* (January 18). Within the brief (fewer than 100) pages of that report and most clearly in the concluding essay by the Program's director, Emmanuel G. Mesthene, one can discern some of the important threads of belief which bind together much current writing on the social implications of technology. Mesthene's essay is worth extended analysis because these beliefs are of interest in themselves and, of greater importance, because they form the basis not of a new but of a newly aggressive right-wing ideology in this country, an ideology whose growing importance was accurately measured by the magnitude of the *Times*'s news report.

. . . Mesthene believes there are two distinct problems in technology's relation to society, a positive one of taking full advantage of the opportunities it offers and the negative one of avoiding unfortunate consequences which flow from the exploitation of those opportunities. Positive opportunities may be missed because the costs of technological development outweigh likely benefits (e.g., Herman Kahn's "Doomsday Machine"). Mesthene seems convinced, however, that a more important case is that in which

> . . . technology lies fallow because existing social structures are in-adequate to exploit the opportunities it offers. This is revealed clearly in the examination of institutional failure in the ghetto car-ried on by [the Program]. . . .

His diagnosis of these problems is generous in the extreme:

> All these factors combine to dilute what may be otherwise a genuine desire to apply our best knowledge and adequate resources to the resolution of urban tensions and the eradication of poverty in the nation.

Moreover, because government and the media ". . . are not yet equipped for the massive task of public education that is needed . . . " if we are to exploit technology more fully, many technological opportu-nities are lost because of the lack of public support. This too is a problem primarily of "institutional innovation."

Mesthene believes that institutional innovation is no less important in combating the negative effects of technology. Individuals or individ-ual firms which decide to develop new technologies normally do not take "adequate account" of their likely social benefits or costs. His critique is anticapitalist in spirit, but lacks bite, for he goes on to add that

> . . . [most of the negative] consequences of technology that are causing concern at the present time—pollution of the environment, potential damage to the ecology of the planet, occupational and social dislocations, threats to the privacy and political significance of the individual, social and psychological malaise—are *negative exter-nalities of this kind.* They are with us in large measure because it has not been anybody's explicit business to foresee and anticipate them. [Italics added.]

Mesthene's abstract analysis and its equally abstract diagnosis in favor of "institutional innovation" place him in a curious and, for us, instructive position. If existing social structures are inadequate to ex-ploit technology's full potential, or if, on the other hand, so-called negative externalities assail us because it is nobody's business to foresee and anticipate them, doesn't this say that we should apply technology to this problem too? That is, we ought to apply and organize the appropriate *organizational* knowledge for the practical purpose of solving the problems of institutional inadequacy and "negative externalities." Hence, in principle, Mesthene is in the position of arguing that the cure for technology's problems, whether positive or negative, is still more technology. This is the first theme of the technological school of writers and its ultimate First Principle.

Technology, in their view, is a self-correcting system. Temporary oversight or "negative externalities" will and should be corrected by technological means. Attempts to restrict the free play of technological innovation are, in the nature of the case, self-defeating. Technological innovation exhibits a distinct tendency to work for the general welfare in the long run. *Laissez innover!*

I have so far deliberately refrained from going into any greater detail than does Mesthene on the empirical character of contemporary technology for it is important to bring out the force of the principle of *laissez innover* in its full generality. Many writers on technology appear to deny in their definition of the subject—organized knowledge for practical purposes—that contemporary technology exhibits distinct trends which can be identified or projected. Others, like Mesthene, appear to accept these trends, but then blunt the conclusion by attributing to technology so much flexibility and "scientific" purity that it becomes an abstraction infinitely malleable in behalf of good, pacific, just, and egalitarian purposes. Thus the analogy to the laissez-faire principle of another time is quite justified. Just as the market or the free play of competition provided in theory the optimum long-run solution for virtually every aspect of virtually every social and economic problem, so too does the free play of technology, according to its writers. Only if technology or innovation (or some other synonym) is allowed the freest possible reign, they believe, will the maximum social good be realized.

What reasons do they give to believe that the principle of *laissez innover* will normally function for the benefit of mankind rather than, say, merely for the belief of the immediate practitioners of technology, their managerial cronies, and for the profits accruing to their corporations? As Mesthene and other writers of his school are aware, this is a very real problem, for they all believe that the normal tendency of technology is, and ought to be, the increasing concentration of decision-making power in the hands of larger and larger scientific-technical bureaucracies. *In principle,* their solution is relatively simple, though not often explicitly stated.[1]

Their argument goes as follows: the men and women who are elevated by technology into commanding positions within various decision-making bureaucracies exhibit no generalized drive for power such as characterized, say, the landed gentry of preindustrial Europe or the capitalist entrepreneur of the last century. For their social and institutional position and its supporting culture as well are defined solely by the fact that these men are problem solvers. (Organized knowledge for practical purposes again.) That is, they gain advantage and reward only to the extent that they can bring specific technical knowledge to bear on the solution of specific technical problems. Any more general drive for power would undercut the bases of their usefulness and legitimacy.

Moreover their specific training and professional commitment to solving technical problems creates a bias against ideologies in general which inhibits any attempts to formulate a justifying ideology for the group. Consequently, they do not constitute a class and have no general interests antagonistic to those of their problem-beset clients. We may refer to all of this as the disinterested character of the scientific-technical decision-maker, or, more briefly and cynically, as the principle of the Altruistic Bureaucrat. . . .

This combination of guileless optimism with scientific tough-mindedness might seem to be no more than an eccentric delusion were the American technology it supports not moving in directions that are strongly antidemocratic. To show why this is so we must examine more closely Mesthene's seemingly innocuous distinction between technology's positive opportunities and its "negative externalities." In order to do this I will make use of an example drawn from the very frontier of American technology, the war in Vietnam.

II

At least two fundamentally different bombing programs [have been] carried out in South Vietnam. There are fairly conventional attacks against targets which consist of identified enemy troops, fortifications, medical centers, vessels, and so forth. The other program is quite different and, at least since March 1968, infinitely more important. With some oversimplification it can be described as follows:

Intelligence data is gathered from all kinds of sources, of all degrees of reliability, on all manner of subjects, and fed into a computer complex located, I believe, at Bien Hoa. From this data and using mathematical models developed for the purpose, the computer then assigns probabilities to a range of potential targets, probabilities which represent the likelihood that the latter contain enemy forces or supplies. These potential targets might include: a canal-river crossing known to be used occasionally by the NLF; a section of trail which would have to be used to attack such and such an American base, now overdue for attack; a square mile of plain rumored to contain enemy troops; a mountainside from which camp fire smoke was seen rising. Again using models developed for the purpose, the computer divides pre-programmed levels of bombardment among these potential targets which have the highest probability of containing actual targets. Following the raids, data provided by further reconnaissance is fed into the computer and conclusions are drawn (usually optimistic ones) on the effectiveness of the raids. This estimate of effectiveness then becomes part of the data governing current and future operations, and so on.

Two features must be noted regarding this program's features, which are superficially hinted at but fundamentally obscured by Mesthene's distinction between the abstractions of positive opportunity and "negative externality." First, when considered from the standpoint of its planners, the bombing program is extraordinarily rational, for it creates previously unavailable "opportunities" to pursue their goals in Vietnam. It would make no sense to bomb South Vietnam simply at random, and no serious person or air force general would care to mount the effort to do so. So the system employed in Vietnam significantly reduces, though it does not eliminate, that randomness. That canal-river crossing which is bombed at least once every eleven days or so is a very poor target compared to an NLF battalion observed in a village. But it is an infinitely more promising target than would be selected by throwing a dart at a grid map of South Vietnam. In addition to bombing the battalion, why not bomb the canal crossing to the frequency and extent that it *might* be used by enemy troops?

Even when we take into account the crudity of the mathematical models and the consequent slapstick way in which poor information is evaluated, it is a "good" program. No single raid will definitely kill an enemy soldier but a whole series of them increases the "opportunity" to kill a calculable number of them (as well, of course, as a calculable but not calculated number of nonsoldiers). This is the most rational bombing system to follow if American lives are very expensive and American weapons and Vietnamese lives very cheap. Which, of course, is the case.

Secondly, however, considered from the standpoint of goals and values not programmed in by its designers, the bombing program is incredibly irrational. In Mesthene's terms, these "negative externalities" would include, in the present case, the lives and well-being of various Vietnamese as well as the feelings and opinions of some less important Americans. Significantly, this exclusion of the interests of people not among the managerial class is based quite as much on the so-called technical means being employed as on the political goals of the system. In the particular case of the Vietnamese bombing system, the political goals of the bombing system clearly exclude the interests of certain Vietnamese. After all, the victims of the bombardment are communists or their supporters, they are our enemies, they resist U.S. intervention. In short, their interests are fully antagonistic to the goals of the program and simply must be excluded from consideration. The technical reasons for this exclusion require explanation, being less familiar and more important, especially in the light of Mesthene's belief in the malleability of technological systems.

Advanced technological systems such as those employed in the bombardment of South Vietnam make use not only of extremely complex and expensive equipment but, quite as important, of large numbers of relatively scarce and expensive-to-train technicians. They have immense capital costs; a thousand aircraft of a very advanced type, literally hundreds of thousands of spare parts, enormous stocks of rockets, bombs, shells and bullets, in addition to tens of thousands of technical specialists: pilots, bombardiers, navigators, radar operators, computer programmers, accountants, engineers, electronic and mechanical technicians, to name only a few. In short, they are "capital intensive."

Moreover, the coordination of this immense mass of esoteric equipment and its operators in the most effective possible way depends upon an extremely highly developed technique both in the employment of each piece of equipment by a specific team of operators and in the management of the program itself. Of course, all large organizations standardize their operating procedures, but it is peculiar to advanced technological systems that their operating procedures embody a very high degree of information drawn from the physical sciences, while their managerial procedures are equally dependent on information drawn from the social sciences. We may describe this situation by saying that advanced technological systems are both "technique intensive" and "management intensive."

It should be clear, moreover, even to the most casual observer that such intensive use of capital, technique, and management spills over into almost every area touched by the technological system in question. An attack program delivering 330,000 tons of munitions more or less selectively to several thousand different targets monthly would be an anomaly if forced to rely on sporadic intelligence data, erratic maintenance systems, or a fluctuating and unpredictable supply of heavy bombs, rockets, jet fuel, and napalm tanks. Thus it is precisely because the bombing program requires an intensive use of capital, technique, and management that the same properties are normally transferred to the intelligence, maintenance, supply, coordination and training systems which support it. Accordingly, each of these supporting systems is subject to sharp pressures to improve and rationalize the performance of its machines and men, the reliability of its techniques, and the efficiency and sensitivity of the management controls under which it operates. Within integrated technical systems, higher levels of technology drive out lower, and the normal tendency is to integrate systems.

From this perverse Gresham's Law of Technology follow some of the main social and organizational characteristics of contemporary technological systems: the radical increase in the scale and complexity of operations that they demand and encourage; the rapid and widespread diffusion of technology to new areas; the great diversity of activities which can be directed by central management; an increase in the ambition of management's goals; and, as a corollary, especially to the last, growing resistance to the influence of so-called negative externalities.

Complex technological systems are extraordinarily resistant to intervention by persons or problems operating outside or below their managing groups, and this is so regardless of the "politics" of a given situation. Technology creates its own politics. The point of such advanced systems is to minimize the incidence of personal or social behavior which is erratic or otherwise not easily classified, of tools and equipment with poor performance, of improvisory techniques, and of unresponsiveness to central management. . . .

To define technology so abstractly that it obscures these observable characteristics of contemporary technology—as Mesthene and his school have done—makes no sense. It makes even less sense to claim some magical malleability for something as undefined as "institutional innovation." Technology, in its concrete, empirical meaning, refers fundamentally to systems of rationalized control over large groups of men, events, and machines by small groups of technically skilled men operating through organizational hierarchy. The latent "opportunities" provided by that control and its ability to filter out discordant "nega-tive externalities" are, of course, best illustrated by extreme cases. Hence the most instructive and accurate example should be of a tech-nology able to suppress the humanity of its rank-and-file and to commit genocide as a by-product of its rationality. The Vietnam bombing pro-gram fits technology to a "T."

* * *

IV

Among the conventional explanations for the rise and spread of the democratic ethos in Europe and North America in the seventeenth, eighteenth, and nineteenth centuries, the destruction of the gap in political culture between the mass of the population and that of the ruling classes is extremely important. There are several sides to this explanation. For example, it is often argued that the invention of the printing press and the spread of Protestant Christianity encouraged a significant growth in popular literacy. . . . The dating of these develop-ments is, in the nature of the case, somewhat imprecise. But certainly by the middle of the eighteenth century, at least in Britain and North America, the literacy of the population was sufficient to support a variety of newspapers and periodicals not only in the large cities but in the smaller provincial towns as well. . . . Common townsmen had closed at least one of the cultural gaps between themselves and the aristocracy of the larger cities.

Similarly, it is often argued that with the expansion and improve-ment of road and postal systems, the spread of new tools and tech-niques, the growth in the number and variety of merchants, the consequent invigoration of town life, and other numerous and famil-iar related developments, the social experiences of larger numbers of people became richer, more varied, and similar in fact to those of the ruling class. This last, the growth in similarity of the social experi-ences of the upper and lower classes, is especially important. Social skills and experiences which underlay the monopoly of the upper classes over the processes of law and government were spreading to important segments of the lower orders of society. For carrying on trade, managing a commercial—not a subsistence—farm, participat-ing in a vestry or workingmen's guild, or working in an up-to-date manufactory or business, unlike the relatively narrow existence of the medieval serf or artisan, were experiences which contributed to what I would call the social rationality of the lower orders.

Activities which demand frequent intercourse with strangers, accurate calculation of near means and distant ends, and a willingness to devise collective ways of resolving novel and unexpected problems demand and reward a more discriminating attention to the realities and deficiencies of social life, and provide thereby a rich variety of social experiences analogous to those of the governing classes. As a result not only were the processes of law and government, formerly treated with semireligious veneration, becoming demystified but, equally important, a population was being fitted out with sufficient skills and interests to contest their control. Still another gap between the political cultures of the upper and lower ends of the social spectrum was being closed.

The same period also witnessed a growth in the organized means of popular expression. . . .

These same developments were also reflected in the spread of egalitarian and republican doctrines such as those of Richard Price and Thomas Paine, which pointed up the arbitrary character of what had heretofore been considered the rights of the higher orders of society, and thus provided the popular ideological base which helped to define the legitimate lower-class demands.

This description by no means does justice to the richness and variety of the historical process underlying the rise and spread of what has come to be called the democratic ethos. But it does, I hope, isolate some of the important structural elements and, moreover, it enables us to illuminate some important ways in which the new technology, celebrated by Mesthene and his associates for its potential contributions to democracy, contributes instead to the erosion of that same democratic ethos. For if, in an earlier time, the gap between the political cultures of the higher and lower orders of society was being widely attacked and closed, this no longer appears to be the case. On the contrary, I am persuaded that the direction has been reversed and that we now observe evidence of a growing separation between ruling and lower-class culture in America, a separation which is particularly enhanced by the rapid growth of technology and the spreading influence of its *laissez innover* ideologues.

Certainly, there has been a decline in popular literacy, that is to say, in those aspects of literacy which bear on an understanding of the political and social character of the new technology. Not one person in a hundred is even aware of, much less understands, the nature of technologically highly advanced systems such as are used in the Vietnam bombing program. . . .

Secondly, the social organization of this new technology, by systematically denying to the general population experiences which are analogous to those of its higher management, contributes very heavily to the growth of social irrationality in our society. For example, modern technological organization defines the roles and values of its members, not vice versa. An engineer or a sociologist is one who does all those things but only those things called for by the "table of organization" and the "job description" used by his employer. Professionals who seek self-realization through creative and autonomous behavior without regard to the defined goals, needs, and channels of their respective departments have no more place in a large corporation or government agency than squeamish soldiers in the army. . . .

However, those at the top of technology's most advanced organizations hardly suffer the same experience. For reasons which are clearly related to the principle of the Altruistic Bureaucracy the psychology of an individual's fulfillment through work has been incorporated into management ideology. As the pages of *Fortune, Time,* or *Business Week* . . . serve to show, the higher levels of business and government are staffed by men and women who spend killing hours looking after the economic welfare and national security of the rest of us. The rewards of this life are said to be very few: the love of money would be demeaning and, anyway, taxes are said to take most of it; its sacrifices are many, for failure brings economic depression to the masses or gains for communism as well as disgrace to the erring managers. Even the essential high-mindedness or altruism of our managers earns no reward, for the public is distracted, fickle, and, on occasion, vengeful. . . . Hence for these "real revolutionaries of our time," as Walt Rostow has called them, self-fulfillment through work and discipline is the only reward. The managerial process is seen as an expression of the vital personalities of our leaders and the right to it an inalienable right of the national elite.

In addition to all this, their lonely and unrewarding eminence in the face of crushing responsibility, etc., tends to create an air of mystification around technology's managers. . . .

It seems fundamental to the social organization of modern technology that the quality of the social experience of the lower orders of society declines as the level of technology grows no less than does their literacy. And, of course, this process feeds on itself, for with the consequent decline in the real effectiveness and usefulness of local and other forms of organization open to easy and direct popular influence their vitality declines still further, and the cycle is repeated.

The normal life of men and women in the lower and, I think, middle levels of American society now seems cut off from those experiences in which near social means and distant social ends are balanced and rebalanced, adjusted and readjusted. But it is from such widespread experience with effective balancing and adjusting that social rationality derives. To the degree that it is lacking, social irrationality becomes the norm, and social paranoia a recurring phenomenon. . . .

Mesthene himself recognizes that such "negative externalities" are on the increase. His list includes ". . . pollution of the environment, potential damage to the ecology of the planet, occupational and social dislocations, threats to the privacy and political significance of the individual, social and psychological malaise. . . . " Minor matters all, however, when compared to the marvelous opportunities *laissez innover* holds out to us: more GNP, continued free world leadership, supersonic transports, urban renewal on a regional basis, institutional innovation, and the millennial promises of his school.

This brings us finally to the ideologies and doctrines of technology and their relation to what I have argued is a growing gap in political culture between the lower and upper classes in American society. Even more fundamentally than the principles of *laissez innover* and the altruistic bureaucrat, technology in its very definition as the organization of knowledge for practical purposes assumes that the primary and really creative role in the social processes consequent on technological change is reserved for a scientific and technical elite, the elite which presumably discovers and organizes that knowledge. But if the scientific and technical elite and their indispensable managerial cronies are the really creative (and hardworking and altruistic) elements in American society, what is this but to say that the common mass of men are essentially drags on the social weal? This is precisely the implication which is drawn by the *laissez innover* school. Consider the following quotations from an article which appeared in *The New Republic* in December 1967, written by Zbigniew Brzezinski, one of the intellectual leaders of the school.

Brzezinski is describing a nightmare which he calls the "technetronic society" (the word like the concept is a pastiche of technology and electronics). This society will be characterized, he argues, by the application of ". . . the principle of equal opportunity for all but . . . special opportunity for the singularly talented few." It will thus combine ". . . continued *respect* for the popular will with an increasing *role* in the key decision-making institutions of individuals with special intellectual and scientific attainments." (Italics added.) Naturally, "The educational and social systems [will make] it increasingly attractive and easy for those meritocratic few to develop to the fullest of their special potential."

However, while it will be ". . . necessary to require everyone at a sufficiently responsible post to take, say, two years of [scientific and technical] retraining every ten years . . . ," the rest of us can develop a new ". . . interest in the cultural and humanistic aspects of life, *in addition to purely hedonistic preoccupations.*" (Italics added.) The latter, he is careful to point out, "would serve as a social valve, reducing tensions and political frustration."

Is it not fair to ask how much *respect* we carefree pleasure lovers and culture consumers will get from the hard-working bureaucrats, going to night school two years in every ten, while working like beavers in the "key decision-making institutions"? The altruism of our bureaucrats has a heavy load to bear.

Stripped of their euphemisms these are simply arguments which enhance the social legitimacy of the interests of new technical and scientific elites and detract from the interests of the rest of us. . . .

As has already been made clear, the *laissez innover* school accepts as inevitable and desirable the centralizing tendencies of technology's social organization, and they accept as well the mystification which comes to surround the management process. Thus equality of opportunity, as they understand it, has precious little to do with creating a more egalitarian society. On the contrary, it functions as an indispensable feature of the highly stratified society they envision for the future. For in their society of meritocratic hierarchy, equality of opportunity assures that talented young meritocrats (the word is no uglier than the social system it refers to) will be able to climb into the "key decision-making" slots reserved for trained talent, and thus generate the success of the new society, and its cohesion against popular "tensions and political frustration."

The structures which formerly guaranteed the rule of wealth, age, and family will not be destroyed (or at least not totally so). They will be firmed up and rationalized by the perpetual addition of trained (and, of course, acculturated) talent. In technologically advanced societies, equality of opportunity functions as a hierarchical principle, in opposition to the egalitarian social goals it pretends to serve. To the extent that is has already become the kind of "equality" we seek to institute in our society, it is one of the main factors contributing to the widening gap between the cultures of upper- and lower-class America.

V

. . . *Laissez innover* is now the premier ideology of the technological impulse in American society, which is to say, of the institutions which monopolize and profit from advanced technology and of the social classes which find in the free exploitation of *their* technology the most likely guarantee of their power, status, and wealth.

This said, it is important to stress both the significance and limitations of what has in fact been said. Here Mesthene's distinction between the positive opportunities and negative "externalities" inherent in technological change is pivotal; for everything else which I've argued follows inferentially from the actual social meaning of that distinction. As my analysis of the Vietnam bombing program suggested, those technological effects which are sought after as positive opportunities and those which are dismissed as negative externalities are decisively influenced by the fact that this distinction between positive and negative within advanced technological organizations tends to be made among the planners and managers themselves. Within these groups there are, as was pointed out, extremely powerful organizational, hierarchical, doctrinal, and other *"technical"* factors, which tend by design to filter out "irrational" demands from below, substituting for them the "rational" demands of technology itself. As a result, technological rationality is as socially neutral today as market rationality was a century ago. . . .

This analysis lends some weight (though perhaps no more than that) to a number of wide-ranging and unorthodox conclusions about American society today and the directions in which it is tending. . . .

First, and most important, technology should be considered as an institutional system, not more and certainly not less. Mesthene's definition of the subject is inadequate, for it obscures the systematic and decisive social changes, especially their political and cultural tendencies, that follow the widespread application of advanced technological systems. At the same time, technology is less than a social system per se, though it has many elements of a social system, viz., an elite, a group of linked institutions, an ethos, and so forth. Perhaps the best summary statement of the case resides in an analogy—with all the vagueness and imprecision attendant on such things: today's technology stands in relation to today's capitalism as, a century ago, the latter stood to the free market capitalism of the time. . . .

A second major hypothesis would argue that the most important dimension of advanced technological institutions is the social one, that is, the institutions are agencies of highly centralized and intensive social control. Technology conquers nature, as the saying goes. But to do so it must first conquer man. More precisely, it demands a very high degree of control over the training, mobility, and skills of the work force. The absence (or decline) of direct controls or of coercion should not serve to obscure from our view the reality and intensity of the social controls which are employed (such as the internalized belief in equality of opportunity, indebtedness through credit, advertising, selective service channeling, and so on).

Advanced technology has created a vast increase in occupational specialties, many of them requiring many, many years of highly specialized training. It must motivate this training. It has made ever more complex and "rational" the ways in which these occupational specialties are combined in our economic and social life. It must win passivity and obedience to this complex activity. Formerly, technical rationality had been employed only to organize the production of rather simple physical objects, for example, aerial bombs. Now technical rationality is increasingly employed to organize all of the processes necessary to the utilization of physical objects, such as bombing systems. For this reason it seems a mistake to argue that we are in a "postindustrial" age, a concept favored by the *laissez innover* school. On the contrary, the rapid spread of technical rationality in organizational and economic life and, hence, into social life is more aptly described as a second and much more intensive phase of the industrial revolution. One might reasonably suspect that it will create analogous social problems.

Accordingly, a third major hypothesis would argue that there are very profound social antagonisms or contradictions not less sharp or fundamental than those ascribed by Marx to the development of nineteenth-century industrial society. The general form of the contradictions might be described as follows: a society characterized by the employment of advanced technology requires an ever more socially disciplined population, yet retains an ever declining capacity to enforce the required discipline. . . .

These are brief and, I believe, barely adequate reviews of extremely complex hypotheses. But, in outline, each of these contradictions appears to bear on roughly the same group of the American population, a technological underclass. If we assume this to be the case, a fourth hypothesis would follow, namely that technology is creating the basis for new and sharp class conflict in our society. That is, technology is creating its own working and managing classes just as earlier industrialization created its working and owning classes. Perhaps this suggests a return to the kind of class-based politics which characterized the U.S. in the last quarter of the nineteenth century, rather than the somewhat more ambiguous politics which was a feature of the second quarter of this century. I am inclined to think that this is the case, though I confess the evidence for it is as yet inadequate.

This leads to a final hypothesis, namely that *laissez innover* should be frankly recognized as a conservative or right-wing ideology. . . .

The point of this final hypothesis is not primarily to reimpress the language of European politics on the American scene. Rather it is to summarize the fact that many of the forces in American life hostile to the democratic ethos have enrolled under the banner of *laissez innover*. Merely to grasp this is already to take the first step toward a politics of radical reconstruction and against the malaise, irrationality, powerlessness, and official violence that characterize American life today.

NOTE

1. For a more complete statement of the argument which follows, see Suzanne Keller, *Beyond the Ruling Class* (New York: Random House, 1963).

8. Technology and the Tragic View

SAMUEL C. FLORMAN

During the more than twenty-five years since their publication, the
Mesthene and McDermott selections have come to be regarded by many
scholars as a classic debate between pro- and antitechnology positions.
To treat a subject as complex as technology and its effects on society in
such simplistic, for-or-against terms is ultimately less than satisfying,
however. Samuel Florman's insightful essay "Technology and the Tragic
View," taken from his book Blaming Technology, suggests another
approach. Florman draws on the classical Greek concept of tragedy to
develop a new perspective on technology. In the tragic view of life, says
Florman,

> [it] is man's destiny to die, to be defeated by the forces of the universe.
> But in challenging his destiny, in being brave, determined, ambitious,
> resourceful, the tragic hero shows to what heights a human being can
> soar. This is an inspiration to the rest of us. After witnessing a tragedy
> we feel good, because the magnificence of the human spirit has been
> demonstrated.

The tragic view accepts responsibility but does not seek to cast blame. It
challenges us to do, with caution, what needs to be done, and to consider
at the same time the consequences of not acting. Florman's view is
ultimately an affirmation of the value of technology in human life,
tempered by a recognition of its limits in sustaining human happiness. It
is a uniquely constructive approach to thinking about technology and
society and a fitting note on which to close the first section of this book.

Samuel C. Florman, author of The Existential Pleasures of Engineering, is a practicing engineer and vice president of Kreisler Borg

Source: From Blaming Technology: The Irrational Search for Scapegoats by Samuel C.
Florman. Copyright © 1981 by Samuel C. Florman. Reprinted with permission from St.
Martin's Press, Inc.

Florman Construction Company in Scarsdale, New York. His more than one hundred articles dealing with the relationship of technology to general culture have appeared in professional journals and popular magazines. Florman, born in New York in 1925, is a fellow of the American Society of Civil Engineers and a member of the New York Academy of Sciences. He holds a bachelor's degree and a civil engineer's degree from Dartmouth College and an M.A. in English literature from Columbia University.

The blaming of technology starts with the making of myths—most importantly, the myth of the technological imperative and the myth of the technocratic elite. In spite of the injunctions of common sense, and contrary to the evidence at hand, the myths flourish.

False premises are followed by confused deductions—a maligning of the scientific view; the assertion that small is beautiful; the mistake about job enrichment; an excessive zeal for government regulation; the hostility of feminists toward engineering; and the wishful thinking of the Club of Rome. These in turn are followed by distracted rejoinders from the technological community, culminating in the bizarre exaltation of engineering ethics.

In all of this it is difficult to determine how much is simple misunderstanding and how much is willful evasion of the truth, a refusal to face up to the harsh realities that underlie life, not only in a technological age, but in every age since the beginning of civilization.

Out of the confusion has come a dialogue of sorts, shaped around views that are deemed "pro"-technology or "anti," "optimistic" or "pessimistic." I believe that we should be thinking in different terms altogether.

House & Garden magazine, in celebration of the American Bicentennial, devoted its July 1976 issue to the topic "American Know-How." The editors invited me to contribute an article, and enticed by the opportunity to address a new audience, plus the offer of a handsome fee, I accepted. We agreed that the title of my piece would be "Technology and the Human Adventure," and I thereupon embarked on a strange adventure of my own.

I thought that it would be appropriate to begin my Bicentennial-inspired essay with a discussion of technology in the time of the Founding Fathers, so I went to the library and immersed myself in the works of Benjamin Franklin, surely the most famous technologist of America's early days. Remembering stories from my childhood about Ben

Franklin the clever tinkerer, I expected to find a pleasant recounting of inventions and successful experiments, a cheering tale of technological triumphs. I found such a tale, to be sure, but along with it I found a record of calamities *caused by* the technological advances of his day.

In several letters and essays, Franklin expressed concern about fire, an ever-threatening scourge in Colonial times. Efficient sawmills made it possible to build frame houses, more versatile and economical than log cabins—but less fire-resistant. Advances in transport made it possible for people to crowd these frame houses together in cities. Cleverly conceived fireplaces, stoves, lamps, and warming pans made life more comfortable, but contributed to the likelihood of catastrophic fires in which many lives were lost.

To deal with this problem, Franklin recommended architectural modifications to make houses more fireproof. He proposed the licensing and supervision of chimney sweeps and the establishment of volunteer fire companies, well supplied and trained in the science of firefighting. As is well known, he invented the lightening rod. In other words, he proposed technological ways of coping with the unpleasant consequences of technology. He applied Yankee ingenuity to solve problems arising out of Yankee ingenuity.

In Franklin's writings I found other examples of technological advances that brought with them unanticipated problems. Lead poisoning was a peril. Contaminated rum was discovered coming from distilleries where lead parts had been substituted for wood in the distilling apparatus. Drinking water collected from lead-coated roofs was also making people seriously ill.

The advancing techniques of medical science were often a mixed blessing, as they are today. Early methods of vaccination for smallpox, for example, entailed the danger of the vaccinated person dying from the artificially induced disease. (In a particularly poignant article, Franklin was at pains to point out that his four-year-old son's death from smallpox was attributable to the boy's *not* having been vaccinated and did not result, as rumor had it, from vaccination itself.)

After a while, I put aside the writings of Franklin and turned my attention to American know-how in the nineteenth century. I became engrossed in the story of the early days of steamboat transport. This important step forward in American technology was far from being the unsullied triumph that it appears to be in our popular histories.

Manufacturers of the earliest high-pressure steam engines often used materials of inferior quality. They were slow to recognize the weakening of boiler shells caused by rivet holes, and the danger of using wrought-iron shells together with cast iron heads that had a different coefficient of expansion. Safety valve openings were often not properly proportioned, and gauges had a tendency to malfunction. Even well-designed equipment quickly became defective through the effects of corrosion and sediment. On top of it all, competition for prestige led to racing between boats, and during a race the usual practice was to tie down the safety valve so that excessive steam pressure would not be relieved.

From 1825 to 1830, 42 recorded explosions killed upward of 270 persons. When, in 1830, an explosion aboard the *Helen McGregor* near Memphis killed more than 50 passengers, public outrage forced the federal government to take action. Funds were granted to the Franklin Institute of Philadelphia to purchase apparatus needed to conduct experiments on steam boilers. This was a notable event, the first technological research grant made by the federal government.

The institute made a comprehensive report in 1838, but it was not until 14 years later that a workable bill was passed by Congress providing at least minimal safeguards for the citizenry. Today we may wonder why the process took so long, but at the time Congress was still uncertain about its right, under the interstate commerce provision of the Constitution, to control the activities of individual entrepreneurs.

When I turned from steamboats to railroads I found another long-forgotten story of catastrophe. Not only were there problems with the trains themselves, but the roadbeds, and particularly the bridges, made even the shortest train journey a hazardous adventure. In the late 1860s more than 25 American bridges were collapsing each year, with appalling loss of life. In 1873 the American Society of Civil Engineers set up a special commission to address the problem, and eventually the safety of our bridges came to be taken for granted.

The more I researched the history of American know-how, the more I perceived that practically every technological advance had unexpected and unwanted side effects. Along with each triumph of mechanical genius came an inevitable portion of death and destruction. Instead of becoming discouraged, however, our forebears seemed to be resolute in confronting the adverse consequences of their own inventiveness. I was impressed by this pattern of progress/setback/renewed-creative-effort. It seemed to have a special message for our day, and I made it the theme of my essay for *House & Garden*.

No matter how many articles one has published, and no matter how much one likes the article most recently submitted, waiting to hear from an editor is an anxious experience. In this case, as it turned out, I had reason to be apprehensive. I soon heard from one of the editors who, although she tried to be encouraging, was obviously distressed. "We liked the part about tenacity and ingenuity," she said, "but, oh dear, *all those disasters*—they are so depressing."

I need not go into the details of what follows: the rewriting, the telephone conferences, the rewriting—the gradual elimination of accidents and casualty statistics, and a subtle change in emphasis. I retreated, with some honor intact I like to believe, until the article was deemed to be suitably upbeat.

I should have known that the Bicentennial issue of *House & Garden* was not the forum in which to consider the dark complexities of technological change. My piece was to appear side by side with such articles as "A House That Has Everything," "Live Longer, Look Younger," and "Everything's Coming Up Roses" (devoted to a review of Gloria Vanderbilt's latest designs).

In the United States today magazines like *House & Garden* speak for those, and to those, who are optimistic about technology. Through technology we get better dishwashers, permanent-press blouses, and rust-proof lawn furniture. "Better living through chemistry," the old Du Pont commercial used to say. Not only is *House & Garden* optimistic, that is, hopeful, about technology; it is cheerfully optimistic. There is no room in its pages for failure, or even for struggle, and in this view it speaks for many Americans, perhaps a majority. This is the lesson I learned—or I should say, relearned—in the Bicentennial year.

Much has been written about the shallow optimism of the United States: about life viewed as a Horatio Alger success story or as a romantic movie with a happy ending. This optimism is less widespread than it used to be, particularly as it relates to technology. Talk of nuclear warfare and a poisoned environment tends to dampen one's enthusiasm. Yet optimistic materialism remains a powerful force in American life. The poll-takers tell us that people believe technology is, on balance, beneficial. And we all know a lot of people who, even at this troublesome moment in history, define happiness in terms of their ability to accumulate new gadgets. The business community, anxious to sell merchandise, spares no expense in promoting a gleeful consumerism.

Side by side with what I have come to think of as *House & Garden* optimism, there is a mood that we might call *New York Review of Books* pessimism. Our intellectual journals are full of gloomy tracts that depict a society debased by technology. Our health is being ruined, according to this view, our landscape despoiled, and our social institutions laid waste. We are forced to do demeaning work and consume unwanted products. We are being dehumanized. This is happening because a technological demon has escaped from human control or, in a slightly different version, because evil technocrats are leading us astray.

It is clear that in recent years the resoluteness exhibited by Benjamin Franklin, and other Americans of similarly robust character, has been largely displaced by a foolish optimism on the one hand and an abject pessimism on the other. These two opposing outlooks are actually manifestations of the same defect in the American character. One is the obverse, the "flip side," of the other. Both reflect a flaw that I can best describe as immaturity.

A young child is optimistic, naively assuming that his needs can always be satisfied and that his parents have it within their power to "make things right." A child frustrated becomes petulant. With the onset of puberty a morose sense of disillusionment is apt to take hold. Sulky pessimism is something we associate with the teenager.

It is not surprising that many inhabitants of the United States, a rich nation with seemingly boundless frontiers, should have evinced a childish optimism, and declared their faith in technology, endowing it with the reassuring power of a parent—also regarding it with the love of a child for a favorite toy. It then follows that technological setbacks would be greeted by some with the naive assumption that all would turn out for the best and by others with peevish declarations of despair. Intellectuals have been in the forefront of this childish display, but every segment of society has been caught up in it. Technologists themselves have not been immune. In the speeches of nineteenth-century engineers, we find bombastic promises that make us blush. Today the profession is torn between a blustering optimism and a confused guilt.

The past 50 years have seen many hopes dashed, but we can see in retrospect that they were unrealistic hopes. We simply cannot make use of coal without killing miners and polluting the air. Neither can we manufacture solar panels without worker fatalities and environmental degradation. (We assume that it will be less than with coal, but we are not sure.) We cannot build highways or canals or airports without despoiling the landscape. Not only have we learned that environmental dangers are inherent in every technological advance, but we find that we are fated to be dissatisfied with much of what we produce because our tastes keep changing. The sparkling, humming, paved metropolises of science fiction—even if they could be realized—are not, after all, the home to which humankind aspires. It seems that many people find such an environment "alienating." There can never be a technologically based Utopia because we discover belatedly that we cannot agree on what form that Utopia might take.

To express our disillusionment we have invented a new word: "trade-off." It is an ugly word, totally without grace, but it signifies, I believe, the beginning of maturity for American society.

It is important to remember that our disappointments have not been limited to technology. (This is a fact that the antitechnologists usually choose to ignore.) Wonderful dreams attended the birth of the New Deal, and later the founding of the United Nations, yet we have since awakened to face unyielding economic and political difficulties. Socialism has been discredited, as was laissez-faire capitalism before it. We have been bitterly disappointed by the labor movement, the educational establishment, efforts at crime prevention, the ministrations of psychiatry, and most recently by the abortive experiments of the so-called counterculture. We have come face to face with *limits* that we had presumed to hope might not exist.

Those of us who have lived through the past 50 years have passed personally from youthful presumptuousness to mature skepticism at the very moment that American society has been going through the same transition. We have to be careful not to define the popular mood in terms of our personal sentiments, but I do not think I am doing that when I observe the multiple disenchantments of our time. We also have to be careful not to deprecate youthful enthusiasm, which is a force for good, along with immaturity, which is tolerable only in the young.

It can be argued that there was for a while good reason to hold out hope for Utopia, since modern science and technology appeared to be completely new factors in human existence. But now that they have been given a fair trial, we perceive their inherent limitations. The human condition is the human condition still.

To persist in saying that we are optimistic or pessimistic about technology is to acknowledge that we will not grow up.

I suggest that an appropriate response to our new wisdom is neither optimism nor pessimism, but rather the espousal of an attitude that has traditionally been associated with men and women of noble character—the tragic view of life.

As a student in high school, and later in college, I found it difficult to comprehend what my teachers told me about comedy and tragedy. Comedy, they said, expresses despair. When there is no hope, we make jokes. We depict people as puny, ridiculous creatures. We laugh to keep from crying.

Tragedy, on the other hand, is uplifting. It depicts heroes wrestling with fate. It is man's destiny to die, to be defeated by the forces of the universe. But in challenging his destiny, in being brave, determined, ambitious, resourceful, the tragic hero shows to what heights a human being can soar. This is an inspiration to the rest of us. After witnessing a tragedy we feel good, because the magnificence of the human spirit has been demonstrated. Tragic drama is an affirmation of the value of life.

Students pay lip service to this theory and give the expected answers in examinations. But sometimes the idea seems to fly in the face of reason. How can we say we feel better after Oedipus puts out his eyes, or Othello kills his beloved wife and commits suicide, than we do after laughing heartily over a bedroom farce?

Yet this concept, which is so hard to grasp in the classroom, where students are young and the environment is serene, rings true in the world where mature people wrestle with burdensome problems.

I do not intend to preach a message of stoicism. The tragic view is not to be confused with world-weary resignation. As Moses Hadas, a great classical scholar of a generation ago, wrote about the Greek tragedians: "Their gloom is not fatalistic pessimism but an adult confrontation of reality, and their emphasis is not on the grimness of life but on the capacity of great figures to adequate themselves to it."[1]

It is not an accident that tragic drama flourished in societies that were dynamic: Periclean Athens, Elizabethan England, and the France of Louis XIV. For tragedy speaks of ambition, effort, and unquenchable spirit. Technological creativity is one manifestation of this spirit, and it is only a dyspeptic antihumanist who can feel otherwise. Even the Greeks, who for a while placed technologists low on the social scale, recognized the glory of creative engineering. Prometheus is one of the quintessential tragic heroes. In viewing technology through a tragic prism we are at once exalted by its accomplishments and sobered by its limitations. We thus ally ourselves with the spirit of great ages past.

The fate of Prometheus, as well as that of most tragic heroes, is associated with the concept of *hubris*, "overweening pride." Yet pride, which in drama invariably leads to a fall, is not considered sinful by the great tragedians. It is an essential element of humanity's greatness. It is what inspires heroes to confront the universe, to challenge the status quo. Prometheus defied Zeus and brought technological knowledge to the human race. Prometheus was a revolutionary. So were Gutenberg, Watt, Edison, and Ford. Technology is revolutionary. Therefore, hostility toward technology is antirevolutionary, which is to say it is reactionary. This charge is currently being leveled against environmentalists and other enemies of technology. Since antitechnologists are traditionally "liberal" in their attitudes, the idea that they are reactionary confronts us with a paradox.

The tragic view does not shrink from paradox; it teaches us to live with ambiguity. It is at once revolutionary and cautionary. *Hubris,* as revealed in tragic drama, is an essential element of creativity; it is also a tragic flaw that contributes to the failure of human enterprise. Without effort, however, and daring, we are nothing. Walter Kerr has spoken of "tragedy's commitment to freedom, to the unflinching exploration of the possible." "At the heart of tragedy," he writes, "feeding it energy, stands godlike man passionately desiring a state of affairs more perfect than any that now exists."[2]

This description of the tragic hero well serves, in my opinion, as a definition of the questing technologist.

An aspect of the tragic view that particularly appeals to me is its reluctance to place blame. Those people who hold pessimistic views about technology are forever reproaching others, if not individual engineers, then the "technocratic establishment," the "megastate," "the pentagon of power," or some equally amorphous entity. Everywhere they look they see evil intent.

There is evil in the world, of course, but most of our disappointments with technology come when decent people are trying to act constructively. "The essentially tragic fact," says Hegel, "is not so much the war of good with evil as the war of good with good."

Pesticides serve to keep millions of poor people from starving. To use pesticides is good; to oppose them when they create havoc in the food chain is also good. To drill for oil and to transport it across the oceans is good, since petroleum provides life-saving chemicals and heat for homes. To prevent oil spills is also good. Nuclear energy is good, as is the attempt to eliminate radioactivity. To seek safety is a worthy goal; but in a world of limited resources, the pursuit of economy is also worthy. We are constantly accusing each other of villainy when we should be consulting together on how best to solve our common problems.

Although the tragic view shuns blame, it does not shirk responsibil-

ity. "The fault, dear Brutus, is not in our stars, but in ourselves. . . . "
We are accountable for what we do or, more often, for what we neglect
to do. The most shameful feature of the antitechnological creed is that
it so often fails to consider the consequences of not taking action. The
lives lost or wasted that might have been saved by exploiting our
resources are the responsibility of those who counsel inaction. The
tragic view is consistent with good citizenship. It advocates making the
most of our opportunities; it challenges us to do the work that needs
doing.

Life, it may be said, is not a play. Yet we are constantly talking about
roles—role-playing, role models, and so forth. It is a primordial urge to
want to play one's part. The outlook I advocate sees value in many
different people playing many different parts. A vital society, like a
meaningful drama, feeds on diversity. Each participant contributes to
the body social: scientist, engineer, farmer, craftsman, laborer, politi-
cian, jurist, teacher, artist, merchant, entertainer. . . . The pro-
growth industrialist and the environmentalist are both needed, and in
a strange way they need each other.

Out of conflict comes resolution; out of variety comes health. This is
the lesson of the natural world. It is the moral of ecological balance; it
is also the moral of great drama. We cannot but admire Caesar, Brutus,
and Antony all together. So should we applaud the guardians of our
wilderness, even as we applaud the creators of dams and paper mills. I
am a builder, but I feel for those who are afraid of building, and I
admire those who want to endow all building with grace.

George Steiner, in *The Death of Tragedy* (1961), claimed that the
tragic spirit was rendered impotent by Christianity's promise of salvation. But I do not think that most people today are thinking in terms of
salvation. They are thinking of doing the best they can in a world that
promises neither damnation nor transcendent victories, but instead
confronts us with both perils and opportunities for achievement. In
such a world the tragic spirit is very much alive. Neither optimism nor
pessimism is a worthy alternative to this noble spirit.

We use words to communicate, but sometimes they are not as precise
as we pretend, and then we confuse ourselves and each other. "Optimism," "pessimism," "tragic view"—these are mere sounds or scratches
on paper. The way we feel is not adequately defined by such sounds or
scratches. René Dubos used to write a column for *The American Scholar*
that he called "The Despairing Optimist." I seem to recall that he once
gave his reasons for not calling it "The Hopeful Pessimist," although I

cannot remember what they were. What really counts, I suppose, is not what we say, or even what we feel, but what we want to do.

By saying that I espouse the tragic view of technology I mean to ally myself with those who, aware of the dangers and without foolish illusions about what can be accomplished, still want to move on, actively seeking to realize our constantly changing vision of a more satisfactory society. I mean to oppose those who would evade harsh truths by intoning platitudes. I particularly mean to challenge those who enjoy the benefits of technology but refuse to accept responsibility for its consequences.

Earlier in this essay I mentioned the problems I encountered in preparing an article for *House & Garden,* and I would like to close by quoting the last few lines from that much-rewritten opus. The prose is somewhat florid, but please remember that it was written in celebration of the American Bicentennial:

> For all our apprehensions, we have no choice but to press ahead. We must do so, first, in the name of compassion. By turning our backs on technological change, we would be expressing our satisfaction with current world levels of hunger, disease, and privation. Further, we must press ahead in the name of human adventure. Without experimentation and change our existence would be a dull business. We simply cannot stop while there are masses to feed and diseases to conquer, seas to explore and heavens to survey.

The editors of *Home & Garden* thought I was being optimistic. I knew that I was being tragic, but I did not argue the point.

NOTES

1. Moses Hadas, *A History of Greek Literature* (New York: Columbia University Press, 1950), p. 75.
2. Walter Kerr, *Tragedy and Comedy* (New York: Simon and Schuster, 1967), p. 107.

Part II
FORECASTING, ASSESSING, AND CONTROLLING THE IMPACTS OF TECHNOLOGY

Few contemporary observers, including those who would cheerfully accept the label "technological optimist," dispute the need for society to develop better ways of forecasting, assessing, and controlling the impacts of technology. In recent years, growing concerns over environmental deterioration, the long-term global impacts of human activities, social dislocations and inequities, and health and safety risks posed by new technologies have led researchers, policymakers, and activists in many nations to begin thinking about new modes of technological planning and decision making. These concerns and the responses to them constitute the focus of the chapters in this section.

In the opening piece, journalist Herb Brody looks at predictions of commercial technology development and explains why these prognostications are so often wrong. The following chapter provides a startling example of such foresight, or rather its lack. In it, historian Paul Ceruzzi conducts a kind of retrospective technology assessment, looking at early expectations of the utilization and impact of computers and finding them remarkably myopic.

Joseph Morone and Edward Woodhouse, in a reading new to this volume, examine the problem of social management of risky technologies and suggest some approaches that may help society avoid technological catastrophes. Next, a rather different perspective on assessing technologies and shaping them to social purposes, one that focuses on the role of women, is presented in Corlann Gee Bush's article. Wrapping up Part II, sociologist Allan Mazur examines technological controversies in which experts disagree, suggesting that they are part of the normal flow of political activity and that they serve an important purpose in our democratic system.

As a footnote to the readings in this section it is worth noting that the termination of the congressional Office of Technology Assessment (OTA) in 1995 by the new Republican majority in Congress should not be interpreted as a signal that technology assessment has outlived its usefulness. Notwithstanding its origins in the early 1970s as a technological "early warning system" for Congress, OTA quickly evolved into a policy research organization—a congressional service agency focused exclusively on scientific and technological issues. It died because it failed to build a strong enough bipartisan support base to withstand the budgetary and ideological winds of the 1990s, not because political leaders felt that assessing the impacts of technology on society is less important now than it has been in recent decades.

9.   Great Expectations: Why Technology Predictions Go Awry

HERB BRODY

Predicting the course of technology development is essential to assessing the potential impacts of the technology on society and, ultimately, to controlling the technology. Yet forecasts of technology development are often wildly inaccurate. Technologies expected to have enormous potential disappear with hardly a trace, while others, largely unheralded, come to play highly significant roles in society.

Why technology predictions go wrong is the subject of this article, which comes from Technology Review, *a monthly magazine published at MIT. The author, Herb Brody, is a journalist with a degree in physics and a career interest in science, technology, and society. Brody writes that a critical look at technology predictions requires that one know who is making the predictions and what their motives are. Market surveys can be poor guides by which to judge the prospects of truly revolutionary technologies. Furthermore, the course of one technology can be strongly influenced by developments in other fields.*

Herb Brody has been senior editor at Technology Review *since 1990. Prior to joining* Technology Review, *he served as managing editor at* Laser Focus, *associate managing editor at* High Technology, *and senior editor at* PC/Computing. *Among his writings are articles on factory automation in Japan, on "smart house" technology, and on the use of personal computers by disabled persons.*

Imagine a world where solar cells and nuclear fusion provide megawatts of pollution-free electricity, the average factory bristles with sophisticated robots, videotex terminals rival the TV set for attention in most households, automobiles run on batteries, and computers manipulate information in the form of light waves rather than electric pulses.

That's the world of today as envisioned by technological forecasters only a few years ago. In reality, of course, manufacturers have bought only a modest quantity of robots, videotex sputtered [but the Internet didn't!—Ed.], solar electricity remains too expensive for all but a handful of applications, and electric cars and nuclear fusion seem, as always, to be at least a decade from practicality. Along the way, other technologies have sneaked into prominence while hardly registering a blip on prognosticators' early-warning radars.

No one can be blamed for not predicting the future with pinpoint accuracy. But ever since the first steamboats were derided as "Fulton's folly," experts' technology forecasts have amounted to a chronicle of wildly missed cues and squandered opportunities. The rapid replacement of vacuum tubes by semiconductors surprised even those close to the industry. Meanwhile, after years of hype, developers of voice recognition have yet to deliver a system that can handle more than a few dozen words of continuous speech.

Such errors in prediction lead to technological fads that can disrupt the orderly allocation of R&D resources. Hyped technologies "become very glamorous for a while, and a lot of good people rush into the field," says Robert Lucky, executive director for communications science research at Bell Labs. Even after a glamour technology's difficulties become clearer, scientists often stay in the field anyway. The result is a pool of scientists whose mission has vanished.

But erroneous forecasts can affect more than individual careers. The evaluation of a technology's future success plays an ever larger part in determining the nation's research agenda as universities depend more heavily on industry for research funding. Government standard-setting agencies have a particularly large stake in knowing how technology will move. Before regulators can set a standard for high-definition television, for example, they need to decide whether HDTV is likely to be piped into homes via fiber optics, in which case the video signal could be much richer than if broadcast over already crowded airwaves.

Are faulty forecasts inevitable? To an extent, perhaps. But certain kinds of mistakes tend to recur. By looking at these patterns, companies and policymakers may be able to judge more intelligently the course of today's embryonic technologies and allocate resources more productively.

MISLEADING MISSIONARIES

Rosy predictions often originate with people who have a financial stake in a new technology. And since their bullish statements of technical potential are often misleadingly packaged as precise market forecasts, unwary businesses and investors often suffer.

To develop something new, you have to believe in it. Developers must convince others of the bright prospects as well. An entrepreneur needs financiers. Scientists in a large corporation need advocates high enough in the hierarchy to allocate funds. And government-supported researchers at universities and national labs have an obvious incentive to overstate their progress and understate the problems that lie ahead: the better the chances for success, the more money an agency is willing to shell out. The result is a climate of raised, and sometimes un-realistic, expectations.

Technological breakthroughs can especially skew the vision of nor-mally level-headed planners. "There's a lot of optimism and specula-tion," says Ian Wilson, a senior management consultant at SRI Interna-tional. "Then the problem turns out to be complex." The development of high-temperature superconductors provides a recent example. After intense press coverage that proclaimed the imminent coming of levi-tated trains, intercontinental power transmission, and superfast elec-tronic switches, the novel superconductors turned out to fall far short of immediate commercial practicality. Says Wilson: "History should have drummed this lesson into our heads by now, but we keep making the same mistake."

Similarly, researchers working on nuclear fusion have kept up a steady barrage of "breakthrough" reports since the mid-1970s. But de-spite bursts of progress, the magic break-even point—at which a reac-tor produces more energy than it consumes—remains elusive.

Market-forecasting firms feed the tendency to overestimate a technol-ogy's near-term promise. Companies such as Dataquest, Frost & Sulli-van, and Business Communications regularly publish reports analyzing the business potential of existing and emerging technologies. Over the past decade, outfits like these have foretold billion-dollar markets for artificial intelligence, videotex, and virtually every other new technol-ogy that laboratories have reported.

Followers of these rosy visions have met a sorry fate. Numerous firms started up in the early 1980s, for example, to mine the supposedly large and growing market for industrial robots. Many received financing from venture capital firms hungry for short-term paybacks. But manu-facturers either declined to modernize or opted for conventional auto-mation instead, stranding the robotics startups. Artificial intelligence went through a similar cycle. Glowing pronouncements of AI's poten-tial to reshape computing inspired a wave of new ventures. When the technology stalled getting out of the laboratory, the budding AI "indus-try" crashed.

One reason for the consistently inflated predictions is that market researchers survey the wrong people. Typically they ask the companies who make a technology how much of it they are selling and how much they expect to sell. It's clear why vendors are a favorite source: they are easy to identify, and their business plans often incorporate sales fore-casts with just the sort of numerical estimates that excite researchers.

But such polls will almost always be skewed. Technology companies—particularly entrepreneurial firms in an emerging field—tend to exhibit an almost missionary zeal about their endeavors. "Vendors believe their own propaganda," says Steven Weissman, senior research editor at BIS CAP International, a Waltham, Mass., firm that follows the computer graphics industry. "You can almost never trust their numbers." Market researchers can create more realistic projections by studying a new tech-nology's potential buyers, who have less of a stake in its success. But the universe of customers is much larger and harder to reach than the uni-verse of sellers. Cost-conscious market researchers usually don't bother.

Weissman cites the advent of CD-ROMs, or compact disc read-only memories. Each CD-ROM disc, identical in appearance to audio CDs, can hold almost 700 megabytes—enough for 275,000 pages of text, or thousands of images. In the mid-1980s, market researchers proclaimed that CD-ROMs were inevitable companions to large numbers of per-sonal computers. But computer users preferred magnetic hard disks, which, unlike CD-ROMs, are erasable and which retrieve data much more rapidly. The CD-ROM advantage in memory capacity has dimin-ished greatly, too, as computer memories have grown from the 10 megabytes common in 1984 to 100 megabytes or more on some of today's machines.

The media have acted as willing accomplices in disseminating over-blown forecasts. The business press, always on the lookout for the next hot technology trend, seizes on an inflated forecast as the basis for a news story. Analysts who go out on a limb with a forecast are cast as "the experts" and cited repeatedly.

Once printed in a reputable publication, forecasts take on a life of their own, and other publications quote them as authoritative. The myth of a huge robot market, for example, grew in large part out of statements by Prudential-Bache vice-president Laura Conigliaro, whose figures echoed throughout dozens of newspaper and magazine stories. Conigliaro's erroneous assumptions about manufacturers' needs took on the weight of truth in the retelling.

Such optimistic predictions find a receptive audience in business. They satisfy a need of companies and investors to identify potentially profitable technological breakthroughs. And those within a company who advocate pursuing a certain technology will pointedly seek data that back up their hunches. Even companies that have already committed themselves to a technology value the reassurance of an enthusiastic report from an independent firm. "The higher my numbers, the more reports I can sell," admits Weissman.

SELLING TODAY'S TECHNOLOGY SHORT

Technological forecasts tend to go astray partly because they underestimate the possibilities for advances in existing technology. "Theoretically, it's been possible for the past 25 years for computers to eliminate photographic film," says Alexander MacLachlan, senior vice-president for R&D at DuPont. But, he points out, continuing chemical refinements have kept silver-halide film in the center of the picture despite a strong challenge from electronic imaging media.

Chip manufacture is another case in point. Today the production of virtually all integrated circuits involves optical lithography, in which light is projected through a mask onto a silicon wafer to form an intricate pattern of transistors and interconnecting wires. To pack more transistors on a chip—and hence increase its computational power or memory capacity—the lithography process must be able to create higher-resolution images.

During the 1970s, conventional wisdom held that optical lithography would soon run out of steam. "People believed there were fundamental reasons that optics would be unable to produce chips with feature dimensions smaller than about 1 micron," says Marc Brodsky, director of technical planning for IBM's Research Division. (A micron is a millionth of a meter.)

The optical process, it was widely thought, would give way to lithography using x-rays, which have much shorter wavelengths than light and so, in principle, can be focused more sharply. But instead of winding down, optical lithography has undergone continuous refinement.

And while chips keep getting more densely packed, semiconductor makers have yet to abandon their tried and true techniques. Optical lithography is expected to carry chips down to feature dimensions of a third or a quarter of a micron, says Brodsky.

An exotic class of electronic switches called Josephson junctions offers a classic case of an entrenched technology outpacing upstart challengers. IBM worked for 15 years on these devices, which use an ultrathin layer of superconducting material to achieve switching speeds far faster than the silicon transistors available during the 1970s. But the junctions' advantage slowly dissipated as silicon chips got faster and faster. "Nobody thought that the improvements in silicon would last so long," explains Brodsky. By the time IBM researchers had overcome the major problems of Josephson junctions, advances in silicon had erased the once compelling need for the new technology.

Another electronic revolution continually postponed is the advent of gallium-arsenide chips. These high-speed devices are already widely used in microwave and opto-electronic equipment, as well as in some space and military systems, where gallium arsenide's ability to function despite exposure to nuclear radiation is prized. During the 1980s, bright prospects for gallium-arsenide chips spurred the formation of a miniature industry, largely with money from venture capital firms hoping to score big by backing the next Intel. But gallium-arsenide developers have so far failed to overtake swiftly improving silicon. "We've been hearing for 10 years that silicon is running out of gas," says Bell Labs' Lucky. "People forget that there are always an army of people working on improving an old technology and only a handful of people working on a new technology."

UNDERESTIMATING THE REVOLUTIONS

Although consumer technology is probably the thorniest arena in which to prophesy, one enduring caveat is that consumers are unwilling to spend a lot of money on something only slightly superior to what they already have, especially if it's also less convenient. Quadraphonic sound, for example, forced audiophiles to rearrange their living rooms to provide seating in the small zone of optimal quadraphonic effect—and for all that, the music sounded only a little bit better. Small wonder the technology quickly disappeared.

In another misreading of consumers' appetites, high hopes abounded in the early 1980s among purveyors of home information services. One forecast in 1980 estimated that 5 percent of all U.S. households would be hooked into videotex by 1985. Consumers would supposedly relish the ability to go on-line with their computers and shop, browse through encyclopedias, and read the latest news, weather, and stock market summaries.

Knight-Ridder spent $60 million setting up a videotex service that never made money and was ultimately abandoned. Technology per se was not the problem—all the necessary computer and telecommunication power existed. Instead, videotex marketers badly misunderstood how people want to use their home computers. There are more efficient ways to do just about everything that videotex provides. Newspapers, printed encyclopedias, and shopping catalogs are superior for most people's purposes. Unlike businesses that deal with financial markets and other late-breaking news, consumers at home rarely demand up-to-the-minute information on anything.

While companies like RCA and Knight-Ridder have lost money by backing a loser, other businesses have made the opposite mistake: failing to pursue a winner. A common mistake is the tendency to evaluate emerging technologies as if they were direct replacements for something familiar. "New things are viewed in the clothing of the old," says James Utterback, a professor of engineering at MIT's Sloan School of Management. The difficulty with this, says Utterback, is that "old things are optimized for what they do already."

When the transistor came out of Bell Labs in the late 1940s, for example, its main use was thought to be as an electronic amplifier in radios. Few saw the potential of the new devices to replace vacuum tubes in digital computers, which were still in their infancy. At the time, many believed that these primitive computers were so potent that the world would never need more than a few dozen of them. Of course, it has been in computing that transistors—first as discrete components, later as devices on integrated circuit chips—have had their most profound impact.

Any truly revolutionary technology defies easy prediction. Until the late 1970s, for example, computer designers focused on building ever bigger and more powerful machines. Few foresaw what has turned out to be the technology's defining trend: the evolution toward personal machines using packaged software. "If you asked computer users in 1970 what they wanted, they'd have probably wished that Cobol [a programming language for mainframe computers] was a little easier to use," says Michael Rappa, a professor of management at the Sloan School. "The idea of a desktop computer with a graphical interface like that of the Macintosh would have seemed idiotic."

Indeed, virtually all the elements of today's personal computers existed in the mid-1970s. IBM reportedly concluded, from its study of what computer users said they wanted, that PCs would appeal only to a small group of hobbyists. Xerox, too, forfeited its headstart in desktop computers because management thought the market was too small to be worth the company's while.

One reason for such problems is that the commercial success of a new technology often depends on factors outside the control of its developer. Videodisc players, for example, failed to take root as a popular consumer appliance despite enthusiastic predictions in the late 1970s. Although RCA invested heavily to market its SelectaVision videodisc system, which played prerecorded programs, sales never approached the mass-market RCA had counted on. In 1989 the company abandoned the product and took a $175 million write-off.

Why did the videodisc fail while the videocassette recorder has enjoyed spectacular success? Certainly the VCR's ability to record programs for later viewing provided an obvious advantage. But millions of people who buy VCRs never figure out how to program the machines to record; they rent movies instead. In fact, the VCR owes much of its success to the advent of video stores. Just as Lotus 1-2-3 gave businesspeople a reason to buy a personal computer, the availability of a large number of movies at low cost justified the purchase of a VCR. RCA, by contrast, counted on videodisc owners to build up a library of discs that they would view repeatedly, even though most people want to watch a movie only once.

If retailers had had the idea of renting out videodiscs 10 years ago, disc players would probably now be much more prevalent in homes. After all, the players entered the market substantially cheaper than VCRs, which were then priced in the luxury range. And the picture from a videodisc is markedly superior to that from a tape.

COPING WITH UNCERTAINTY

Given that technology forecasting is a precarious science, some experts advise organizations not to pay much attention to predictions. "The illusion of knowing what's going to happen is worse than not knowing," contends MIT's Utterback. Rather than basing their strategy on flawed visions of the future, he says, managers and policymakers should make sure their organizations are agile enough to respond to technological changes as they occur.

Yet organizations with a financial interest in knowing the future of technology need not abandon prediction altogether, says SRI's Wilson. "Not everything is uncertain." He suggests a middle ground. Rather than predicting a single outcome, a forecaster should paint several scenarios of the future, each hinging on different assumptions. These scenarios, says Wilson, would suggest a "portfolio of technologies that will bound the envelope of uncertainty." Decision makers can speculate on the potential impact of each scenario on the organization's goals and on the technologies needed to reach those goals.

The fate of many energy technologies, for example, depends on the strength of the international environmental movement, particularly the effort to stem emissions of greenhouse gases. Nonfossil energy, notably solar and nuclear, will probably receive more urgent attention—and greater government funding—if the greenhouse warnings hold up to scrutiny. Organizations, says Wilson, should continually tune their planning as current events render scenarios more or less likely.

The uneven record of past predictions suggests a few other guidelines for avoiding costly and embarrassing mistakes:

- Watch developments in related fields. Ten years ago, laser printers cost $10,000 or more. The technology for inexpensive laser printers came not from the printer industry but from Canon, which developed the laser units for photocopiers.
- Discount predictions based on information from vested interests. Investors, management, and the media repeatedly disregard this seemingly obvious advice. Interested parties include not only the companies that stand to make money from a technology but also scientists whose funding grows and wanes with the level of public excitement and who are also distant from the marketplace. "Researchers can be pretty naive about what's a good idea," says Phil Brodsky, Monsanto's director of corporate research.
- Expect existing technologies to continue improving. And don't expect people to abandon what they have for something new that is 10 percent better.
- Beware of predictions based on simple trend extrapolation. Telling the future by looking at the past assumes that conditions remain

constant. This is like driving a car by looking in the rear-view mirror.

- Distinguish between technological forecasts and market predictions. It is one thing to say that new gallium-arsenide chips will be able to operate faster than silicon chips—that is a matter of physics and engineering. But to predict that the market for gallium-arsenide chips will reach a given level in a given year hinges on many hard-to-predict factors, such as the difficulty of handling the material in mass production, the demand for computers, and progress in competing technologies. People often deduce market impact from technology predictions without giving these other factors their due.

- Give innovation time to diffuse. Truly innovative technologies typically take 10 to 25 years to enter widespread use. This is true even for computer and telecommunications technologies that seem to have come out of nowhere: fax machines first appeared in the 1940s, and fiber optics have been around since the 1960s.

- Pay attention to the infrastructure on which a technology's success depends. Lee De Forest invented the vacuum tube in 1906, but radio broadcasting did not begin until 1921; developing techniques to mass-produce the tubes reliably took 15 years. Any attempt today to figure when TVs will be hanging on living room walls must begin with an analysis of the technology available for making flat-panel displays—slim imaging devices that weigh less and consume less power than the cathode-ray tubes used in conventional TVs and computer terminals.

There's certainly plenty of material for would-be technology seers to practice on. Neural-network computers, shirt-pocket telephones, hypermedia, computer-generated virtual realities, intelligent highway systems—all have their enthusiasts who tell of impending "revolutions." Which promises will be fulfilled and which broken?

Following the principles outlined here will help bring some order to the confusing torrent of technology predictions. But at bottom, prophesy is still a gamble. Bell Labs' Lucky likens technology development to the manipulations of a Ouija board. "Everybody's got their hands on it," he says, "but it always feels like somebody else is moving it."

10.   An Unforeseen Revolution: Computers and Expectations, 1935–1985

PAUL CERUZZI

A good illustration of some of Herb Brody's points is contained in Paul Ceruzzi's essay, which looks back at the early forecasts of the societal impact of computers. Ceruzzi writes that the computer pioneers of the 1940s assumed that perhaps half a dozen of the new machines would serve the world's needs for the foreseeable future! As Joseph Corn notes in the introduction to his book where this reading was first published (Imagining Tomorrow), *this "dazzling failure of prophecy" is explained by the fact that most of the computer pioneers were physicists who viewed the new devices as equipment for their experiments and found it hard to imagine that their inventions might be applied to entirely different fields, such as payroll processing and graphic arts. The impact of computers on society has been staggering, and it is sobering to realize how little of this impact was anticipated by those most responsible for developing and introducing this new technology.*

Paul Ceruzzi is a curator in the Department of Space History at the National Air and Space Museum in Washington, D.C. He is the author of Beyond the Limits: Flight Enters the Computer Age *(Cambridge, MA: MIT Press, 1989).*

The "computer revolution" is here. Computers seem to be everywhere: at work, at play, and in all sorts of places in between. There are perhaps half a million large computers in use in America today [in 1986], 7 or 8 million personal computers, 5 million programmable calculators, and

Source: From *Imagining Tomorrow: History, Technology and the American Future*, edited by Joseph J. Corn (Cambridge, MA: MIT Press, 1986). Reprinted by permission.

millions of dedicated microprocessors built into other machines of every description.

The changes these machines are bringing to society are profound, if not revolutionary. And, like many previous revolutions, the computer revolution is happening very quickly. The computer as defined today did not exist in 1950. Before World War II, the word *computer* meant a human being who worked at a desk with a calculating machine, or something built by a physics professor to solve a particular problem, used once or twice, and then retired to a basement storeroom. Modern computers—machines that do a wide variety of things, many having little to do with mathematics or physics—emerged after World War II from the work of a dozen or so individuals in England, Germany, and the United States. The "revolution," however one may define it, began only when their work became better known and appreciated.

The computer age dawned in the United States in the summer of 1944, when a Harvard physics instructor named Howard Aiken publicly unveiled a giant electromechanical machine called the Mark I. At the same time, in Philadelphia, J. Presper Eckert, Jr., a young electrical engineer, and John Mauchly, a physicist, were building the ENIAC, which, when completed in 1945, was the world's first machine to do numerical computing with electronic rather than mechanical switches.

Computing also got underway in Europe during the war. In 1943 the British built an electronic machine that allowed them to decode intercepted German radio messages. They built several copies of this so-called Colossus, and by the late 1940s general-purpose computers were being built at a number of British institutions. In Germany, Konrad Zuse, an engineer, was building computers out of used telephone equipment. One of them, the Z4, survived the war and had a long and productive life at the Federal Technical Institute in Zurich.

These machines were the ancestors of today's computers. They were among the first machines to have the ability to carry out any sequence of arithmetic operations, keep track of what they had just done, and adjust their actions accordingly. But machines that only solve esoteric physics problems or replace a few human workers, as those computers did, do not a revolution make. The computer pioneers did not foresee their creations as doing much more than that. They had no glimmering of how thoroughly the computer would permeate modern life. The computer's inventors saw a market restricted to few scientific, military, or large-scale business applications. For them, a computer was akin to a wind tunnel: a vital and necessary piece of apparatus, but one whose expense and size limited it to a few installations.

For example, when Howard Aiken heard of the plans of Eckert and Mauchly to produce and market a more elegant version of the ENIAC, he was skeptical. He felt they would never sell more than a few of them, and he stated that four or five electronic digital computers would satisfy all the country's computing needs.[1] In Britain in 1951, the physicist Douglas Hartree remarked: "We have a computer here in Cambridge; there is one in Manchester and one at the [National Physical Laboratory]. I suppose there ought to be one in Scotland, but that's about all."[2] Similar statements appear again and again in the folklore of computing.[3] This perception clearly dominated early discussions about the future of the new technology.[4] At least two other American computer pioneers, Edmund Berkeley and John V. Atanasoff, also recall hearing estimates that fewer than ten computers would satisfy all of America's computing needs.[5]

By 1951 about half a dozen electronic computers were running, and in May of that year companies in the United States and England began producing them for commercial customers. Eckert and Mauchly's dream became the UNIVAC—a commercial electronic machine that for a while was a synonym for *computer,* as *Scotch Tape* is for cellophane tape or *Thermos* is for vacuum bottles. It was the star of CBS's television coverage of the 1952 presidential election, when it predicted, with only a few percent of the vote gathered, Eisenhower's landslide victory over Adlai Stevenson. With this election, Americans in large numbers suddenly became aware of this new and marvelous device. Projects got underway at universities and government agencies across the United States and Europe to build computers. Clearly, there was a demand for more than just a few of the large-scale machines.

But not many more. The UNIVAC was large and expensive, and its market was limited to places like the U.S. Census Bureau, military installations, and a few large industries. (Only the fledgling aerospace industry seemed to have an insatiable appetite for those costly machines in the early years.) Nonetheless, UNIVAC and its peers set the stage for computing's next giant leap, from one-of-a-kind special projects built at universities to mass-produced products designed for the world of commercial and business data processing, banking, sales, routine accounting, and inventory control.

Yet, despite the publicity accorded the UNIVAC, skepticism prevailed. The manufacturers were by no means sure of how many computers they could sell. Like the inventors before them, the manufacturers felt that only a few commercial computers would saturate the market. For example, an internal IBM study regarding the potential market for a computer called the Tape Processing Machine (a prototype of which had been completed by 1951) estimated that there was a market for no more than 25 machines of its size.[6] Two years later, IBM developed a smaller computer for business use, the Model 650, which was designed to rent for $3,000 a month—far less than the going price for large computers like the UNIVAC, but nonetheless a lot more than IBM charged for its other office equipment. When it was announced in 1953, those who were backing the project optimistically foresaw a market for 250 machines. They had to convince others in the IBM organization that this figure was not inflated.[7]

As it turned out, businesses snapped up the 650 by the thousands. It became the Model T of computers, and its success helped establish IBM as the dominant computer manufacturer it is today. The idea finally caught on that a private company could manufacture and sell computers—of modest power, and at lower prices than the first monsters—in large quantities. The 650 established the notion of the computer as a machine for business as well as for science, and its success showed that the low estimates of how many computers the world needed were wrong.

Why the inventors and the first commercial manufacturers underestimated the computer's potential market by a wide margin is an interesting question for followers of the computer industry and for historians of modern technology. There is no single cause that accounts for the misperception. Rather, three factors contributed to the erroneous picture of the computer's future: a mistaken feeling that computers were fragile and unreliable; the institutional biases of those who shaped policies toward computer use in the early days; and an almost universal failure, even among the computer pioneers themselves, to understand the very nature of computing (how one got a computer to do work, and just how much work it could do).

It was widely believed that computers were unreliable because their vacuum-tube circuits were so prone to failure. Large numbers of computers would not be built and sold, it was believed, because their unreliability made them totally unsuitable for routine use in a small business or factory. (Tubes failed so frequently they were plugged into sockets, to make it easy to replace them. Other electronic components were more reliable and so were soldered in place.) Eckert and Mauchly's ENIAC had 18,000 vacuum tubes. Other electronic computers got by with fewer, but they all had many more than most other electronic equipment of the day. The ENIAC was a room-sized Leviathan whose tubes generated a lot of heat and used great quantities of Philadelphia's electric power. Tube failures were indeed a serious problem, for if even one tube blew out during a computation it might render the whole machine inoperative. Since tubes are most likely to blow out within a few minutes after being switched on, the ENIAC's power was left on all the time, whether it was performing a computation or not.

Howard Aiken was especially wary of computers that used thousands of vacuum tubes as their switching elements. His Mark I, an electromechanical machine using parts taken from standard IBM accounting equipment of the day, was much slower but more rugged than computers that used vacuum tubes. Aiken felt that the higher speeds vacuum tubes offered did not offset their tendency to burn out. He reluctantly designed computers that used vacuum tubes, but he always kept the numbers of tubes to a minimum and used electromechanical relays wherever he could.[8] Not everyone shared Aiken's wariness, but his arguments against using vacuum-tube circuits were taken seriously by many other computer designers, especially those whose own computer projects were shaped by the policies of Aiken's Harvard laboratory.

That leads to the next reason for the low estimates: Scientists controlled the early development of the computer, and they steered post-war computing projects away from machines and applications that might have a mass market. Howard Aiken, John von Neumann, and Douglas Hartree were physicists or mathematicians, members of a scientific elite. For the most part, they were little concerned with the mundane payroll and accounting problems that business faced every day. Such problems involved little in the way of higher mathematics, and their solutions contributed little to the advancement of scientific knowledge. Scientists perceived their own place in society as an important one but did not imagine that the world would need many more men like themselves. Because their own needs were satisfied by a few powerful computers, they could not imagine a need for many more such machines. Even at IBM, where commercial applications took precedence, scientists shaped the perceptions of the new invention. In the early 1950s the mathematician John von Neumann was a part-time consultant to the company, where he played no little role in shaping expectations for the new technology.

The perception of a modest and limited future for electronic computing came, most of all, from misunderstandings of its very nature. The pioneers did not really understand how humans would interact with machines that worked at the speed of light, and they were far too modest in their assessments of what their inventions could really do. They felt they had made a breakthrough in numerical calculating, but they missed seeing that the breakthrough was in fact a much bigger one. Computing turned out to encompass far more than just doing complicated sequences of arithmetic. But just how much more was not apparent until much later, when other people gained familiarity with computers. A few examples of objections raised to computer projects in the early days will make this clear.

When Howard Aiken first proposed building an automatic computer, in 1937, his colleagues at Harvard objected. Such a machine, they said, would lie idle most of the time, because it would soon do all the work required of it. They were clearly thinking of his proposed machine in terms of a piece of experimental apparatus constructed by a physicist; after the experiment is performed and the results gathered, such an apparatus has no further use and is then either stored or dismantled so that the parts can be reused for new experiments. Aiken's proposed "automatic calculating machine," as he called it in 1937, was perceived that way. After he had used it to perform the calculations he wanted it to perform, would the machine be good for anything else? Probably not. No one had built computers before. One could not propose building one just to see what it would look like; a researcher had to demonstrate the need for a machine with which he could solve a specific problem that was otherwise insoluble. Even if he could show that only with the aid of a computer could he solve the problem, that did not necessarily justify its cost.[9]

Later on, when the much faster electronic computers appeared, this argument surfaced again. Mechanical computers had proved their worth, but some felt that electronic computers worked so fast that they would spew out results much faster than human beings could assimilate them. Once again, the expensive computer would lie idle, while its human operators pondered over the results of a few minutes of its activity. Even if enough work was found to keep an electronic computer busy, some felt that the work could not be fed into the machine rapidly enough to keep its internal circuits busy.[10]

Finally, it was noted that humans had to program a computer before it could do any work. Those programs took the form of long lists of arcane symbols punched into strips of paper tape. For the first electronic computers, it was mostly mathematicians who prepared those tapes. If someone wanted to use the computer to solve a problem, he was allotted some time during which he had complete control over the machine; he wrote the program, fed it into the computer, ran it, and took out the results. By the early 1950s, computing installations saw the need for a staff of mathematicians and programmers to assist the person who wanted a problem solved, since few users would be expected to know the details of programming each specific machine. That meant that every computer installation would require the services of skilled mathematicians, and there would never be enough of them to keep more than a few machines busy. R. F. Clippinger discussed this problem at a meeting of the American Mathematical Society in 1950, stating: "In order to operate the modern computing machine for maximum output, a staff of perhaps twenty mathematicians of varying degrees of training are required. There is currently such a shortage of persons trained for this work, that machines are not working full time."[11] Clippinger forecast a need for 2,000 such persons by 1960, implying that there would be a mere 100 computers in operation by then.

These perceptions, which lay behind the widely held belief that computers would never find more than a limited (though important) market in the industrialized world, came mainly from looking at the new invention strictly in the context of what it was replacing: calculating machines and their human operators. That context was what limited the pioneers' vision.

Whenever a new technology is born, few see its ultimate place in society. The inventors of radio did not foresee its use for broadcasting entertainment, sports, and news; they saw it as a telegraph without wires. The early builders of automobiles did not see an age of "automobility"; they saw a "horseless carriage." Likewise, the computer's inventors perceived its role in future society in terms of the functions it was specifically replacing in contemporary society. The predictions that they made about potential applications for the new invention had to come from the context of "computing" that they knew. Though they recognized the electronic computer's novelty, they did not see how it would permit operations fundamentally different from those performed by human computers.

Before there were digital computers, a mathematician solved a complex computational problem by first recasting it into a set of simpler problems, usually involving only the four ordinary operations of arithmetic—addition, subtraction, multiplication, and division. Then he would take this set of more elementary problems to human computers, who would do the arithmetic with the aid of mechanical desktop calculators. He would supply these persons with the initial input data, books of logarithmic and trigonometric tables, paper on which to record intermediate results, and instructions on how to proceed. Depending on the computer's mathematical skill, the instructions would be more or less detailed. An unskilled computer had to be told, for example, that the product of two negative numbers is a positive number; someone with more mathematical training might need only a general outline of the computation.[12]

The inventors of the first digital computers saw their machines as direct replacements for this system of humans, calculators, tables, pencils and paper, and instructions. We know this because many early experts on automatic computing used the human computing process as the standard against which the new electronic computers were compared. Writers of early textbooks on "automatic computing" started with the time a calculator took to multiply two ten-digit numbers. To that time they added the times for the other operations: writing and copying intermediate results, consulting tables, and keying in input values. Although a skilled operator could multiply two numbers in 10 or 12 seconds, in an 8-hour day he or she could be expected to perform only 400 such operations, each of which required about 72 seconds.[13] The first electronic computers could multiply two ten-digit decimal numbers in about 0.003 second; they could copy and read internally stored numbers even faster. Not only that, they never had to take a coffee break, stop for a meal, or sleep; they could compute as long as their circuits were working.

Right away, these speeds radically altered the context of the arguments that electronic components were too unreliable to be used in more than a few computers. It was true that tubes were unreliable, and that the failure of even one during a calculation might vitiate the results. But the measure of reliability was the number of operations between failures, not the absolute number of hours the machine was in service. In terms of the number of elementary operations it could do before a tube failed, a machine such as ENIAC turned out to be quite reliable after all. If it could be kept running for even one hour without a tube failure, during that hour it could do more arithmetic than the supposedly more reliable mechanical calculators could do in weeks. Eventually the ENIAC's operators were able to keep it running for more than 20 hours a day, 7 days a week. Computers were reliable enough long before the introduction of the transistor provided a smaller and more rugged alternative to the vacuum tube.

So an electronic computer like the ENIAC could do the equivalent of about 30 million elementary operations in a day—the equivalent of the work of 75,000 humans. By that standard, five or six computers of the ENIAC's speed and size could do the work of 400,000 humans. However, measuring electronic computing power by comparing it with that of humans makes no sense. It is like measuring the output of a steam engine in "horsepower." For a 1- or 2-horsepower engine the comparison is appropriate, but it would be impossible to replace a locomotive with an equivalent number of horses. So it is with computing power. But the human measure was the only one the pioneers knew. Recall that between 1945 and 1950 the ENIAC was the only working electronic computer in the United States. At its public dedication in February 1946, Arthur Burks demonstrated the machine's powers to the press by having it add a number to itself over and over again—an operation that reporters could easily visualize in terms of human abilities. Cables were plugged in, switches set, and a few numbers keyed in. Burks then said to the audience, "I am now going to add 5,000 numbers together," and pushed a button on the machine. The ENIAC took about a second to add the numbers.[14]

Almost from the day the first digital computers began working, they seldom lay idle. As long as they were in working order, they were busy, even long after they had done the computations for which they were built.

As electronic computers were fundamentally different from the human computers they replaced, they were also different from special-purpose pieces of experimental apparatus. The reason was that the computer, unlike other experimental apparatus, was programmable. That is, the computer itself was not just "a machine," but at any moment it was one of an almost infinite number of machines, depending on what its program told it to do. The ENIAC's users programmed it by plugging in cables from one part of the machine to another (an idea borrowed from telephone switchboards). This rewiring essentially changed it into a new machine for each new problem it solved. Other early computers got their instructions from punched strips of paper tape; the holes in the tape set switches in the machine, which amounted to the same kind of rewiring effected by the ENIAC's plug-boards. Feeding the computer a new strip of paper tape transformed it into a completely different device that could do something entirely different from what its designers had intended it to do. Howard Aiken designed his Automatic Sequence Controlled Calculator to compute tables of mathematical functions, and that it did reliably for many years. But in between that work it also solved problems in hydrodynamics, nuclear physics, and even economics.[15]

The computer, by virtue of its programmability, is not a machine like a printing press or a player piano—devices which are configured to perform a specific function.[16] By the classical definition, a machine is a set of devices configured to perform a specific function: one employs motors, levers, gears, and wire to print newspapers; another uses motors, levers, gears, and wire to play a prerecorded song. A computer is also made by configuring a set of devices, but its function is not implied by that configuration. It acquires its function only when someone programs it. Before that time it is an abstract machine, one that can do "anything." (It can even be made to print a newspaper or play a tune.) To many people accustomed to the machines of the Industrial Revolution, a machine having such general capabilities seemed absurd, like a toaster that could sew buttons on a shirt. But the computer was just such a device; it could do many things its designers never anticipated.

The computer pioneers understood the concept of the computer as a general-purpose machine, but only in the narrow sense of its ability to solve a wide range of mathematical problems. Largely because of their institutional backgrounds, they did not anticipate that many of the applications computers would find would require the sorting and retrieval of non-numeric data. Yet outside the scientific and university milieu, especially after 1950, it was just such work in industry and business that underlay the early expansion of the computer industry. Owing to the fact that the first computers did not do business work, the misunderstanding persisted that anything done by a computer was somehow more "mathematical" or precise than that same work, done by other means. Howard Aiken probably never fully understood that a computer could not only be programmed to do different mathematical problems but could also do problems having little to do with mathematics. In 1956 he made the following statement: ". . . if it should ever turn out that the basic logics of a machine designed for the numerical solution of differential equations coincide with the logics of a machine intended to make bills for a department store, I would regard this as the most amazing coincidence that I have ever encountered."[17] But the logical design of modern computers for scientific work in fact coincides with the logical design of computers for business purposes. It is a "coincidence," all right, but one fully intended by today's computer designers.

The question remained whether electronic computers worked too fast for humans to feed work into them. Engineers and computer designers met the problem of imbalance of speeds head-on, by technical advances at both the input and output stages of computing. To feed in programs and data, they developed magnetic tapes and disks instead of tedious plugboard wiring or slow paper tape. For displaying the results of a computation, high-speed line printers, plotters, and video terminals replaced the slow and cumbersome electric typewriters and card punches used by the first machines.

Still, the sheer bulk of the computer's output threatened to inundate the humans who ultimately wanted to use it. But that was not a fatal fault, owing (again) to the computer's programmability. Even if in the course of a computation a machine handles millions of numbers, it need not present them all as its output. The humans who use the computer need only a few numbers, which the computer's program itself can select and deliver. The program may not only direct the machine to solve a problem; it also may tell the machine to select only the "important" part of the answer and suppress the rest.

Ultimately, the spread of the computer beyond physics labs and large government agencies depended on whether people could write programs that would solve different types of problems and that would make efficient use of the high internal speed of electronic circuits. That challenge was not met by simply training and hiring armies of programmers (although sometimes it must have seemed that way). It was met by taking advantage of the computer's ability to store its programs internally. By transforming the programming into an activity that did not require mathematical training, computer designers exploited a property of the machine itself to sidestep the shortage of mathematically trained programmers.

Although the computer pioneers recognized the need for internal program storage, they did not at first see that such a feature would have such a significant effect on the nature of programming. The idea of storing both the program and data in the same internal memory grew from the realization that the high speed at which a computer could do arithmetic made sense only if it got its instructions at an equally high speed. The plugboard method used with the ENIAC got instructions to the machine quickly but made programming awkward and slow for humans. In 1944 Eckert proposed a successor to the ENIAC (eventually called the EDVAC) whose program would be supplied not by plugboards but by instructions stored on a high-speed magnetic disk or drum.

In the summer of 1944, John von Neumann first learned (by chance) of the ENIAC project, and within a few months he had grasped that giant machine's fundamentals—and its deficiencies, which Eckert and Mauchly hoped to remedy with their next computer. Von Neumann then began to develop a general theory of computing that would influence computer design to the present day.[18] In a 1945 report on the progress of the EDVAC he stated clearly the concept of the stored program and how a computer might be organized around it.[19] Von Neumann was not the only one to do that, but it was mainly from his report and others following it that many modern notions of how best to design a computer originated.

For von Neumann, programming a digital computer never seemed to be much of an intellectual challenge; once a problem was stated in mathematical terms, the "programming" was done. The actual writing of the binary codes that got a computer to carry out that program was an activity he called coding, and from his writings it is clear that he regarded the relationship of coding to programming as similar to that of typing to writing. That "coding" would be as difficult as it turned out to be, and that there could emerge a profession devoted to that task, seems not to have occurred to him. That was due in part to von Neumann's tremendous mental abilities and in part to the fact that the problems that interested him (such as long-range weather forecasting and complicated aspects of fluid dynamics[20]) required programs that were short relative to the time the computer took to digest the numbers. Von Neumann and Herman Goldstine developed a method (still used today) of representing programs by flow charts. However, such charts could not be fed directly into a machine. Humans still had to do that, and for those who lacked von Neumann's mental abilities the job remained as difficult as ever.

The intermediate step of casting a problem in the form of a flow chart, whatever its benefits, did not meet the challenge of making it easy for nonspecialists to program a computer. A more enduring method came from reconsidering, once again, the fact that the computer stored its program internally.

In his reports on the EDVAC, von Neumann had noted the fact that the computer could perform arithmetic on (and thus modify) its instructions as if they were data, since both were stored in the same physical device.[21] Therefore, the computer could give itself new orders. Von Neumann saw this as a way of getting a computer with a modest memory capacity to generate the longer sequences of instructions needed to solve complex problems. For von Neumann, that was a way of condensing the code and saving space.

However, von Neumann did not see that the output of a computer program could be, rather than numerical information, another program. That idea seemed preposterous at first, but once implemented it meant that users could write computer programs without having to be skilled mathematicians. Programs could take on forms resembling English and other natural languages. Computers then would translate these programs into long complex sequences of ones and zeroes, which would set their internal switches. One even could program a computer by simply selecting from a "menu" of commands (as at an automated bank teller) or by paddles and buttons (as on a computerized video game). A person need not even be literate to program.

That innovation, the development of computer programs that translated commands simple for humans to learn into commands the computer needs to know, broke through the last barrier to the spread of the new invention.[22] Of course, the widespread use of computers today owes a lot to the technical revolution that has made the circuits so much smaller and cheaper. But today's computers-on-a-chip, like the "giant brains" before them, would never have found a large market had a way not been found to program them. When the low-cost, mass-produced integrated circuits are combined with programming languages and applications packages (such as those for word processing) that are fairly easy for the novice to grasp, all limits to the spread of computing seem to drop away. Predictions of the numbers of computers that will be in operation in the future become meaningless.

What of the computer pioneers' early predictions? They could not foresee the programming developments that would spread computer technology beyond anything imaginable in the 1940s. Today, students with pocket calculators solve the mathematical problems that prompted the pioneers of that era to build the first computers. Furthermore, general-purpose machines are now doing things, such as word processing and game playing, that no one then would have thought appropriate for a computer. The pioneers did recognize that they were creating a new type of machine, a device that could do more than one thing depending on its programming. It was this understanding that prompted their notion that a computer could do "anything." Paradoxically, the claim was more prophetic than they could ever have known. Its implications have given us the unforeseen computer revolution amid which we are living.

NOTES

1. Harold Bergstein, "An Interview with Eckert and Mauchly," *Datamation* 8, no. 4 (1962), pp. 25–30.
2. Simon Lavington, *Early British Computers* (Bedford, Mass.: Digital Press, 1980), p. 104.
3. See, for example, John Wells, The Origins of the Computer Industry: A Case Study in Radical Technological Change, Ph.D. Diss., Yale University, 1978, pp. 93, 96, 119; Robert N. Noyce, "Microelectronics," *Scientific American* 237 (September 1977), p. 674; Edmund C. Berkeley, "Sense and Nonsense about Computers and Their Applications," in proceedings of World Computer Pioneer Conference, Llandudno, Wales, 1970, also in Phillip J. Davis and Reuben Hersh (eds.), *The Mathematical Experience* (New York: Houghton Mifflin, 1981).
4. See, for example, the proceedings of two early conferences: *Symposium on Large-Scale Digital Calculating Machinery*, Annals of Harvard University Computation Laboratory, vol. 16, 1949; *The Moore School Lectures: Theory and Techniques for Design of Electronic Digital Computers*, lectures given at Moore School of Electrical Engineering, University of Pennsylvania, 1946 (Cambridge, Mass.: MIT Press, 1986).
5. Georgia G. Mollenhoff, "John V. Atanasoff, DP Pioneer," *Computerworld* 8, no. 11 (1974), pp. 1, 13.
6. Byron E. Phelps, *The Beginnings of Electronic Computation*, IBM Corporation Technical Report TR-00.2259, Poughkeepsie, N.Y., 1971, p. 19.
7. Cuthbert C. Hurd, "Early IBM Computers: Edited Testimony," *Annals of the History of Computing* 3 (1981), pp. 163–182.
8. Anthony Oettinger, "Howard Aiken," *Communications ACM* 5 (1962), pp. 298–299, 352.
9. Henry Tropp, "The Effervescent Years: A Retrospective," *IEEE Spectrum* 11 (February 1974), pp. 70–81.
10. For an example of this argument, and a refutation of it, see John von Neumann, *Collected Works*, vol. 5 (Oxford: Pergamon, 1961), pp. 182, 236.
11. R. F. Clippinger, "Mathematical Requirements for the Personnel of a Computing Laboratory," *American Mathematical Monthly* 57 (1950), p. 439; Edmund Berkeley, *Giant Brains, or Machines that Think* (New York: Wiley, 1949), pp. 108–109.
12. Ralph J. Slutz, "Memories of the Bureau of Standards SEAC," in N. Metropolis, J. Howlett, and G. Rota (eds.), *A History of Computing in the Twentieth Century* (New York: Academic, 1980), pp. 471–477.
13. In a typical computing installation of the 1930s, humans worked, with mechanical calculators that could perform the four elementary operations of arithmetic, on decimal numbers having up to ten digits, taking a few seconds per operation. Although the machines were powered by electric

motors, the arithmetic itself was always done by mechanical parts–gears, wheels, racks, and levers. The machines were sophisticated and complex, and they were not cheap; good ones cost hundreds of dollars. For a survey of early mechanical calculators and early computers, see Francis J. Murray, *Mathematical Machines,* vol. 1: *Digital Computers* (New York: Columbia University Press, 1961); see also Engineering Research Associates, *High-Speed Computing Devices* (New York: McGraw-Hill, 1950; Cambridge, Mass.: MIT Press, 1984).

14. Quoted in Nancy Stern, *From ENIAC to UNIVAC* (Bedford, Mass.: Digital Press, 1981), p. 87.
15. Oettinger, "Howard Aiken."
16. Abbott Payson Usher, *A History of Mechanical Inventions,* second edition (Cambridge, Mass.: Harvard University Press, 1966), p. 117.
17. Howard Aiken, "The Future of Automatic Computing Machinery," in *Elektronische Rechenmaschinen und Informationsverarbeitung,* proceedings of a symposium, published in *Nachrichtentechnische Fachberichte* no. 4 (Braunschweig: Vieweg, 1956), pp. 32–34.
18. Herman H. Goldstine, *The Computer From Pascal to von Neumann* (Princeton University Press, 1972), p. 182.
19. Von Neumann's "First Draft of a Report on the EDVAC" was circulated in typescript for many years. It was not meant to be published, but it nonetheless had an influence on nearly every subsequent computer design. The complete text has been published for the first time as an appendix to Nancy Stern's *From ENIAC to UNIVAC* (note 14).
20. Von Neumann, *Collected Works,* vol. 5, pp. 182, 236.
21. Martin Campbell-Kelley, "Programming the EDVAC," *Annals of the History of Computing* 2 (1980), p. 15.
22. For a discussion of the concept of high-level programming languages and how they evolved, see H. Wexelblatt (ed.), *History of Programming Languages* (New York: Academic, 1981), especially the papers on FORTRAN, BASIC, and ALGOL.

11. Strategies for Regulating Risky Technologies

JOSEPH G. MORONE
EDWARD J. WOODHOUSE

Controlling technologies in modern society means understanding and dealing with their risks. Some of the most beneficial technologies available to society bring with them risks of potential catastrophes. Is it possible to manage such technologies in ways that avert the catastrophic consequences of accident or misuse? In this chapter, taken from the conclusion of their book, Averting Catastrophe, *Joseph Morone and Edward Woodhouse examine the strategies that have evolved in the United States for dealing with major risks in five areas: toxic chemicals, nuclear power, recombinant DNA research, threats to the ozone layer, and the greenhouse effect.*

In particular, the authors focus their attention on the problems that remain unsolved by existing risk management techniques. "Can we do better?" they ask. Their analysis leads them to look at how society answers the question "How safe is safe enough?" To improve the current system, they propose a strategic approach, including four steps: (1) attack egregious risks; (2) seek alternatives that transcend or circumvent risks; (3) develop research strategies to reduce key uncertainties; and (4) be prepared to learn from error, rather than expecting to find all the answers in advance.

Joseph Morone is with the Corporate Research and Development Department of the General Electric Company in Schenectady, New York. He served as an Industrial Research Institute fellow at the White House Office of Science and Technology Policy in 1984. Edward (Ned)

Source: Chapter 8 from *Averting Catastrophe: Strategies for Regulating Risky Technologies*, pp. 150–202. Copyright © 1986 by The Regents of the University of California. Permission granted by The Regents of the University of California and the University of California Press.

Woodhouse is a political scientist who teaches in the interdisciplinary Department of Science and Technology Studies at Rensselaer Polytechnic Institute (RPI) in Troy, New York.

HOW SAFE IS SAFE ENOUGH?

As regulators have developed strategies for coping with potential catastrophes, these very strategies have created a new and sometimes more perplexing problem: when is the catastrophe-aversion system good enough? The question arises because the strategies can be implemented in any number of ways, from very rigorously to very loosely. One such strategy (proceed cautiously) clearly illustrates the problem: How cautious is cautiously? How cautious should be the implementation of initial precautions: very rigorous (early rDNA), moderate (U.S. fluorocarbons), or mild (scrutiny of new chemicals)? No matter how strictly catastrophe-aversion strategies have been applied, they can always be applied even more rigorously—even to the point of an outright ban. The problem would be easily resolved were it not for the fact that the precautions are costly, and each degree of rigor brings additional costs.

Among the cases we have reviewed, the problem of weighing the benefits of additional safeguards against the costs is most apparent in the case of nuclear power, and, indeed, is a central element in contemporary nuclear policy debates. Since containment cannot be guaranteed, emphasis is placed on preventing malfunctions and mistakes from triggering serious mishaps. Since prevention requires that all serious possibilities be taken into account, there always seems to be one more set of precautions—and expenses—that perhaps ought to be undertaken. The increase in precautions (and expenses) brings with it the issue of whether additional precautions are worth the cost. Increasingly, new questions are raised: When have we gone far enough in attempting to avoid potential catastrophe? When are reactors safe enough?

The same problem arises, to varying degrees, in our other cases. For toxic chemicals, is it reasonable to focus regulatory concern on the fifty most potentially hazardous substances? Why not twenty, or one hundred, or even five hundred? How many different types of tests should be required, conducted in how many species of laboratory animals? And how dangerous should a chemical be in order to restrict or prohibit its use? The Delaney amendment specifies a zero tolerance for any food additive that causes cancer in animals. But this requirement clearly is too cautious.[1] As toxicologists become able to detect chemicals at the parts per billion or trillion level, virtually all foods will be found to contain traces of something objectionable. The problem thus is similar to the one faced by nuclear regulators when they recognized that containment no longer could be guaranteed in large reactors. In both cases, the zero risk option—no risk of radiation exposures, no trace of a toxic compound—becomes impossible to achieve. Some risk is unavoidable, so the issue is then how much risk is acceptable.

For the greenhouse effect, the same question arises in a somewhat different form. It is not "How far should we go in reducing the risks?" but "How far should we reduce the uncertainties before beginning to reduce the risks?" A potential for catastrophic changes in climate due to greenhouse gases is undeniable, but there remain major uncertainties about the timing and magnitude of the risks as well as the costs associated with the regulatory options. At what point does the likelihood of catastrophe outweigh the costs of action? Possibly this point has not yet been reached, for most scientists and policymakers are still waiting for clarification of the greenhouse threat before deciding whether action is necessary. But this stance rests as much on judgment as on science.

If there is a difference between the greenhouse problem and the nuclear power and toxic substances cases, it is that the nondecision on the greenhouse effect has not been subject to as much controversy. Nevertheless, the potential for yet another open-ended, contentious debate is present, as illustrated by the simultaneous (though coincidental) release in late 1983 of two reports on the greenhouse effect. A National Academy of Sciences report concluded that any responses to the threat should await a reduction in the uncertainties. An EPA report—quickly disavowed by the Reagan administration—concluded the reverse: uncertainties notwithstanding, action should be taken soon. The issue, while not yet a full-blown controversy, looms ahead: how uncertain is uncertain enough?

Another way to put this is, "How much safety should we purchase?" Because people disagree about how much they are willing to spend to reduce risks to health and to the environment, political battles and compromises over safety expenditures are inevitable. This topic is inherently controversial, so it is no surprise that there are long-running, fiercely contested debates. When we consider not just the potential for catastrophe and strategies for avoiding it but also the issue of how stringently to employ those strategies to achieve a sensible balance of costs and benefits, the task facing regulators becomes much more demanding. Whereas regulators seem to be learning to handle the catastrophe-aversion problem, they are having a much harder time with the "How safe?" question.

Setting a Safety Goal

In the early 1980s, the Nuclear Regulatory Commission made an explicit attempt to resolve the "How safe?" question for nuclear power plants. We review that attempt here to illustrate the nature of the problem and the reasons that it has proven so difficult to resolve. The lessons that can be drawn from this example apply to most risky technologies.

The notion of explicitly addressing the "How safe?" issue emerged well before Three Mile Island, but the accident provided a strong impetus. Several post-accident analyses recommended that the NRC explicitly identify a safety goal—a level of risk at which reactors would be safe enough. Establishing such a goal, advocates believed, would end the interminable debates over whether reactors should be made safer. What quickly became apparent, however, was that establishing a stopping point was far easier to recommend than to achieve. To establish a safety goal, regulators would have to resolve two complex and politically sensitive issues. First, what is an acceptable risk of death and injury? And second, how should regulators determine whether reactors actually pose such an acceptably low risk?

Identifying an Acceptable Level of Risk.   A commonly proposed solution to the first problem is to make the acceptable level of risk for nuclear reactors comparable to the risks associated with other technologies.[2] If society has accommodated these other technologies, the argument goes, it is reasonable to assume that society accepts the associated risks.

This approach has proven to be problematic. To begin with, an already accepted technology that bears comparison with nuclear energy is yet to be found. To illustrate this problem, consider the risks of driving an automobile. One can drive a large car or a small one; one can drive cautiously or recklessly, soberly or drunkenly, with seat belts or without. In contrast, the risks of nuclear power are not as much within the individual's control; the only option an individual has is to move farther away from a reactor. The nature of the hazard associated with these two technologies differs also. Automobiles produce many fatalities through numerous independent events; a serious reactor accident might provide many fatalities from a single event. So does it make sense to compare automobiles with nuclear reactors? Some say "yes"—a death is a death. Others say "no"—high-probability, low-consequence risks that are partially subject to individual control are fundamentally different from low-probability, high-consequence risks over which the individual has no control.

A possible way to overcome this difficulty is to compare the risks from other sources of electricity with those from nuclear power. But this leads to a new problem: how to measure those risks. The hazards of coal are well known—air pollution, acid rain, and possible overheating of the earth's atmosphere—but the level of risk is uncertain.[3] Also in dispute is the range of risks that should be included in such a comparison. Should the risks of mining and transportation be included? What about the risks of waste disposal and sabotage? If the risks of nuclear power are compared to burning oil, what about the risks of a cutoff of oil from the Mideast or the chances of being involved in a war in the Mideast?

Using this comparative approach to define an acceptable level of risk for nuclear power also poses other problems. It assumes that society, after reasoned evaluation, actually has accepted the risks associated with these technologies. Judging from the controversies surrounding air pollution, acid rain, the greenhouse effect, and the health and safety of miners, millions of people do not accept the levels of risk currently posed by coal burning. Moreover, in cases where people seem to accept high risks for an activity that easily could be made significantly safer (such as driving a car), the implicit rejection of precautions that lower risks might not be rational. (Indeed, it is hard to see how the refusal to fasten seat belts can be anything but irrational.) Should the irrational standards that society applies to driving or other unnecessarily risky activities also be applied to nuclear power?

In spite of these problems, the NRC proposed a safety goal in February 1982, after about a year and a half of deliberations. Reactors would be considered safe enough when, among other requirements:

1. The risk to the population near the reactor of being killed in a reactor accident should not exceed 0.1 percent of the risk of being killed in any kind of an accident; and

2. The risk to the population living within fifty miles of the plant of eventually dying from cancer as a result of a reactor accident should not exceed 0.1 percent of the risk of dying from any other causes.[4]

When first proposed, the second of these goals set off a flurry of controversy because 0.1 percent of the cancer rate for a fifty-mile radius would amount to an average of three cancer fatalities per reactor per year. This would be a total of 13,500 deaths over the next thirty years in an industry comprised of 150 reactors—a figure critics argued was too high. The NRC could have responded to this criticism by revising the second goal, but this would have triggered criticism from proponents of nuclear power, who would have argued that the goal was too strict compared with other risks that society accepts. Thus, both parts of the safety goal have remained as originally drafted.

Verifying That the Safety Goal Has Been Met

If, despite the difficulties, an acceptable level of risk could be agreed on by a majority of policymakers, regulators then would have to determine whether the goal actually has been met. To evaluate this, regulators must know the level of safety achieved by the various safety strategies: they must have the right facts. For nuclear regulators, such a task is even more difficult than identifying an acceptable risk level. The NRC recognized this, and announced that because of "the sizeable uncertainties . . . and gaps in the data base" regarding actual safety levels, the two goals would serve as "aiming points or numerical benchmarks," not as stopping points.[5]

As an illustration of factual uncertainties, consider the first NRC goal concerning the risks of being promptly killed by a reactor accident. Five people in ten thousand are killed by some kind of an accident each year. For a reactor with two hundred people living within a mile, the NRC's goal implies that the annual probability of an accident killing one person should be no more than one in ten thousand.[6] The probability (per reactor per year) of accidents in which ten people are killed should be no more than one in 100 thousand, for one hundred deaths no more likely than one in a million, and so on.

These probabilities are minuscule; they are reassuring because they suggest that the NRC expects serious accidents to be extremely rare. But precisely because the probabilities are so small, it would take hundreds of years for an industry of one hundred reactors to accumulate enough experience to show that reactors satisfy the safety goal. Unless actual probabilities are much higher than those deemed acceptable, experience cannot help in determining whether the risks associated with reactors are as low as stipulated by the safety goal.

The only alternative to learning from experience for determining whether the actual probability of reactor accidents satisfies the safety goal is to use analytic techniques such as fault tree analysis; this, in fact, is how advocates of the safety goal propose to proceed. In fault tree analysis, the analyst attempts to identify all the possible sequences of errors and malfunctions that could lead to serious accidents. For each sequence, the probabilities of each of the errors and malfunctions must be estimated, and from these individual probabilities a probability estimate for the entire sequence is derived. Assuming that the various sequences of events are independent, the analyst then totals the probabilities of each of the sequences. This sum represents the probability estimate of a serious accident.

The key to this form of analysis is that the analyst does not attempt to estimate the probability of a serious accident directly. Because such events have never occurred, there is no data upon which to base an estimate. Instead, the analyst focuses on the sequence of events that would lead to the accident. Unlike the accident itself, the individual events in each of the sequences are relatively common—not only in reactors, but also in a variety of industrial enterprises. For example, the nuclear industry has decades of experience with control rod mechanisms (which control the rate of chain reaction in the reactor), so it is possible to develop fairly reliable estimates of the likelihood that the mechanisms will fail. Similarly, from experience with the nuclear and other industries, it is possible to estimate the probability of power failures, pipe failures, pump failures, and so on.

Unfortunately, fault tree analysis is subject to the same uncertainties that have plagued nuclear regulators since the mid-1960s. What if the analysis fails to identify all the possible sequences of malfunctions that could lead to a serious accident? What if safety systems presumed to be independent actually are vulnerable to common faults? What if the probability of inherently uncertain problems, such as operator errors and terrorist attacks, have been underestimated? The regulator must thus confront yet another dilemma: the only practical method for determining the actual probability of reactor accidents is to use analytic techniques, but such techniques are subject to considerable uncertainties. At best, analysis can result in estimates of probabilities. The only way to verify these estimates is through experience, but experience, because of the very low probabilities, is of little help.

Determining whether reactors satisfy a safety goal is further complicated by the fact that the consequences of core melts are uncertain. One of the effects of the mid-1960s shift to a prevention philosophy was that all research about the behavior of core melts halted. The argument was that since core melts were to be prevented, there was no need to study them. As a result, little is now known about the consequences of serious mishaps with the core. Among many other unknowns, it is not clear what portion of the radioactive fission products would actually escape from the reactor in a core melt.[7] It is also not known what would happen to the core if it melted entirely. Its heat might dissipate on the containment floor and the core solidify there instead of melting through. If the molten core came in contact with water accumulated on the containment floor, would it set off a steam explosion? If so, would it be powerful enough to rupture containment?

These and many other aspects of meltdowns are of critical importance in determining the actual risks of reactors, and all of these aspects are uncertain. If the actual risks of reactors are uncertain, then regulators cannot determine with confidence whether the risks associated with reactors are acceptable. They might agree on the levels of risk that would be acceptable, but because of the uncertain probabilities and consequences of reactor accidents, regulators do not know whether the actual risks are as low as required.

In general, most of the proponents of a safety goal failed to appreciate the significance of existing factual uncertainties. They urged the NRC to establish a goal, and (although this step was controversial) the NRC did so. Yet the goal had no impact on the nuclear regulatory process, which currently remains as open-ended and contentious as ever. A safety goal is of little value unless partisans in a dispute can recognize when the goal has been achieved. Ironically, such recognition is prevented by the very uncertainties that made the regulatory process open-ended and led to the demand for a safety goal.

Discussion

Establishing safety goals is the epitome of the analytic approach to risk analysis that dominates professional thinking. Various forms of risk–benefit analysis and cost–benefit analysis are often presented by risk professionals as ways to settle disagreements.[8] To be workable, however, all such analytic methods must confront the same two obstacles that the NRC faced in setting the safety goal: value uncertainties ("And how much for your grandmother?")[9] and factual uncertainties. No analytic methods surmount these obstacles.

With toxic chemicals, for example, regulators are confronted with essentially the same two questions as they are in the nuclear controversy. First, how large a risk of cancer is acceptable? Is it one in one thousand, one in ten thousand, one in a million? And then there is the factual uncertainty that arises with attempts to establish the actual level of risk. Chemical testing typically is performed with animals. Are the responses in animals comparable to responses in humans? Are the dosage rates comparable? Is it even possible to define the human population at risk? As with nuclear power, these questions are answered largely with assumptions, and to a considerable extent, the assumptions are untestable in practice.

For example, the pesticide EDB produces cancer in animals exposed to high dosage levels. We make the conservative assumption that it will do so in humans as well, but we may never be able to test this assumption. Obviously, we would not want to use humans as test subjects, so the only alternative is to use epidemiological evidence—to study exposed populations. Such study is problematic, however, since the population that might be tested for low exposures to EDB might simultaneously have been exposed to so many other possible sources of cancer that it becomes impossible to link cause and effect. We can identify the presence of EDB and, in some cases, levels of exposure to it, but in practice we are not able to establish a close relationship to the possible effects.

It becomes apparent, on reflection, that there is something inappropriate about applying an analytic solution to risk disputes. Even putting aside all the practical difficulties of verifying when a safety goal has actually been achieved, the idea that conflicts of value can be reduced to a formula is at odds with the way that real people and real societies actually function. Of course, as discussed, society does in a certain sense implicitly accept various levels of risk from technologies now in use. But most people find it morally offensive to plan explicitly for the number of deaths and injuries that will be acceptable. Thus, it is not surprising that there is reluctance to directly confront the "How safe?" problem, even if it is "rational" to do so.

Moreover, while governments frequently deal with value-charged social issues (such as abortion), these issues rarely are quickly resolved. Instead, elements of the issue will repeatedly show up on governmental agendas over decades, producing compromises that gradually evolve with changing social mores. Analytic strategies rarely, if ever, affect these outcomes, and there is little reason to expect that the value side of risk disputes will depart from this pattern. To be workable, theories of risk management must be compatible with how society's value decisions actually are made.

In principle, establishing a safety target is a perfectly rational approach to improving the catastrophe-aversion system. In practice, it requires that touchy and politically charged value judgments be backed up by factual judgments that are difficult or impossible to verify. So analysis inevitably falls short, as illustrated in the cases studied. While analysis often can be useful as an adjunct, it rarely is a substitute for judgment and strategy.

A STRATEGIC APPROACH TO IMPROVED RISK MANAGEMENT

To improve the efficiency and effectiveness of the catastrophe-aversion system, we must adopt a more strategic approach. We see four promising steps.

1. Attack egregious risks—those clearly worse than others even after allowing for uncertainties;
2. Seek and employ alternatives that transcend or circumvent risks;
3. Develop carefully prioritized research strategies to reduce key uncertainties;
4. Be actively prepared to learn from error, rather than naively expecting to fully analyze risks in advance or passively waiting for feedback to emerge.

Attack Egregious Risks

Because resources always are limited, society is forced to set priorities. Dollars spent to avert catastrophes are not available for social services. Money spent to avert one type of catastrophe is not available for averting other types. Priority setting can be done in a relatively systematic manner, or it can be done haphazardly. Priorities are now set haphazardly; we are grossly inconsistent in our attempts to reduce various kinds of risks.

For example, chemical wastes are many times greater in volume than radioactive wastes, and some are actually longer lived. Yet they tend to be buried in insecure landfills near the surface of the earth, rather than in the deep geological repositories being designed for radioactive wastes; and EPA regulations require that hazardous wastes be contained for only thirty years, compared with ten thousand years for radioactive wastes. And why do we worry about some exposures to radioactivity and not about others? In the mid-1970s, Swedish scientists examining newer housing found high levels of radon from concrete containing alum shale, high in radium. Researchers subsequently have found alarming levels of indoor radon in many parts of the United States; even average homes expose occupants to a cancer risk greater than that posed by most dangerous chemicals. Remedial steps to reduce radon risks are available, but regulation had not been initiated by 1986.[10]

This inconsistency in our approach to risks is by no means an exception, as we can see from the following data.[11]

Safety Measure	Estimated Cost per Life Saved
Cancer screening programs	$10,000+
Mobile cardiac emergency units	$30,000
Smoke detectors	$50,000+
Seat belts	$80,000
Emergency core cooling system	$100,000
Scrubbers to remove sulfur dioxide from coal-fired power plants	$100,000+
Auto safety improvements, 1966–1970	$130,000
Highway safety programs	$140,000
Kidney dialysis treatment units	$200,000
Automobile air bags	$320,000
Proposed upholstered furniture flammability standard	$500,000
Proposed EPA drinking water regulations	$2.5 million
Reactor containment building	$4 million
EPA vinyl chloride regulations	$4 million
OSHA coke fume regulations	$4.5 million
On-site radioactive waste treatment system for nuclear power plants	$10 million
OSHA benzene regulations	$300 million
Hydrogen recombiners for nuclear reactors	$3 billion

This study shows that certain design changes in nuclear reactors would cost as much as $3 billion per life saved, whereas additional highway safety could be achieved for as little as $140,000 per life. Other analyses have resulted in somewhat different estimates, but it is clear that there is a vast discrepancy concerning funds spent to save lives from various threats.

Focusing political attention on the overall costs of averting risks would help balance such gross discrepancies. One course would be to establish a government agency or congressional committee with authority to set priorities for risk reduction. A more realistic option would make total expenditures subject to a unified congressional authorization procedure. Currently, competing proposals for risk abatement do not confront one another. New safety procedures required by the NRC for electric utilities that use nuclear power in no way impinge on the amount spent for highway safety, nor does either of these expenditures influence expenditures for testing and regulation of chemicals. The result is that safety proposals are not compared with each other, so neither government nor the media nor the public is forced to think about comparative risks.

Factual uncertainties prevent precise comparisons among risks, but precise comparisons often are not needed. There are such gross discrepancies in our approaches to different risks that much can be done to reduce these risks without having to confront the intractable uncertainties. Compared to attacking egregious risks that have been relatively unattended, making precise comparisons among risks that already are regulated seems like fine tuning. While it might be nice to make precise comparisons and resolve the "How safe?" debate, doing so is not as important as attacking the egregious risks. Unfortunately, such fine tuning preoccupies professional risk assessors, regulators, and political activists and results in a waste of time and energy.

Transcend or Circumvent Risks

A second strategic approach would take advantage of risk-reduction opportunities that circumvent troublesome risks. The greenhouse issue provides a good illustration. Virtually all attention devoted to this problem has focused on carbon dioxide emissions from combustion of fossil fuels. Yet fossil fuels are considered fundamental to contemporary life, and the costs of significant reductions in their use could be severe; so there is widespread reluctance to take any action without a much better understanding of the risks. The net effect is that we wait and debate whether the risk is real enough to warrant action. Until the uncertainties are reduced, there is no rational basis for resolving the debate.

But there may be an alternative. Carbon dioxide is not the only contributor to the greenhouse problem. Other gases, such as nitrous oxide, are also major factors. It is conceivable that emissions of these other gases might be easier to control and might thereby offer an opportunity to at least delay or reduce the magnitude of the greenhouse effect. The 1983 NAS and EPA studies make note of this possibility but do not analyze it in any detail.[12] By early 1986 little sustained attention had been paid to the policy options potentially available for reducing non-CO_2 greenhouse gases.

Similarly, discussions of the options for combating the greenhouse effect have focused on costly restrictions on the use of high carbon fuels, but it may be possible to achieve at least some of the benefits of such restrictions through a much less costly combination of partial solutions. This combination of solutions might include partial reforestation, plus research on crop strains better adapted to dry climates, plus partial restrictions on only the highest carbon fuels.

Another means of circumventing uncertainties about a risk is to develop a method of offsetting the risk. Quite inadvertently, the ozone threat eased when it was found that low-flying airplanes emit chemicals that help produce ozone. Could a similar approach be pursued deliberately for some technological risks? In the greenhouse case, deliberate injection of sulfur dioxide or dust into the atmosphere might result in temporary cooling similar to that achieved naturally by volcanic dust. Deliberate intervention on such a scale might pose more environmental danger than the original problem, but careful analysis of this possibility is surely warranted.

The case of nuclear power provides another possible approach to circumventing risks and uncertainties about risks. Interest is growing in the notion of inherently (or passively) safe reactors—reactors for which there is no credible event or sequence of events that could lead to a meltdown. The reactor concepts now receiving the most attention include small high temperature gas cooled reactors and the PIUS reactor (a light water reactor with the core immersed in a pool of borated water).[13] Preliminary analyses indicate that these reactors are effectively catastrophe proof. Even if the control systems and cooling systems fail, the reactors will still shut themselves down.

Skeptics argue that the concept of inherent safety probably cannot be translated into practice, and that such reactors in any case would not be economical. But in the history of commercial power reactors there has never before been a deliberate attempt to build an inherently safe reactor, and some analysts believe that these new reactors can provide, if not "walk away" safety, at least substantially reduced risks. If this is true, these new reactor concepts provide the opportunity to short circuit much of the "How safe?" debate for nuclear power plants. If it can be shown that such reactors are resistant to core melts in all credible accident scenarios, then many of the open-ended and contentious safety arguments could be avoided. While we do not know whether inherently safe reactors will prove feasible, and while there are other controversial aspects of the nuclear fuel cycle (particularly waste disposal), nonetheless, the possibility that reactors could approach inherent safety is well worth considering. Resistance to this concept apparently is due more to organizational inertia than to sound technical arguments. Thus, in spite of the fact that the concept of inherent safety has been in existence for thirty years, society has been subjected to a bitter and expensive political battle, that a more strategic approach to this topic might have circumvented.

A very different approach to transcending factual uncertainties is to compromise. When policymakers are at an impasse over how safe a technology is or should be, it may at times be possible to reach a solution that does not depend on the resolution of the uncertainties. This strategy is already used, but it is not employed consciously enough or often enough. Because each opportunity for creative compromise necessarily is unique, there can be no standard operating procedure. However, examples of the advantages of compromise abound.

For example, the Natural Resources Defense Council, EPA, and affected industries have reached several judicially mediated agreements that have accomplished most of the limited progress made to date against toxic water pollutants.[14] Another example is the negotiated approach to testing of priority chemicals adopted in 1980 by EPA toward the chemical industry. The possibility of creative compromise was not envisioned by the framers of the Toxic Substances Control Act, but neither was it prohibited. Numerous protracted analysis-based hearings and judicial challenges thereby have been avoided, and judging from the limited results available to date, testing appears to be proceeding fairly rapidly and satisfactorily.

Had compromises and tradeoffs been the basis for setting standards throughout the toxic substances field, many more standards could have been established than actually have been.[15] Then they could have been modified as obvious shortcomings were recognized. Of course, compromise agreements can be very unsatisfying to parties on either side of the issue who believe they know the truth about the risks of a given endeavor. But, by observing past controversies where there was under- or overreaction to possible risks, there is a fair prospect that all parties to future controversies gradually will become more realistic.

Reduce Uncertainties: Focused Research

A third option for strengthening the catastrophe-aversion system is to create research and development programs focused explicitly on reducing key factual uncertainties. This seems an obvious approach, yet it has not been pursued systematically in any major area of technological risk except for recombinant DNA. Of course, regulatory agencies have research and development (R&D) programs that investigate safety issues, but priorities ordinarily are not well defined and research tends to be ill matched to actual regulatory debates.

The greenhouse case again provides a good illustration, particularly since the uncertainties associated with it are so widely recognized as being at the heart of the debate about whether or not action is required. The NAS report could not have been more explicit about the importance of the uncertainties to the greenhouse debate:

> Given the extent and character of the uncertainty in each segment of the argument—emissions, concentrations, climatic effects, environmental and societal impacts—a balanced program of research, both basic and applied, is called for, with appropriate attention to more significant uncertainties and potentially more serious problems.[16]

Yet as clearly as the report recognizes the importance of the factual uncertainties, it fails to develop a strategy for dealing with them. It merely cites a long list of uncertainties that requires attention. The NRC listed over one hundred recommendations, ranging from economic and energy simulation models for predicting long-term CO_2 emissions, to modeling and data collection on cloudiness, to the effects of climate on agricultural pests.

Certainly answers to all of these questions would be interesting and perhaps useful; but, just as certainly, answers to some of them would be more important than answers to others. What are the truly critical uncertainties? What kinds of information would make the biggest differences in deciding whether or not to take action? As R&D proceeds and information is gained, are there key warning signals for which we should watch? What would be necessary to convince us that we should not wait any longer? Policymakers and policy analysts need a strategy for selectively and intelligently identifying, tracking, and reducing key uncertainties.

A similar problem arises in the case of nuclear power. In principle, nuclear regulators should systematically identify the central remaining safety uncertainties—the issues that will continue to lead to new requirements for regulations. Regulators should then devise a deliberate R&D agenda to address such uncertainties. A prime example is uncertainty about the behavior of the reactor core once it begins to melt. Clearly, this lies at the heart of the entire nuclear debate, since the major threat to the public results from core melts. Yet, as we discussed earlier, virtually no research was performed on core melts in the 1960s and 1970s.

Information and research resulting from the experience of Three Mile Island now have called into question some of the basic assumptions about core melts. For example, if the TMI core had melted entirely, according to the Kemeny Commission it probably would have solidified on the containment floor.[17] Even the nuclear industry had assumed that a melted core would have gone through the floor. Moreover, it appears that there were a variety of ways in which the core melt could have been stopped. Prior to the accident, the common assumption was that core melts could not be stopped once underway. Also overestimated, according to some recent studies, is the amount of radioactive material predicted to escape in a serious reactor accident: prior assumptions may have been ten to one thousand times too pessimistic.[18]

If such revised ideas about reactor accidents were to be widely accepted, they would have a substantial effect on the perceived risks of reactor accidents. But all such analyses are subject to dispute. To the extent feasible, therefore, it clearly makes sense to invest in research and development that will narrow the range of credible dispute— without waiting for the equivalent of a TMI accident. As with the greenhouse effect, what is needed is a systematic review of prevailing uncertainties and an R&D program devised to strategically address them. The uncertainties that make the biggest difference must be identified, those that can be significantly reduced by R&D must be selected, and an R&D program focused on these uncertainties must then be undertaken. In other words, a much better job can be done of using analysis in support of strategy.

Reduce Uncertainty: Improve Learning from Error

As noted previously, learning from error has been an important component of the strategies deployed against risky technologies. But learning from error could be better used as a focused strategy for reducing uncertainties about risk. As such, it would constitute a fourth strategic approach for improving the efficiency and effectiveness of the catastrophe-aversion system.

The nuclear power case again offers a good illustration of the need to prepare actively for learning from error. Suppose that a design flaw is discovered in a reactor built ten years ago for a California utility company. Ideally, the flaw would be reported to the Nuclear Regulatory Commission. The NRC would then devise a correction, identify all other reactors with similar design flaws, and order all of them to institute the correction. In actual operation, the process is far more complicated and the outcome far less assured.

To begin with, in any given year the NRC receives thousands of reports about minor reactor mishaps and flaws. The agency must have a method of sifting this mass of information and identifying the problems that are truly significant. This is by no means a straightforward task, as exemplified by the flaw that triggered the Three Mile Island accident. A similar problem had been identified at the Davis-Besse reactor several years earlier, but the information that was sent to the NRC apparently was obscured by the mass of other data received by the agency. Several studies of the TMI accident noted this unfortunate oversight, and concluded that the NRC and the nuclear industry lacked an adequate mechanism for monitoring feedback. In response, the nuclear industry established an institute for the express purpose of collecting, analyzing, and disseminating information about reactor incidents. This action represents a significant advance in nuclear decision makers' ability to learn from experience.

Even with a well-structured feedback mechanism, there are still other obstacles to learning from experience. One such obstacle arises from the contentious nature of current U.S. regulatory environments, which can actually create disincentives to learning. Given the adversarial nature of the nuclear regulatory environment, many in the nuclear industry believe that they will only hurt themselves if they propose safety improvements in reactor designs. They fear that opponents of nuclear power will use such safety proposals to argue that existing reactors are not safe enough, and that regulators will then force the industry to make the change on existing reactors, not just on new ones. This would add another round of costly retrofits.

Another obstacle to learning from experience can arise from the nature of the industry. For example, the nuclear industry is comprised of several vendors who over the years have sold several different generations of two different types of reactors to several dozen unrelated utility companies. Furthermore, even reactors of the same generation have been partially custom designed to better suit the particular site for which they were intended. This resulting nonuniformity of reactor design is a significant barrier to learning from experience, because lessons learned with one reactor are not readily applicable to others.

The design flaw uncovered at our hypothetical California utility's ten-year-old reactor probably can be generalized to the few reactors of the same generation (unless the flaw was associated with some site-specific variation of the basic design). It is less likely to apply to reactors built by the same vendor but of different generations, much less likely to apply to reactors of the same general type made by other vendors, and extremely unlikely to apply to other reactor types. Furthermore, lessons gained from experience in maintaining and operating reactors are also hard to generalize. Since reactors are owned by independent utilities, the experience of one utility in operating its reactor is not easily communicated to other utilities. In many respects, therefore, each utility must go through an independent learning cycle.

There also are significant barriers to learning about most toxic chemicals. The large number of such chemicals, the vast variety of uses and sites, and the esoteric nature of the feedback make the task of monitoring and learning from experience extraordinarily difficult. Yet the EPA's tight budget and the limited resources of major environmental groups means that routine monitoring will not get the attention that is given to other more pressing needs. What a good system for such monitoring would be is in itself a major research task, but just obtaining reliable information on production volumes, uses, and exposures would be a good place to start.

The point, then, is that active preparation is required to promote learning from experience. The institutional arrangements in the regulatory system must be devised from the outset with a deliberate concern for facilitating learning from error. In the nuclear power case, the ideal might be a single reactor vendor, selling a single, standardized type of reactor to a single customer. The French nuclear system comes close to this pattern.[19]

CONCLUSION

In summary, there are at least four promising avenues for applying risk-reduction strategies more effectively. The first strategy is to make an overall comparison of risks and to focus on those that clearly are disproportionate. The second is to transcend or circumvent risks and uncertainties by employing creative compromise, making technical corrections, and paying attention to easier opportunities for risk reduction. The third strategy is to identify key uncertainties and focus research on them. The fourth is to prepare from the outset to learn from error; partly this requires design of appropriate institutions, but partly it is an attitudinal matter of embracing error as an opportunity to learn. Finally, implicit throughout this study is a fifth avenue for improvement: by better understanding the repertoire of strategies available for regulating risky technologies, those who want to reduce technological risks should be able to take aim at their task more consciously, more systematically, and therefore more efficiently.

Of these, the first strategy probably deserves most attention. Attacking egregious risks offers simultaneously an opportunity to improve safety and to improve cost effectiveness. As an example, consider the 1984 Bhopal, India, chemical plant disaster.[20] The accident occurred when:

A poorly trained maintenance worker let a small amount of water into a chemical storage tank while cleaning a piece of equipment;

A supervisor delayed action for approximately one hour after a leak was reported because he did not think it significant and wanted to wait until after a tea break;

Apparently as an economy measure, the cooling unit for the storage tank had been turned off, which allowed a dangerous chemical reaction to occur much more quickly;

Although gauges indicated a dangerous pressure buildup, they were ignored because "the equipment frequently malfunctioned";

When the tank burst and the chemical was released, a water spray designed to help neutralize the chemical could not do so because the pumps were too small for the task;

The safety equipment that should have burned off the dangerous gas was out of service for repair and anyway was designed to accommodate only small leaks;

The spare tank into which the methyl isocyanate (MIC) was to be pumped in the event of an accident was full, contrary to Union Carbide requirements;

Workers ran away from the plant in panic instead of transporting nearby residents in the buses parked on the lot for evacuation purposes;

The tanks were larger than Union Carbide regulations specified, hence they held more of the dangerous chemical than anticipated;

The tanks were 75 percent filled, even though Union Carbide regulations specified 50 percent as the desirable level, so that pressure in the tank built more quickly and the overall magnitude of the accident was greater.

The length of this list of errors is reminiscent of the Three Mile Island accident. The difference between the two incidents is that TMI had catastrophe-aversion systems that prevented serious health effects, while at least two thousand died in Bhopal and nearly two hundred thousand were injured. Even though the U.S. chemical industry is largely self-regulated, most domestic plants employ relatively sophisticated safety tactics that use many of the strategies of the catastrophe-aversion system. Still, questions remain about how effectively these strategies have been implemented.[21] For example, a 1985 chemical plant accident in Institute, West Virginia, while minor in its effects, revealed a startling series of "failures in management, operations, and equipment."[22]

The Bhopal and Institute incidents suggest that, relative to other risks, safety issues in chemical manufacturing deserve more governmental attention than they previously have received. In addition to whatever changes are warranted at U.S. chemical plants, special attention should be paid to the process of managing risk at many overseas plants owned by U.S. firms. If the practices at the Bhopal plant were typical, safety strategies abroad are haphazard. While the Bhopal incident has led to a fundamental review of safety procedures in chemical plants worldwide, it should hardly have required a catastrophe to reveal such a vast category of hazard. This oversight demonstrates that some entire categories of risk may not yet be taken into account by the catastrophe-aversion system.

The catastrophe-aversion system likewise was not applied, until recently, to hazardous waste in the United States. State and federal laws made no special provisions for toxic waste prior to the 1970s; there were no requirements for initial precautions, or for conservatism in the amounts of waste that were generated. Systematic testing for underground contamination was not required, and waste sites were not monitored for potential problems. It is a tribute to the resilience of the ecosystem that after-the-fact cleanup now in progress has a good chance of keeping damage from past dumping below catastrophic levels. The next step is to find ways of limiting the generation of new wastes.

What does all this add up to? In our view, society's standard operating procedure should be as follows:

First, apply each of the catastrophe-aversion strategies in as many areas of risk as possible;

After this has been accomplished, proceed with more detailed inquiry, debate, and action on particular risks.

To pursue detailed debates on a risk for which a catastrophe-aversion system already is operative, continuing to protect against smaller and smaller components of that risk, is likely to be a misallocation of resources until the full range of potential catastrophes from civilian technologies has been guarded against. The "How safe?" questions that have become so much the focus of concern are matters of fine tuning; they may be important in the long run, but they are relatively minor compared to the major risks that still remain unaddressed.

Concluding Note

The highly respected social critic Lewis Mumford claimed in 1970 that "The professional bodies that should have been monitoring our technology . . . have been criminally negligent in anticipating or even reporting what has actually been taking place." Mumford also said that technological society is "a purely mechanical system whose processes can neither be retarded nor redirected nor halted, that has no internal mechanism for warning of defects or correcting them."[23] French sociologist Jacques Ellul likewise asserted that the technological

> system does not have one of the characteristics generally regarded as essential for a system: feedback. . . . [Therefore] the technological system does not tend to modify itself when it develops nuisances or obstructions. . . . [H]ence it causes the increase of irrationalities.[24]

Reflecting on different experiences several decades earlier, Albert Schweitzer thought he perceived that "Man has lost the capacity to foresee and forestall. He will end by destroying the earth."[25]

Although one of us began this investigation extremely pessimistic and the other was hardly an optimist, we conclude that Mumford, Ellul, Schweitzer, and many others have underestimated the resilience both of society and of the ecosystem. We found a sensible set of tactics for protecting against the potentially catastrophic consequences of errors. We found a complex and increasingly sophisticated process for monitoring and reporting potential errors. And we found that a fair amount of remedial action was taken on the basis of such monitoring (though not always the right kind of action or enough action, in our judgment).

Certainly not everyone would consider averting catastrophe to be a very great accomplishment. Most citizens no doubt believe that an affluent technological society ought to aim for a much greater degree of safety than just averting catastrophes. Many industry executives and engineers as well as taxpayers and consumers also no doubt believe that sufficient safety could be achieved at a lower cost. We agree with both. But wanting risk regulation to be more efficient or more effective is very different from being caught up in an irrational system that is leading to catastrophic destruction. We are glad—and somewhat surprised—to be able to come down on the optimistic side of that distinction.

Finally, what are the implications of our analysis for environmentally conscious business executives, scientists, journalists, activists, and public officials? Is it a signal for such individuals to relax their efforts? We do not intend that interpretation. The actions taken by concerned groups and individuals are an important component of the catastrophe-aversion system described in these pages. To relax the vigilance of those who monitor errors and seek their correction would be to change the system we have described. Quick reaction, some-times even overreaction, is a key ingredient in that part of regulating risky technologies that relies on trial and error. So to interpret these results as justifying a reduction of efforts would be a gross misreading of our message.

Instead, we must redirect some of our concern and attention. Envi-ronmental groups should examine whether they could contribute more to overall safety by focusing greater attention on egregious risks that have not yet been brought under the umbrella of the catastrophe-aversion system—instead of focusing primarily on risks that already are partially protected against. The Union of Concerned Scientists, for example, devotes extended attention to analyses of nuclear plant safety but has contributed almost nothing on the dangers of coal combustion, international standards for chemical plants, or toxic waste generation—egregious risks that have not been taken into ac-count by catastrophe-aversion strategies. Regardless of whether con-temporary nuclear reactors are safe enough, there is no question that they have been intensively subjected to the restraints of the cata-strophe-aversion system. We doubt that much more safety will be produced by further debate of the sort that paralyzed nuclear policy making during the 1970s and 1980s. In general, we believe it is time for a more strategic allocation of the (always limited) resources avail-able for risk reduction.

Our main message, however, has been that the United States has done much better at averting health and safety catastrophes than most people realize, considering the vast scope and magnitude of the threats posed by the inventiveness of science and industry in the twentieth century. Careful examination of the strategies evolved to cope with threats from toxic chemicals, nuclear power, recombinant DNA, ozone depletion, and the greenhouse effect suggests that we have a reasonably reliable system for discovering and analyzing poten-tial catastrophes. And, to date, enough preventive actions have been taken to avoid the worst consequences. How much further improve-ment will be achieved depends largely on whether those groups and individuals concerned with health and safety can manage to win the political battles necessary to extend and refine the strategies now being used. Because we have a long way to go in the overall process of learning to manage technology wisely, recognizing and appreciating the strengths of our catastrophe-aversion system may give us the inspiration to envision the next steps.

NOTES

1. On the recent criticisms, see Marjorie Sun, "Food Dyes Fuel Debate Over Delaney," *Science* 229 (1985): 739–41.
2. There is a large and growing literature on the subject of acceptable risk. An early statement was William W. Lowrance's, *Of Acceptable Risk: Science and the Determination of Safety* (Los Altos, Calif.: William Kaufman, 1976); a [more] recent overview is William W. Lowrance's, *Modern Science and Human Values* (New York: Oxford University Press, 1985). Also see Richard C. Schwing and Walter A. Albers, *Societal Risk Assessment: How Safe Is Safe Enough?* (New York: Plenum, 1980).
3. See, for example, A. V. Cohen and D. K. Pritchard, *Comparative Risks of Electricity Production Systems: A Critical Survey of the Literature*, Health and Safety Executive, Research Paper no. 11 (London: Her Majesty's Stationery Office, 1980).
4. U.S. Nuclear Regulatory Commission, *Safety Goal for Nuclear Power Plants: A Discussion Paper* (Washington, D.C.: U.S. Nuclear Regulatory Commission, 1982).
5. Ibid., xi.
6.

5/10,000	$\times$	0.1%	$\times$	200 persons	=	1/10,000
(overall accident death rate)		(acceptable reactor death rate)		(population near reactor)		(annual probability of one death from reactor accident)

7. For a discussion of the uncertainty and associated controversy surrounding the size of the source term—the amount of fission products that escape in a serious accident—see "Source Terms: The New Reactor Safety Debate," *Science News* 127 (1984): 250–53.
8. For a typical example, see Edmund A. C. Crouch and Richard Wilson, *Risk/Benefit Analysis* (Cambridge: Ballinger, 1982).
9. J. G. U. Adams, ". . . And How Much for Your Grandmother?," reprinted in Steven E. Rhoads, ed., *Valuing Life: Public Policy Dilemmas* (Boulder, Colo.: Westview Press, 1980), 135–46.
10. Anthony V. Nero, Jr., "The Indoor Radon Story," *Technology Review* 89 (January 1986): 28–40.
11. This example is adapted from Table 6, p. 534, in E. P. O'Donnell and J. J. Mauro, "A Cost–Benefit Comparison of Nuclear and Nonnuclear Health and Safety Protective Measures and Regulations," *Nuclear Safety* 20 (1979): 525–40. For a different analysis that makes the same basic point, see Crouch and Wilson, *Risk/Benefit Analysis*.
12. See, for example, the brief reference in NRC, *Changing Climate*, 4.
13. Alvin M. Weinberg and Irving Spiewak, "Inherently Safe Reactors and a Second Nuclear Era," *Science* 224 (1984): 1398–1402.
14. See *NRDC v. Train*, 8 ERC 2120 (D.D.C. 1976) and *NRDC v. Costle*, 12 ERC 1830 (D.D.C. 1979).
15. For a more extended analysis of this issue, see Giandomenico Majone, "Science and Trans-Science in Standard Setting."
16. NRC, *Changing Climate*, 3.

17. President's Commission on the Accident at Three Mile Island, *The Need for Change: The Legacy of TMI* (Washington, D.C.: U.S. Government Printing Office, October 1979), 56.

18. Alvin M. Weinberg et al., "The Second Nuclear Era," research memorandum ORAU/IEA-84-(M) (Oak Ridge, Tenn.: Institute for Energy Analysis, February 1984), 57.

19. And there are other methods to promote learning. For example, one possible benefit of "energy parks," with a number of reactors close together, is that learning could occur via informal contacts among personnel; see Alvin M. Weinberg, "Nuclear Safety and Public Acceptance," presented at the International ENS/ANS Con-Cycles, Brussels, April 30, 1982.

20. On the Bhopal incident, see the special issue of *Chemical and Engineering News* 63 (February 11, 1985), and the investigative reports in *The New York Times,* January 28 through February 3, 1985.

21. Stuart Diamond, "Carbide Asserts String of Errors Caused Gas Leak," *The New York Times,* August 24, 1985, 1.

22. For an overview of chemical plant safety issues, see Charles Perrow, *Normal Accidents* (New York: Basic Books, 1984), 101–22.

23. Lewis Mumford, *The Pentagon of Power* (New York: Harcourt Brace Jovanovich, 1970), 410.

24. Jacques Ellul, *The Technological System* (New York: Continuum, 1980), 117.

25. Albert Schweitzer, quoted in Rachel Carson, *Silent Spring* (Boston, Mass.: Houghton Mifflin, 1962), v.

12. Women and the Assessment of Technology

CORLANN GEE BUSH

Assessing and controlling technology and its impacts can mean different things to different people. In this selection, taken from Joan Rothschild's book, Machina Ex Dea, *Corlann Gee Bush argues that a feminist perspective can alter the direction of technological change.*

Bush proposes a new definition of technology that includes women and facilitates the understanding of technology's equity dimensions— who wins and who loses from a given development: "Technology is a form of human cultural activity that applies the principles of science and mechanics to the solution of problems. It includes the resources, tools, processes, personnel, and systems developed to perform tasks and create immediate particular, and personal and/or competitive advantages in a given ecological, economic, and social context." It is essential to "unthink" the old myths of technology, writes Bush, in order to use it to promote a more just society.

Corlann Gee Bush is director of affirmative action programs at Montana State University–Bozeman. She has previously served as assistant dean of students at the University of Idaho in Moscow. Bush has been chair of the Committee on Technology of the American Association of University Women. She is the author of Taking Hold of Technology, *and has written and spoken widely on the impact of technology on women's lives.*

"Everything is what it is, what it isn't, and its direct opposite. That technique, so skillfully executed, might help account for the compel-

ling irrationality . . . *double double think is very easy to deal with if we just realize that we have only to double double unthink it.*"
—Dworkin 1974, p. 63.

Although Andrea Dworkin is here analyzing Pauline Reage's literary style in the *Story of O*, her realization that we can "double double unthink" the mind fetters by which patriarchal thought binds women is an especially useful one. For those of us who want to challenge and change female victimization, it is a compelling concept.

SOMETHING ELSE AGAIN

The great strength of the women's movement has always been its twin abilities to unthink the sources of oppression and to use this analysis to create a new and synthesizing vision. Assertiveness is, for example, something else again: a special, learned behavior that does more than merely combine attributes of passivity and aggressiveness. Assertiveness is an unthinking and a transcendence of those common, control-oriented behaviors.[1]

Similarly, in their books *Against Our Will* and *Rape: The Power of Consciousness*, Susan Brownmiller (1974) and Susan Griffin (1979) unthink rape as a crime of passion and rethink it as a crime of violence, insights which led to the establishment of rape crisis and victim advocacy services. But a good feminist shelter home-crisis service is something else again: it is a place where women are responsible for the safety and security of other women, where women teach self-defense and self-esteem to each other. In like manner, women's spirituality is something else again. Indebted both to Mary Daly for unthinking Christianity in *Beyond God the Father* (1973) and *The Church and the Second Sex* (1968) and to witchcraft for rethinking ritual, women's spirituality is more than a synthesis of those insights, it is a transformation of them.

In other words, feminist scholarship and feminist activism proceed not through a sterile, planar dialectic of thesis, antithesis, synthesis, but through a dynamic process of unthinking, rethinking, energizing, and transforming. At its best, feminism creates new life forms out of experiences as common as seawater and insights as electrifying as lightning.

The purpose of this chapter is to suggest that a feminist analysis of technology would be, like assertiveness, something else again. I will raise some of the questions that feminist technology studies should seek to ask, and I will attempt to answer them. Further, I hope to show how scholars, educators, and activists can work together toward a transformation of technological change in our society.

The endeavor is timely not least because books such as this, journal issues, articles, and conferences are increasingly devoting time and energy to the subject or because technologically related political issues such as the antinuclear movement and genetic engineering consume larger and larger amounts of both our news space and our consciousness. The most important reason why feminists must unthink and rethink women's relationship to technology is that the *tech-fix* (Weinberg 1966, p. 6 [reprinted in Part I of this book]) and the public policies on which it is based are no longer working. The tech-fix is the belief that technology can be used to solve all types of problems, even social ones. Belief in progress and the tech-fix has long been used to rationalize inequity: it is only a matter of time until technology extends material benefits to all citizens, regardless of race, sex, class, religion, or nationality.

> Technology has expanded our productive capacity so greatly that even though our distribution is still inefficient, and unfair by Marxian precepts, there is more than enough to go around. Technology has provided a "fix"—greatly expanded production of goods—which enables our capitalistic society to achieve many of the aims of the Marxist social engineer without going through the social revolution Marx viewed as inevitable. Technology has converted the seemingly intractable social problem of *widespread* poverty into a relatively tractable one (Weinberg 1966, p. 7).

While Weinberg himself advocates cooperation among social *and* technical engineers in order to make a "better society, and thereby, a better life, for all of us who are part of society" (Weinberg 1966, p. 10), less conscientious philosophers and politicians have seen in the tech-fix a justification for laissez-faire economics and discriminatory public policy. Despite its claim to the contrary, the tech-fix has not worked well for most women or for people of color; recent analyses of the feminization of poverty, for example, indicate that jobs, which have always provided men with access to material goods, do not get women out of poverty.

> Social welfare programs based on the old male model of poverty do not consider the special nature of women's poverty. One fact that is little understood and rarely reflected in public welfare policy is that

women in poverty are almost invariably productive workers, partici-pating fully in both the paid and the unpaid work force. The inequi-ties of present public policies molded by the traditional economic role of women cannot continue. Locked into poverty by capricious programs designed by and for male policymakers . . . women who are young and poor today are destined to grow old and poor as the years pass. Society cannot continue persisting with the male model of a job automatically lifting a family out of poverty . . . (McKee 1982, p. 36).

As this example illustrates, the traditional social policies for dealing with inequity—*get a job*—and traditional solutions—*produce more efficiently*—have not worked to make a better society for women. Therefore, it is essential that women begin the unthinking of these traditions and the rethinking of new relationships between social and technical engineering.

UNTHINKING TECH-MYTHS

In her poem "To An Old House in America," Adrienne Rich describes the attitude that women should take toward the task of unthinking public policy in regard to technology: "I do not want to simplify/Or: I would simplify/By naming the complexity/It has been made o'er simple all along" (Rich 1975, p. 240). Partly because it is in their best interest to do so and partly because they truly see nothing else, most politicians and technocrats paint the canvas of popular opinion about technology with the broadest possible brushstrokes, rendering it, in pure type, as TOOL, as THREAT, or as TRIUMPH.[2] From each of these assump-tions proceed argument, legislation, public policy, and, ironically, pow-erlessness. In order to develop a feminist critique of technology, we must analyze these assumptions and unthink them, making them sim-pler by naming their complexity.

The belief that technology represents the triumph of human intelli-gence is one of America's most cherished cultural myths; it is also the easiest to understand, analyze, and disprove. Unfortunately, to discuss it is to resort to clichés: "There's nothing wrong that a little good old American ingenuity can't fix"; "That's progress"; or "Progress is our most important product." From such articles of faith in technology stemmed Manifest Destiny, the mechanization of agriculture, the urbanization of rural and nomadic cultures, the concept of the twentieth as the "American Century," and every World's Fair since 1893. That such faith seems naive to a generation that lives with the arms race, acid rain, hazardous waste, and near disasters at nuclear power plants is not to diminish one *byte* either Western culture's faith in the tech-fix or its belief that technological change equals material progress. And, indeed, like all generalizations, this myth is true—at least partially. Technology *has* decreased hardships and suffering while raising standards of health, living, and literacy throughout the industrialized world.

But, not without problems, as nay-sayers are so quick to point out. Those who perceive technology as the ultimate threat to life on the planet look upon it as an iatrogenic disease, one created, like nausea in chemotherapy patients, by the very techniques with which we treat the disease. In this view, toxic wastes, pollution, urban sprawl, increasing rates of skin cancer, even tasteless tomatoes, are all problems created through our desire to control nature through technology. Characterized by their desire to go cold turkey on the addiction to the tech-fix, contemporary critics of technology participate in a myriad of activities and organizations (Zero Population Growth, Friends of the Earth, Sierra Club, the Greenpeace Foundation) and advocate a variety of goals (peace, arms limitation, appropriate technology, etc.). And, once again, their technology-as-threat generalization is true, or at least as true as its opposite number: in truth, no one, until Rachel Carson (1955), paid much attention to the effects of technology on the natural world it tried to control; indeed, technology has created problems as it has set out to solve others.

Fortunately, the inadequacy of such polarized thinking is obvious: technology is neither wholly good nor wholly bad. "It has both positive and negative effects, and it usually has the two *at the same time and in virtue of each other*" (Mesthene 1970, p. 26 [reprinted in Part I of this book]). Every innovation has both positive and negative consequences that pulse through the social fabric like waves through water.

Much harder to unthink is the notion that technologies are merely tools: neither good nor bad but neutral, moral only to the extent that their user is moral. This, of course, is the old saw "guns don't kill people, people kill people" writ large enough to include not only guns and nuclear weapons but also cars, televisions, and computer games. And there is truth here, too. Any given person can use any given gun at any given time either to kill another person for revenge or to shoot a grouse for supper. The gun is the tool through which the shooter accomplishes his or her objectives. However, just as morality is a collective concept, so too are guns. As a class of objects, they comprise a technology that is designed for killing in a way that ice picks, hammers, even knives—all tools that have on occasion been used as weapons—are not. To believe that technologies are neutral tools subject only to the motives and morals of the user is to miss completely their collective significance. Tools and technologies have what I can only describe as *valence*, a bias or "charge" analogous to that of atoms that have lost or gained electrons through ionization. A particular technological system, even an individual tool, has a tendency to interact in similar situations in identifiable and predictable ways. In other words, particular tools or technologies tend to be favored in certain situations, tend to perform in a predictable manner in these situations, and tend to bend other interactions to them. Valence tends to seek out or fit in with certain social norms and to ignore or disturb others.

Jacques Ellul (1964) seems to be identifying something like valence when he describes "the specific weight" with which technique is endowed:

> It is not a kind of neutral matter, with no direction, quality, or structure. It is a power endowed with its own peculiar force. It refracts in its own specific sense the wills which make use of it and the ends proposed for it. Indeed, independently of the objectives that man pretends to assign to any given technical means, that means always conceals in itself a finality which cannot be evaded (pp. 140–41).

While this seems to be overstating the case a bit—valence is not the atom, only one of its attributes—tools and techniques do have tendencies to pull or push behavior in definable ways. Guns, for example, are valenced to violence; the presence of a gun in a given situation raises the level of violence by its presence alone. Television, on the other hand, is valenced to individuation; despite the fact that any number of people may be present in the same room at the same time, there will not be much conversation because the presence of the TV itself pulls against interaction and pushes toward isolation. Similarly, automobiles and microwave ovens are individuating technologies while trains and campfires are accretionary ones.

Unthinking tech-myths and understanding valence also requires greater clarity of definition (Winner 1977, pp. 10–12). Several terms, especially *tool, technique,* and *technology,* are often used interchangeably when, in fact, they describe related but distinguishable phenomena. *Tools* are the implements, gadgets, machines, appliances, and instruments themselves. A hammer is, for example, a tool as is a spoon or an automatic washing machine. *Techniques* are the skills, methods, procedures, and processes that people perform in order to use tools. Carpentry is, therefore, a technique that utilizes hammers, baking is a technique that uses spoons, and laundering a technique that employs washing machines. *Technology* refers to the organized systems of interactions that utilize tools and involve techniques for the performance of tasks and the accomplishment of objectives. Hammers and carpentry are some of the tools and techniques of architectural or building technology. Spoons and baking, washing machines and laundering are some of the tools and techniques of domestic or household technology.

A feminist critique of the public policy debate over technology should, thus, unthink the tripartite myth that sees technology in simple categories as tool, triumph, or threat. In unthinking it, we can simplify it by naming its complexity:

- A tool is not a simple isolated thing but is a member of a class of objects designed for specific purposes.
- Any given use of tools, techniques, or technologies can have both beneficial and detrimental effects at the same time.
- Both use and effect are expressions of a valence or propensity for tools to function in certain ways in certain settings.
- Polarizing the rhetoric about technology enables advocates of particular points of view to gain adherents and power while doing nothing to empower citizens to understand, discuss, and control technology on their own.

"Making it o'er simple all along" has proven an excellent technique for maintaining social control. The assertion that technology is beneficial lulls people into believing that there is nothing wrong that can't be fixed, so they do nothing. Likewise, the technophobia that sees technology as evil frightens people into passivity and they do nothing. The argument that technology is value-free either focuses on the human factor in technology in order to obscure its valence or else concentrates on the autonomy of technology in order to obscure its human control. In all cases, the result is that people feel they can do nothing. In addition, by encouraging people to argue with and blame each other, rhetoric wars draw public attention away from more important questions such as: Who is making technological decisions? On what basis? What will the effects be?

CONTEXT, CONTEXT, WHITHER ART THOU, CONTEXT?

In unthinking the power dynamics of technological decision making, a feminist critique needs to pay special attention to the social messages whispered in women's ears since birth: mother to daughter, "Don't touch that, you'll get dirty"; father to daughter, "Don't worry your pretty little head about it"; teacher to young girl, "It doesn't matter if you can't do math"; woman to woman, "Boy, a man must have designed this."

Each of these statements is talking about a CONTEXT in which technological decisions are made, technical information is conveyed, and technological innovations are adopted. That such social learning is characterized by sex role stereotyping should come as no surprise. What may be surprising is not the depths of women's ignorance—after all, women have, by and large, been encouraged to be ignorant—but the extent to which men in general, inventors, technocrats, even scholars, all share an amazing ignorance about the contexts in which technology operates. There are four:

1. *The design or developmental context* which includes all the decisions, materials, personnel, processes, and systems necessary to create tools and techniques from raw materials.
2. *The user context* which includes all the motivations, intentions, advantages, and adjustments called into play by the use of particular techniques or tools.
3. *The environmental context* that describes nonspecific physical surroundings in which a technology or tool is developed and used.
4. *The cultural context* which includes all the norms, values, myths, aspirations, laws, and interactions of the society of which the tool or technique is a part.

Of these, much more is known about the design or developmental context of technology than about the other three put together. Western culture's collective lack of knowledge about all but the developmental context of technology springs in part from what Langdon Winner calls technological orthodoxy: a "philosophy of sorts" that has seldom been "subject to the light of critical scrutiny" (Winner 1979, p. 75). Standard tenets of technological orthodoxy include:

- That men know best what they themselves have made.
- That the things men make are under their firm control.
- That technologies are neutral: they are simply tools that can be used one way or another; the benefit or harm they bring depends on how men use them (Winner 1979, p. 76).

If one accepts these assumptions, then there is very little to do except study processes of design and invent ever-newer gadgets. The user and environmental contexts become obscured if not invisible, an invisibility further confirmed by the fact that, since the Industrial Revolution, men have been inventors and designers while women have been users and consumers of technology. By and large, men have created, women have accommodated.

The sex role division of labor that characterizes Western societies has ensured that boys and girls have been brought up with different expectations, experiences, and training, a pattern that has undergone remarkably little change since the nineteenth century.

Games for girls were carefully differentiated from boys' amusement. A girl might play with a hoop or swing gently, but the "ruder and more daring gymnastics of boys" were outlawed. Competitive play was also anathema: A "little girl should never be ambitious to swing higher than her companions." Children's board games afforded another insidious method of inculcating masculinity and femininity. On a boys' game board the player moved in an upward spiral, past temptations, obstacles, and reverses until the winner reached a pinnacle of propriety and prestige. A girl's playful enactment of her course in life moved via a circular ever-inward path to the "mansion of happiness," a pastel tableau of mother and child. The dice of popular culture were loaded for both sexes and weighted with domesticity for little women. The doctrine of (separate) spheres was thereby insinuated in the personality of the child early in life and even during the course of play (Ryan 1979, p. 92).

It is difficult to invent a better mousetrap if you're taught to be afraid of mice; it is impossible to dream of becoming an engineer if you're never allowed to get dirty.

As compared to women, men do, indeed, know a great deal about what they would call the "design interface" of technology; they know more about how machines work; they discovered the properties of elements and the principles of science. They know math, they develop cost–benefit risk analyses; they discover, invent, engineer, manufacture, and sell. Collectively, men know almost everything there is to know about the design and development of tools, techniques, and systems; but they understand far less about how their technologies are used—in part because there is less money in understanding than in designing, in part because the burden of adjusting to technological change falls more heavily on women. What is worse, however, is that most men do not know that they do not know anything about women and the user context.

> From the preliminary conceptualization to the final marketing of a product, most decision making about technology is done by men who design, usually subconsciously, a model of the physical world in which they would like to live, using material artifacts which meet the needs of the people—men—they best know. The result [is] technological development based on particular sets of male conditioning, values, and roles . . . (Zimmerman 1981, p. 2).

Ironically, until very recently, most women did not realize that they possessed information of any great significance. With all the cultural attention focused on the activity in the developmental context, it was hard to see beyond the glare of the spotlights into the living rooms and kitchens and laundries where women were working and living out the answers to dozens of unverbalized questions: How am I spending my time here? How is my work different from what I remember my mother doing? Am I really better off? Why does everything seem so out of control? Rephrased, these are the questions that will comprise a feminist assessment of technology: How have women's roles changed as a result of modern technology? Has women's status in society kept pace with the standard of living? Do women today have more opportunities or merely more expectations? What is the relationship of material possessions to personal freedom?

Think for a moment about washing machines. Almost every family in the United States has access to one; across the country, women spend thousands of hours each day in sorting, washing, drying, folding, and ironing clothes. The automatic washing machine has freed women from the pain and toil described so well by Agnes Smedley (1973) in *Daughter of Earth*. But as washing technology has changed, so too has clothing (it gets dirtier faster) and wardrobes (we own more clothes) and even standards of cleanliness (clothes must be whiter than white), children change clothes more often, there are more clothes to wash. Joann Vanek (1974, p. 118), in her work on time spent in housework, asserts that women spent as much time in household related tasks in 1966 as they did in 1926.

More has changed, however, than just standards of cleanliness. Doing laundry used to be a collective enterprise. When I was a child in the late 1940s and early 1950s, my mother and grandmother washed the family's clothes together. My grandmother owned a semiautomatic machine but she lived 45 miles away; my mother had hot water, a large sink, and five children. Every Sunday, we would dump the dirty clothes in a big wicker basket and drive to my grandmother's house where all the womenfolk would spend the afternoon in the basement, talking and laughing as we worked. By evening, the wicker basket would again be full, but this time with neatly folded, clean smelling piles of socks, sheets, towels, and underwear that would have to last us a week. Crisply ironed dresses and slacks, on hangers, waited to be hung, first on those little hooks over the side doors of the car, then in our closets at home.

Nostalgic as these memories are, doing laundry was not romantic. It was exhausting, repetitive work, and neither my mother nor I would trade in our own automatic washers to go back to it (Armitage 1982, pp. 3–6). Yet, during my childhood, laundry was a communal activity, an occasion for gossip, friendship, and bonding. Laundering was hard work, and everyone in the family and in the society knew it and respected us as laborers. Further, having laundry and a day on which to do it was an organizing principle (Monday, washday; Tuesday, iron; Wednesday . . .) around which women allocated their time and resources. And, finally, there was a closure, a sense of completion and accomplishment impossible to achieve today when my sister washes, dries, folds, and irons her family's clothes every day or when I wash only because I have nothing to wear.

Admittedly, this homey digression into soap opera (One Woman's Wash) is a far cry from the design specification and cost–benefit analyses men use to describe and understand the developmental context of washing machines, but it is equally valid for it describes the user context in the user's terms. Analyzing the user context of technological change is a process of collecting thousands and thousands of such stories and rethinking them into an understanding of the effects of technological change on women's lives.[3] From unthinking the developmental context as such and rethinking the user context, it is only a short step to studying the environmental and cultural contexts of technological change. Of these, our knowledge of the environmental context is the better developed, partly because we have given it more serious attention but mostly because environmental studies has been a legitimate career option for men.

While concern about the effects of technology on the natural environment is an idea that can be traced back to de Crevecoeur (1968 [1782]) and James Fenimore Cooper (1832), Rachel Carson (1955, 1961, 1962) is the person most responsible for our current level of ecological awareness and for the scientific rather than aesthetic basis on which it rests. As we learn more about the fragile reciprocity within ecosystems, we begin to unthink the arrogance of our assumption that we are separate from and superior to nature. In an ecosystem, it is never possible to do only *one* thing; for every action there are chain reactions of causes and effects. The continued survival of the world depends upon developing more precise models of the environment so we can predict and prevent actual catastrophe without being immobilized by the risking of it.

Perhaps no one could have foreseen that the aerosol sprays we used to apply everything from paint to anti-perspirant would degrade the earth's ozone layer, but no one seems to have asked. That drums for burying toxic waste would eventually corrode and leak seems so obvious that millions ought to have been able to predict the risk, yet no one seems to have had the desire or the clout to deal with the problem of hazardous waste before it became a crisis. In pursuit of progress, we have been content to ignore the ecological consequences of our technological decisions because, until it was pointed out to us, we did not realize that there *was* an environmental context surrounding the tools we use.

The environmental impact analysis (EIA) has become the most popular means by which governments and industries attempt to predict and assess the ecological impact of technological change. While most EIAs are long, tedious, and nonconfrontive, the idea behind them and much of the work that has gone into them is sound. In her articles on appropriate technology, Judy Smith (1978, 1981) from the Women and Technology Project in Missoula, Montana, has suggested that sex-role impact reports could be used to improve our understanding of the cultural context of technology in much the same way that the EIA has improved our knowledge of the environmental context.

And we do need something, for we know next to nothing about the interactions of culture and technology, having always seen these as separate phenomena. Most people welcome technological change because it is *material*, believing that it makes things better, but it doesn't make them different. They resist social change because it is *social* and personal; it is seen as making things different . . . and worse. The realization that technological change stimulates social change is not one that most people welcome.

Feminists need to unthink this cultural blindness. Because women are idealized as culture carriers, as havens of serenity in a heartless world (Lasch 1977), women are supposed to remain passive while the rest of the culture is allowed, even encouraged, to move rapidly ahead. Women are like the handles of a slingshot whose relatively motionless support enables the elastic and shot to build up energy and to accelerate past them at incredible speeds. The culture measures its progress by women's stasis. When women do try to move, when they try to make changes rather than accommodations, they are accused of selfishness, of me-ness, of weakening the family, of being disloyal to civilization (Rich 1979, pp. 275–310).

However, it is crucial that feminists continue to unthink and rethink the cultural contexts of technology for a reason more significant than our systematic exclusion from it: it is dangerous not to. Technology always enters into the present culture, accepting and exacerbating the existing norms and values. In a society characterized by a sex-role division of labor, any tool or technique—it has valence, remember—will have dramatically different effects on men than on women.

Two examples will serve to illustrate this point. Prior to the acquisition of horses between the late sixteenth and mid-seventeenth centuries, women and dogs were the beasts of burden for Native American tribes on the Great Plains. Mobility was limited by both the topography and the speed at which people and dogs could walk. Physical labor was women's province in Plains culture, but since wealth in those societies was determined by how many dogs a person "owned" and since women owned the dogs, the status of women in pre-equestrian tribes was relatively high—they owned what men considered wealth (Roe 1955, p. 29). Women were central to the economic and social life of their tribes in more than the ownership of the dogs. They controlled the technology of travel and food: they were responsible for the foraging, gathering, and preserving of food for the tribe and, in many cases, determined the time and routes of tribal migration. They had access to important women's societies and played a central part in religious and community celebrations (Liberty 1982, p. 14).

Women's roles in Plains Indian societies changed profoundly and rapidly as horses were acquired and domesticated. In less than two centuries—for some tribes in less than a generation—a new culture evolved. The most immediate changes were technological and economic; horses became the technology for transportation and they were owned by men. Women could still own dogs, but this was no longer the measure of wealth it had been.

With their "currency" debased, women's status slipped further as important economic, social, and religious roles were reassigned to men. As the buffalo became a major source of food and shelter, the value of women's foraging activities decreased. Hunting ranges were expanded, causing more frequent moves with women doing more of the packing up and less of the deciding about when and where to go. As each tribe's hunting range increased, competition for land intensified; and warfare, raiding, and their concomitants for women—rape and slavery—also increased.

Of course, not all the effects were negative. Technologies are substitutes for human labor: horses made women's work easier and more effective. Also, several tribes, including the Blackfeet, allowed a woman to retain ownership of her own horse and saddle. However, a woman was seldom allowed to trade or raid for horses, and her rights to her husband's herd usually ended with his death.

Thus, for Native American women, the horse was a mixed blessing. It eased their burdens and made transportation easier. But it also added new tasks and responsibilities without adding authority over those tasks or increasing autonomy. The opposite was true for men; the horse provided few new tasks and responsibilities—men had always been responsible for hunting, defense, and warfare—but it did enhance these traditional roles, giving men more decision-making authority, more autonomy, and more access to status. Paradoxically, while a woman's absolute status was greatly improved by the changes from dog to horse culture, her status relative to men actually declined. In this manner, horses changed the nature of Native American culture on the high plains, but women and men were affected in profoundly different ways.

A similar phenomenon occurred at the end of the horse farming era in the Palouse region of Idaho and Washington in the United States. During the 1920s, it was common for a farmer to employ 15 to 25 hired men and to use 25 to 44 horses to harvest his crops; farmers and their hands worked back-breaking, twenty-hour days. On the other hand, women also worked long days during harvest, cooking five meals a day for as many as forty people. During the year, women were responsible for a family's food, nutrition, health, safety, and sanitation. Women's work had economic value. Performing their traditional roles as wives, mothers, and homemakers, women were economically crucial to the survival of the labor intensive family farm (Bush 1982). Unfortunately, in the same manner that the horse made a Plains Indian woman's work easier even as it lowered her status relative to men, so too did the conversion from horses to diesel power and electricity ease the farm wife's hardships while it decreased the economic significance of her labor. In both cases, technological innovation had profoundly different consequences for men's and women's work. In both cases, the innovation was coded or valenced in such a way that it loaded the status of men's roles while eroding status for women.

TECHNOLOGY AND EQUITY

Technology is, therefore, an equity issue. Technology has everything to do with who benefits and who suffers, whose opportunities increase and whose decrease, who creates and who accommodates. If women are to transform or "re-valence" technology, we must develop ways to assess the equity implications of technological development and develop strategies for changing social relationships as well as mechanical techniques. To do this, we must have a definition of technology that will allow us to focus on such questions of equity.

Not surprisingly, there are no such empowering definitions in the existing literature. Equity has not been a major concern of either technophobes or technophiles. In fact, most definitions of technology fall short on several counts. The most commonly accessible definitions, those in dictionaries, tell us little: Webster's "the science of the industrial arts" and "science used in a practical way," and the American Heritage Dictionary's "the application of science, especially to industrial and commercial objectives" and "the entire body of methods, and materials used to achieve such objectives" are definitions so abstract as to be meaningless. Other attempts clarify function but lose the crucial connection to science, as in James Burke's (1980, p. 23) "the sum total of all the objects and systems used to produce goods and perform services."

Better definitions connect technology to other categories of human behavior and to human motivation:

A form of cultural activity devoted to the production or transforma-tion of material objects, or the creation or procedural systems, in order to expand the realm of practical human possibility (Hannay & McGinn 1980, p. 27).

On rare occasions, definitions do raise equity questions as in John McDermott's attempt:

Technology, in its concrete, empirical meaning, refers fundamen-tally to systems of rationalized control over large groups of men, events, and machines by small groups of technically skilled men operating through organizational hierarchy (McDermott 1969, p. 29 [reprinted in Part I of this book]).

However, this definition is really defining *technocracy* rather than *tech-nology*. More often, there are romantic definitions that enmesh us in cotton candy:

[Technology's task] is to employ the earth's resources and energy income in such a way as to support all humanity while also enabling all people to enjoy the whole earth, all its historical artifacts and its beautiful places without any man enjoying life around earth at the cost of another (Fuller 1969, p. 348).

While no one could argue with such ideals, Buckminster Fuller leaves us where the boon and bane theorists leave us—confounded by dou-blethink. It is impossible to ask tough questions of such a definition or to examine closely why technology does not now support all humanity equally.

More distressing is the tendency of scholars to use the generic "he/ man" to represent all of humanity. For example, "without one man interfering with the other, without any man enjoying life around the earth at the cost of another" is a statement that completely disregards the fact that, around the earth, men enjoy their lives at *women's* cost. Similarly, statements such as "because of the autonomy of technique, man cannot choose his means any more than his ends" (Ellul 1964, p. 40) and "the roots of the machine's genealogical tree is in the brain of this conceptual man . . . after all it was he who made the machine" (Usher 1954, p. 22) grossly mislead us because they obscure the histori-cal and contemporary roles that women have played in technological development.[4] Worse, they reinforce the most disabling myth of all, the assumption that men and women are affected similarly by and benefit equally from technological change.

Therefore, because of the oversimplification of some definitions and the exclusion of women from others, feminists need to rethink a definition of technology that both includes women and facilitates an equity analysis. Such a definition might be:

Technology is a form of human cultural activity that applies the principles of science and mechanics to the solution of problems. It includes the resources, tools, processes, personnel, and systems developed to perform tasks and create immediate particular, and personal and/or competitive advantages in a given ecological, economic, and social context (Bush in *Taking Hold of Technology* 1981, p. 1).

The chief virtue of this definition is its consideration of advantage; people accept and adopt technology to the extent that they see advantage for themselves and, in competitive situations, disadvantage for others. Thus, an equity analysis of an innovation should focus on benefits and risks within the contexts in which the technology operates. An equity analysis of a technology would examine the following:

The Developmental Context

- the principles of science and mechanics applied by the tool or technique
- the resources, tools, processes, and systems employed to develop it
- the tasks to be performed and the specific problems to be solved

The User Context

- the current tool, technique, or system that will be displaced by its use
- the interplay of this innovation with others that are currently in use
- the immediate personal advantage and competitive advantage created by the use of technology
- the second and third level consequences for individuals

The Environmental Context	· the ecological impact of accepting the technology versus the impact of continuing current techniques
The Cultural Context	· the impact on sex roles · the social system affected · the organization of communities · the economic system involved and the distribution of goods within this system

A specific example will serve to illustrate how an equity analysis might be approached. Refrigeration was "invented" in the 1840s in Apalachicola, Florida, by John Gorrie as a by-product of his work on a cure for malaria (Burke 1980, p. 238). Gorrie's invention was a freezing machine that used a steam-driven piston to compress air in a cylinder that was surrounded by salt water. (As the piston advances, it compresses air in the cylinder; as the piston retracts, the air expands.) An expanding gas draws heat from its surroundings; after several strokes of the piston, the gas has extracted all the heat available from the surrounding brine. If a flow of continuously cold air is then pumped out of the cylinder into the surrounding air, the result is air conditioning; if the air is continuously allowed to cool the brine solution, the brine itself will draw heat from water, causing it to freeze and make ice. If the gas (air) or brine is allowed to circulate in a closed system, heat will be drawn from the surrounding air or matter (food), causing refrigeration.

The Developmental Context

Thus, refrigeration applies the laws of science (especially the properties of gases) and the principles of mechanics (thermodynamics and compression) to perform the tasks of making ice, preserving and freezing food, and cooling air. Refrigeration also solves the problems of retarding food spoilage and coping with heat waves, thereby creating personal advantage. The resources and tools used include a gas, a solution, a source of energy, and a piston-driven compressor.

The developmental context is enormously complex and intercon-
nected; however, a general analysis would include all the supply, manufacture, and distribution systems for the refrigeration units themselves—
everything from the engineers who design the appliances, to the factory
workers who make, inspect, and pack them, to the truckers who transport them, to the clerk who sells them. A truly expansive analysis of the
development context would also include the food production, packing,
and distribution systems required to make available even one box of
frozen peas as well as the artists, designers, paper producers, and advertis-
ers who package the peas and induce us to buy them.

The User Context

Refrigeration has affected our lives in such a myriad of ways that
elaborating on them all would require another paper in itself. Refrigeration has important commercial uses as well as medical ones, and it
would not be overstating the case to assert that there is no aspect of
modern life that has not been affected by refrigeration. Nonetheless, a
more limited analysis of refrigeration as it has affected domestic and
family life in the United States is both revealing and instructive.

To the self-sufficient farm family of the early twentieth century, refrigeration meant release from the food production and preservation chores
that dominated much of men's and women's lives: canning garden
produce to get the family through the winter; butchering, smoking, and
drying meat from farm-raised hogs and cattle; milking cows daily and
churning butter. The advantages of owning a refrigerator in such a situation were immediate and dramatic: food could be preserved for longer
periods of time so there was less spoilage; food could be cooked ahead of
serving time allowing women to spend less time in meal preparation;
freezing produce and meat was a faster, easier, and more sanitary process
than canning or smoking, again, saving women time and improving the
family's health. The refrigerator thus generated positive changes for
women, freeing them from hard, hot work and improving their absolute
status. However, the second and third level effects of refrigeration
technology were not as benign for women as the primary effects.

Since refrigeration kept food fresh for long periods of time, fresh
produce could be shipped across country, thus improving nutrition
nationwide. Food processing and preservation moved out of the home,
and new industries and services paid workers to perform the duties that

had once been almost solely women's domestic responsibility. Within the home, the nature of women's work changed from responsibility for managing food production to responsibility for managing food consumption. Also, farmers stopped growing food for family subsistence and local markets and started growing cash crops for sale on national and international markets. Opportunities for employment shifted from farm labor to industrial labor, and families moved from rural areas to cities and suburbs. Thus, the use of refrigeration changed the work roles of individual women and men and, through them, the economy, the content of work, and the nature of culture and agriculture.

The Environmental Context

An analysis of the environmental context of refrigeration technology would examine the effects of the developmental and user contexts on the environment by asking such questions as: Since refrigeration affects agriculture, what are the ecological effects of cash crop monoculture on, say, soil erosion or the use of pesticides? Since refrigeration retards the growth of bacteria and preserves blood and pharmaceuticals, what are the consequences for disease control? What are the effects of increased transportation of food on energy supplies and air pollution?

The Cultural Context

Finally, an examination of the sex role impact of refrigeration technology would reveal a disparate effect on men and women. In the United States, men have been largely responsible for food production, women for food preservation and preparation. Refrigerators were a valenced technology that affected women's lives by, generally, removing food preservation from their domestic duties and relocating it in the market economy. Women now buy what they once canned. Women's traditional roles have been eroded, as their lives have been made easier. On the other hand, men, who originally had very little to do with food preservation, canning or cooking, now control the processes by which food is manufactured and sold. Men's roles and responsibilities have been loaded and their opportunities increased, although their work has not necessarily been made easier. Refrigeration has, thus, been adopted and diffused throughout a sexist society; we should not be surprised to learn that its effects have been dissimilar and disequitable.

THE GREAT CHAIN OF CAUSATION

Of course, not one of us thinks about the effects of refrigeration on soil erosion or women's status when we open the frige to get a glass of milk. We are gadget-rich and assessment-poor in this society, yet each private act connects us to each other in a great chain of causation. Unfortunately, to think about the consequences of one's actions is to risk becoming immobilized; so the culture teaches us to double think rather than think, and lulls us into believing that individual solutions can work for the collective good.

Of course, we can continue to double think such things only so long as we can foist negative effects and disadvantages off onto someone else: onto women if we are men, onto blacks if we are white, onto youth if we are old, onto the aged if we are young. Equity for others need not concern us as long as *we* are immediately advantaged.

Feminists above all, must give the lie to this rationale, to unthink it; for if the women's movement teaches anything, it is that there can be no individual solutions to collective problems. A feminist transformation of technological thought must include unthinking the old myths of technology as threat or triumph and rethinking the attendant rhetoric. A feminist unthinking of technology should strive for a holistic understanding of the contexts in which it operates and should present an unflinching analysis of its advantages and disadvantages. Above all, a feminist assessment of technology must recognize technology as an equity issue. The challenge to feminists is to transform society in order to make technology equitable and to transform technology in order to make society equitable. A feminist technology should, indeed, be something else again.

NOTES

1. I am indebted for this insight to Betsy Brown and the other students in my seminar "The Future of the Female Principle," University of Idaho, Spring 1982.
2. In *Technological Change: Its Impact on Man and Society* (1970), Emmanuel Mesthene identifies "three unhelpful views about technology: technology as blessing, technology as curse, and technology as unworthy of notice." He does not mention the "technology as neutral tool argument," perhaps because he is one of its leading proponents.
3. Obviously, oral history is the only way that scholars can accumulate this data. Oral historians should ask respondents questions about their acquisition of

and adaptation to household appliances. Such questions might include: "When did you get electricity?"; "What was the first appliance you bought?"; "What was your first washing machine like?"; "How long did it take you to learn how to use it?"; "What was your next machine like?"; "When did you get running water?"; "Are you usually given appliances for presents or do you buy them yourself?", etc.

4. This situation is slowly changing thanks to much good work by Elise Boulding (1976), Patricia Draper (1975), Nancy Tanner and Adrienne Zihlman (1976), and Autumn Stanley (1984, and the volume from which this chapter was taken).

REFERENCES

Armitage, Susan. 1982. Wash on Monday: The housework of farm women in transition. *Plainswoman* VI, 2:3–6.

Boulding, Elise. 1976. *The underside of history: A view of women through time.* Boulder, Colo.: Westview Press.

Brownmiller, Susan. 1974. *Against our will: Men, women and rape.* New York: Simon and Schuster.

Burke, James. 1980. *Connections.* Boston: Little, Brown.

Bush, Corlann Gee. 1982. The barn is his; the house is mine: Agricultural technology and sex roles. *Energy and transport.* Eds. George Daniels and Mark Rose. Beverly Hills, Calif.: Sage Publications, 235–59.

Carson, Rachel. 1955. *The edge of the sea.* Boston: Houghton Mifflin.

Carson, Rachel. 1961. *The sea around us.* New York: Oxford University Press.

Carson, Rachel. 1962. *Silent spring.* Boston: Houghton Mifflin.

Cooper, James Fenimore. 1832. *The pioneer.* Philadelphia: Carey & Lea.

Daly, Mary. 1968. *The church and the second sex.* New York: Harper & Row.

Daly, Mary. 1973. *Beyond God the father: Toward a philosophy of women's liberation.* Boston: Beacon Press.

de Crevecoeur, Michel Guillaume St. Jean. 1968. *Letters from an American farmer: Reprint of 1782 edition.* Magnolia, Mass.: Peter Smith.

Draper, Patricia. 1975. !Kung women: Contrasts in sexual egalitarianism in foraging and sedentary contexts. *Toward an anthropology of women.* Ed. Rayna Reiter. New York: Monthly Review Press: 77–109.

Dworkin, Andrea. 1974. *Woman hating.* New York: E. P. Dutton.

Ellul, Jacques. 1964. *The technological society.* New York: Knopf.

Fuller, R. Buckminster. 1969. *Utopia or oblivion: The prospects for humanity.* New York: Bantam Books.

Griffin, Susan. 1979. *Rape: The power of consciousness.* New York: Harper & Row.

Hannay, N. Bruce; and McGinn, Robert. 1980. The anatomy of modern technology: Prolegomenon to an improved public policy for the social management of technology. *Daedalus* 109, 1:25–53.

Lasch, Christopher. 1977. *Haven in a heartless world: The family besieged.* New York: Basic Books.

Liberty, Margot. 1982. Hell came with horses: Plains Indian women in the equestrian era. *Montana: The Magazine of Western History* 32, 3:10–19.

McDermott, John. 1969. Technology: The opiate of the intellectuals. *New York Review of Books* XVI, 2 (July):25–35.

McKee, Alice. 1982. The feminization of poverty. *Graduate Woman* 76, 4:34–36.

Mesthene, Emmanuel G. 1970. *Technological change: Its impact on man and society.* Cambridge, Mass.: Harvard University Press.

Rich, Adrienne. 1975. *Poems: Selected and new 1950–1974.* New York: W. W. Norton.

Rich, Adrienne. 1979. Disloyal to civilization. *On lies, secrets and silence: Selected prose.* New York: W. W. Norton.

Roe, Frank Gilbert. 1955. *The Indian and the horse.* Norman, Okla.: University of Oklahoma Press.

Ryan, Mary P. 1979. *Womanhood in America: From colonial times to the present.* 2nd. ed. New York: Franklin Watts.

Smedley, Agnes. 1973. *Daughter of earth.* Old Westbury, N.Y.: The Feminist Press.

Smith, Judy. 1978. *Something old, something new, something borrowed, something due: Women and appropriate technology.* Butte, Mont.: National Center for Appropriate Technology.

Smith, Judy. 1981. Women and technology: What is at stake? *Graduate Woman* 75, 1:33–35.

Stanley, Autumn. 1984. *Mothers of invention: Women inventors and innovators through the ages.* Metuchen, N.J.: Scarecrow Press.

Taking hold of technology: Topic guide for 1981–83. 1981. Washington, D.C.: American Association of University Women.

Tanner, Nancy; and Zihlman, Adrienne. 1976. Women in evolution: Part I: Innovation and selection in human origins. *Signs* 1 (Spring):585–608.

Usher, Abbott Payson. 1954. *A history of mechanical inventions.* Cambridge: Harvard University Press.

Vanek, Joann. 1974. Time spent in housework. *Scientific American* 231 (November): 116–20.

Weinberg, Alvin M. 1966. Can technology replace social engineering? *University of Chicago Magazine* 59:6–10.

Winner, Langdon. 1977. *Autonomous technology: Technics-out-of-control as a theme in political thought.* Cambridge: MIT Press.

Winner, Langdon. 1979. The political philosophy of alternative technology. *Technology in Society* 1:75–86.

Zimmerman, Jan. 1981. Introduction. *Future, technology and woman: Proceedings of the conference.* San Diego, Calif.: Women's Studies Department, San Diego State University.

13. Controlling Technology

ALLAN C. MAZUR

Despite impressive developments in the field of risk management such as those described in the article by Morone and Woodhouse, technological controversies persist. Many of these controversies find highly qualified experts taking positions that seem to be in direct opposition to one another. What can a layperson make of such controversies, and how should society deal with them?

Allan Mazur has spent many years studying technological controversies. His paper "Controlling Technology," which is drawn from the last chapter of his book The Dynamics of Technical Controversy *(Communications Press, 1981), concludes that such disputes have a positive function in the political process. Although they do cause their share of problems, they represent, in a democratic society, a valid means of focusing attention on aspects of technological development that would otherwise be neglected in policy-making. What Mazur feels we need are better means of structuring the controversies so that citizens can understand the actual basis on which experts disagree and then make judgments based on their values.*

Allan C. Mazur is a professor of sociology at Syracuse University. He has previously served on the faculties of MIT and Stanford University. Mazur earned an M.S. in engineering from UCLA and worked for several years as an aerospace engineer in industry before obtaining a Ph.D. in sociology from Johns Hopkins University. His articles have appeared in such journals as Science, Minerva, *and* The American Sociological Review.

Public opposition to technology is not new. The contentious histories of the lightning rod and of the Luddites who smashed the machines that were to put them out of work, as well as of vaccination, the telegraph, the railroad, the automobile, the use and control of nuclear power after World War II, fluoridation in the 1950s, and others, tell us that technical controversies have been around a long time, if not in the strength and numbers we see today. They have become more common since 1970 and more salient, particularly the giant controversy over nuclear power.

WHY ARE THERE MORE PROTESTS?

Some analysts believe that these controversies represent a basic loss of confidence in science and scientific reasoning, a loss of faith in the ideal of progress through economic growth and prosperity.[1] But to argue that those who oppose nuclear power are at the same time opposing progress is to accept completely the proponents' claim that nuclear power is an important requirement for progress. This interpretation ignores the fact that most opponents do not believe the industry's forecasts of economic disaster unless we have a nuclear power program. The antinuclear ideology holds that important social benefits can be maintained, without nuclear power, through conservation, the development of renewable energy sources (particularly solar), and by matching power sources to end uses. This ideology does not envision a degradation of life-style but, to the contrary, promotes increases in health, decreases in environmental insult, and a spread of benefits to those portions of the population which do not now enjoy them. This seems closer to a Utopian than an antiprogress position. Whether or not this antinuclear program is realistic, it should be clear that it does not oppose progress and prosperity per se. Similarly, the challengers in most other technical controversies, including those where economic growth is not an issue, usually believe that progress and prosperity will be better served if the technology is not implemented than if it is. For these reasons, I do not believe that technical controversies are caused to any great extent by a disdain for progress.

World War I demonstrated both the wonders and horrors of mechanized warfare, and since then Western attitudes toward science and technology have best been characterized as ambivalent, praising and damning at the same time. While the decades between the world wars showed a great delight with technology, as in the Art Deco movement in architecture and product design, the same period was rich in criticism of science and technology,[2] producing classics like Aldous Huxley's *Brave New World* (1932) and Charlie Chaplin's *Modern Times* (1936). After World War II it was impossible to doubt the power of science-based technology, or to believe that it was necessarily progressive and good. I grew up with the science fiction of that time in which mutant monsters, produced inadvertently by nuclear radiation, were beaten at the end of the movie by a brave young scientist and a beautiful girl who was the daughter or assistant of a kind and wise older scientist. This ambivalence toward science and its products continues into the [present], now more salient because there is more technology in our lives and because we are more interested in it and knowledgeable about it. If one focused exclusively on the pessimistic articles and films produced in the last decade, the accidents involving jumbo jets, nuclear power, chemical storage, and fallen satellites, then a negative impression would be inevitable. On the other hand, if one focused on achievements in space, microprocessors, and medical technology, then the impression would be positive. Most members of the public apparently recognize both good and bad features of technology, combining them into a net view that is positive, if qualified.[3]

Nisbet (1979) believes that it is the intellectuals, if not the majority of the people, who have lost faith in progress and scientific reasoning, and he is no doubt correct to some extent. However, it is also true that some of the leading opponents of particular technologies are themselves prominent and productive scientists, and many of those who oppose one technology are active proponents of another. Furthermore, scientific reasoning is used by *both* sides in a controversy, not just those in favor of the technology. A healthy distrust of scientists appears on both sides too, each doubting the other's experts. Unfortunately, we also see some partisans on each side who reject immediately *anything* that is said by an opposing scientist while accepting uncritically the claims of their own experts. This is not a general skepticism about expertise but simply blatant bias in deciding whom to believe and whom not to. These observations do not seem to me to be consistent with the claim that opposition to a technology basically reflects a loss of faith in science.

I do not believe that the recent increase in technical controversies is properly explained by changing American beliefs about science or progress. A better explanation begins with the recognition that opposition to a technology is a special case of a broader class of political activities usually referred to as "special interest" politics, as opposed to the politics of party identification or patronage.[4] Since the 1960s, there has been a marked increase in the number of groups concerned about particular issues such as peace or arms, schools, pollution, abortion, or electoral reform. Encouraged by legislative actions and administrative rulings, the number of interest-group political-action committees quadrupled between 1972 and 1979.[5] The growth of this form of political activity is not completely understood, but some contributors to growth are apparent. One is growth in the college-educated proportion of the population. (Education is known to be correlated with political participation.)[6] Another is the model provided by popular attempts to improve civil rights, to end the war in Vietnam, to clean up the environment, and to depose a president. Legislative actions, court rulings, and administrative decisions have given new power to challenge groups, for example, the creation of environmental impact statements for federally funded projects, and the opportunities for successful suits based on this requirement. These factors have encouraged a growth in special interest politics which is not limited to the left or right, or to any particular issue content, and it is certainly not limited to problems of technology.

Organizations involved with these activities grew and became established as effective adversaries, as in the cases of Common Cause, the Sierra Club, and the Nader groups. They attracted talented and enthusiastic personnel, cash resources, and sympathetic (or at least interested) mass media, all enhancing their ability to pursue both technical and nontechnical issues.[7] The growth of antitechnology protests need not be explained as a unique phenomenon but as simply one component of the large and growing body of special interest politics.

KNOWLEDGE AND CONTROL

Technical controversies are part of the normal flow of political activity. Therefore it seems reasonable to decide the political issues raised in these controversies in the same way we decide other political issues. *If* we assume that our current political system is an acceptable means for making nonscience policy—a strong assumption—then it ought to be acceptable for making policy regarding science and technology, with little modification. The counterargument to this position is that science and technology are special and therefore must be treated by special means. The particular ways in which they are regarded as special are, first, that they carry their own inertia and so are impervious to social control, and second, that their subject matter is beyond the comprehension of laymen and therefore policy must be made by technical experts. Are either of these points valid?

In *Frankenstein*, Mary Shelley wrote of a scientist who lost control of his discovery and was eventually destroyed by it, an image that is easily transferred to today's atomic scientists. Technology too is often portrayed as a force that pushes forward inexorably, driven by its own momentum, choosing its own direction, and accelerating unperturbed by any human efforts to control it.[8] Perhaps these ideas had more force in the 1950s and 1960s when all "ripe" technologies, whether in warfare, space, or medicine, seemed to be running at full speed. Today things have slowed down, research funds are not so readily available, and we have examples of large technologies that have been stopped in their tracks. We know now that there are two ready levers to control the pace and direction of science and technology: money and government regulation. These are traditional mechanisms which require no special system of political control. We now understand that the usual political bodies—the Executive Branch, Congress, and the courts and state agencies—can increase or decrease the flow of money, and they can tighten or loosen regulatory requirements. We can turn on a war on cancer or turn off space exploration by shifting federal funds, though of course the funding of increased research on, say, cancer does not guarantee that completely satisfactory results will be achieved. We may set standards for the safety of nuclear power plants or the permissible levels of automobile emissions, though there will be instances when these standards are seriously violated. Nonetheless, gross controls exist and are currently in practice.

Our problem is not the absence of powerful controls but rather the failure to exercise properly the controls which are available. This difficulty extends to many policy areas besides science and technology, but in this case the problem is compounded because the material is difficult to understand, the esoteric business of specialists. How should we deal with this problem of specialized knowledge in making technology policy?

It is sometimes claimed that the average citizen is capable of understanding most technical issues that are relevant to policy decisions.[9] I don't believe that anyone who has tried to teach statistics to humanities doctoral students would hold that view for long. The typical voter in a referendum on fluoridation or nuclear power seems to have at best a vague understanding of the underlying scientific disputes which have been discussed during the campaign. The confusion of legislators, when hearing technical experts make apparently inconsistent claims about factual matters, is apparent from congressional hearings.

One could simply ignore technical issues, realizing that controversies are basically political in nature and are rarely settled by the resolution of factual disputes. Some courts follow essentially this course today when hearing technical cases. These come up typically when a regulatory agency has set a standard on, say, effluent of a carcinogen from a chemical plant. The agency is often sued, either by the plant, which may think the regulation too strict, or by environmentalists who think it too lax.[10] In either case, the suit goes to court where conflicting technical claims are heard about "safe" levels of population exposure to the carcinogen. The judges know little of technical matters[11] and usually base their decisions on procedural matters, asking if the regulatory agency followed the proper form and acted within its mandate in setting the standard.[12] Technical policy is thus set without much regard for the substantive issues in dispute.

A more informed approach is to have scientists, who understand the technical issues, settle the policy questions as well. The National Academy of Sciences makes policy recommendations on many technical problems every year, and though these are not binding in law, they have a good deal of stature and frequently become the legal policy of agencies or the Congress. The problem here is that in the process of giving the scientist a strong voice in his arena of competence, we also give him a particularly strong voice in setting policy for the society, an arena in which he has neither special competence nor a public mandate. Scientists have their own axes to grind, and the elite scientists of the Academy have elite axes.

The approach which I prefer is to separate scientific questions of fact from policy questions of value, a separation which seems quite feasible. This is a premise of the "science court" as well as of other proposals for dealing with technical disputes. Whichever of these mechanisms one uses for examining facts, the basic principle is to let the scientists do what they are especially well equipped to do, which is science, while keeping them away from policy decisions, where they have neither special competence nor the public mandate. Once the factual matters are settled, at least provisionally, then the usual procedures of government can be used to make policy decisions, whether by the Congress, the federal agencies, the courts, or by citizens voting in a referendum.

My preference runs counter to the current call for extraordinary public participation in policy decisions regarding science and technology. At one time it was commonly thought that the wisest solutions to technical problems would come from interdisciplinary congresses of scientists, engineers, philosophers, humanists, and clergymen, all pooling their diverse perspectives into a synergistic whole that would somehow be greater than the sum of its parts. In time this view was recognized as elitist (laborers and housewives were never included), and there were new suggestions that the participation of the common man and woman was more important than the involvement of philosophers or clerics. This view has received support from "public interest" groups, though they represent the average citizen about as well as corporate boards represent small stockholders. When President Carter appointed a special commission to investigate the accident at Three Mile Island, he included a "housewife and mother of six" who lived near the site. When the City Council of Cambridge challenged the construction of a laboratory at Harvard for recombinant DNA research, it appointed a much-publicized "citizen court" of eight laypersons from diverse occupations and educational levels,[13] which presumably was a better representative of the city than the City Council. Sweden, the Netherlands, and Austria have made elaborate attempts to involve the average citizen in decision making about nuclear power.[14] All of these procedures run against my suggestion that *normal* political means be used to make science and technology policy (with some modification to accommodate the specialized knowledge problem).

I certainly do not oppose public participation as such, but I do not see why there should be more of it in science and technology policy than in other kinds of policy. We never make a point of bringing housewives and blue collar laborers into formal decisions about the prime interest rate or on whether or not to attack Iran, so why do it when evaluating nuclear power plants and recombinant DNA laboratories? If decision making by government officials is adequate for military and economic policy, then why not for science and technology policy?

THE PROPER FUNCTION OF
TECHNICAL CONTROVERSIES

Technical controversies tend to be regarded as aberrations, as undesirable disruptions which ought to be ignored, disposed of, or avoided altogether. However, there is another viewpoint which emphasizes their positive function.

This viewpoint begins with the assumption that the growth of both scientific knowledge and technological invention is good and ought to be encouraged. At the same time, however, great care must be taken in the distribution of scarce research resources and in the implementation and deployment of innovations. We must minimize waste, health risks, environmental insults, and other costs. These are nearly unassailable assumptions, accepted consensually. It follows that there must be proper roles in the society for people who will advance science and promote new technologies.

It would be nice if the scientist and the engineer were cognizant of, and deeply concerned about, the potential risks, secondary consequences, and other costs of their creations. However, it does not seem realistic to me to expect inventors and promoters of innovation to behave this way. The promoters of thymic irradiation for children thought they were performing a great service, saving many lives; they were not convinced by arguments that "enlarged" thymus was innocuous, nor by early indications that such treatment might be dangerous. Of course, they took care to avoid obvious problems like burns from x-ray, but the promoter's role is not compatible with cautious restraint unless there is an obvious reason for it.

If we are to have social roles which promote potentially hazardous innovations, we should have other social roles, filled by different people, to counterbalance the promoters, to search for the hazards and other costs which the promoters do not see, or ignore, or actively avoid.

Regulatory agencies fulfill some of this function, but neither the routine work setting of a bureaucracy nor the job motivations of career bureaucrats are conducive to an enthusiastic and imaginative attack on the problem. Furthermore, innovations, by their nature, often fall outside of the competence and mandate of preexisting agencies. The congressional Office of Technology Assessment was established to deal with some of these concerns, but it too is a bureaucracy and may not be an appropriate setting for the task. Committees of the National Academy of Sciences have been useful in evaluating some hazards which have already been identified, but if we look at salient cases such as nuclear power or DDT, the Academy has not been particularly effective in locating problems. Furthermore there have been occasional charges that members of Academy committees sometimes have vested interests in the technologies which they are supposed to evaluate.[15]

There have by now been numerous instances when the informal process of social controversy has been more effective in identifying and explicating the risks and benefits of a technology than have been any of the formal means which are supposed to do this. Critics attack with great vigor, stretching their imaginations for all manner of issues with which to score points against their target. Proponents counterattack, producing new analyses and funding new experiments in order to refute the critics. Each side probes and exposes weaknesses in the other side's arguments. As the controversy proceeds, there is a filtering of issues so that some with little substance become ignored while others move to the fore.

The coming and going of issues is particularly clear in the long controversy over nuclear power. I examined polemic literature against nuclear power from the period 1968–72, comparing it to post–Arab oil embargo literature, and could easily discern the shifting pattern of major concerns from one period to another. Here are problems which were perceived as major in 1968–72 and then became minor afterwards: (1) "thermal pollution" of lakes and rivers by waste heat from the reactor; (2) cancers and genetic damage resulting from routine radiation released by normally operating power plants; (3) the Price-Anderson Act, which provided federally subsidized insurance for the nuclear industry as well as limits on liability; and (4) the poor performance of the Atomic Energy Commission (AEC) as a regulatory agency.

Why do some issues fade away while others persist? Perhaps some problems disappeared because they were solved. Early criticism of the AEC's performance—that it could not simultaneously promote and regulate nuclear power—may have been settled by the dissolution of the AEC and its replacement by separate promotional and regulatory agencies. But virtually all the critics of the AEC, and many of its supporters, regarded the reorganization as a superficial shuffle of AEC personnel. Yet the issue of regulatory competence declined in salience, at least until 1979 when it reemerged as a reaction to the accident at Three Mile Island.

The decline of concern about routine radiation emissions from normally operating power plants is usually explained as a result of the stringent revision by the AEC of its radiation standards in 1971. Declining concern with thermal pollution is similarly explained as the result of the increased insistence of the AEC on cooling towers and ponds at that time. But the early 1970s also marked the decline of massive popular interest in the environment, so the passing of these particularly ecological problems may simply reflect waning public concern with the larger issue of the environment. . . . I see no indication that the insurance issue has been settled; it seems simply to have diminished.

New issues have emerged since 1972. One of the most exciting is the problem of safeguarding nuclear material from terrorists who might construct a bomb. Certainly this issue parallels the fascination of the public during the 1970s with terrorist activities, more than it reflects any new facts about bomb construction. After 1974, when India exploded an atomic bomb built from fissionable material produced in a Canadian-supplied reactor, critics brought the nuclear proliferation issue into the controversy.

The Arab oil embargo of 1973 initiated the energy crisis which has been with us in more or less intensity since then. In the ensuing climate of public opinion, anyone who opposed nuclear energy was likely to champion an alternate energy source (particularly small-scale solar), to emphasize the need for energy conservation, and to minimize our need for additional energy. These issues are prominent in the polemic literature after 1973 though they were virtually nonexistent before then.

What of the more persistent issues? There are two such issues, that of catastrophic nuclear reactor accidents, and of disposal of long-term radioactive wastes. These continue as major concerns of the critical literature throughout the controversy. Perhaps the test of time separates wheat from chaff, and these are the "real" issues. Yet even in the persistent concern with reactor accidents, we see a shift in emphasis from year to year: first emphasis was laid on human error, then in 1971–74 on the unproven emergency core-cooling system, then on the credibility of the "Rasmussen Report,"[16] which purports to estimate the probability and severity of various kinds of reactor accidents, and most recently, after Three Mile Island, back to human error and regulatory competence.

Precisely the same kind of shifting occurs in promotional rhetoric. The nuclear industry has extolled atomic power for 25 years, but its major advantage has changed repeatedly. In the beginning it was touted as the electricity that would be too cheap to meter, which proved to be an exaggeration. With the rise of environmental concern in the 1960s, it became the clean source of electricity. By 1970, failing to win the support of environmentalists, the industry supported a number of studies showing that nuclear power was superior to fossil fuel generation in risk–benefit tradeoffs. After the oil embargo of 1973, nuclear power was the only proven energy source that would end our dependence on Arab oil.

New issues are continually thought up by partisans, and others are thrust at us by events such as the Arab oil embargo or the accident at Three Mile Island. Some issues quickly fall of their own weight, such as the claim that nuclear power facilities consume more energy than they produce; other issues never gain any following, such as the claim that increasing radiation levels have caused mental deficiencies in children, thus accounting for the decade-long decline in academic achievement test scores.

It would be a mistake to view this process as nothing more than the shifting of popular concern from one faddish issue to another. It is true that some unimportant issues gain an inordinate amount of attention for awhile, but they usually drop from concern once the popular mood changes, unless they can sustain interest on their intrinsic merits. In the meantime, intelligent people on both sides of the controversy search enthusiastically for new problems, diligently preparing charges and rebuttals, testing the strength of their arguments in open debate. In this manner the controversy functions as a funnel, bringing diverse problems together, and it has the potential to act as a sieve which separates important concerns from those without real merit.

One should have no illusions that this process will lead to a settlement of the controversy. We sometimes hear people speak as if a controversy is a set of n issues such that, if each were solved, the controversy would end. That is not the nature of political debate. The alignment is primary and the various rationales for it—the substantive issues—are often secondary, coming and going while the basic alignment persists. But that need not be a problem for the policymaker who understands the dynamics of controversy and who recognizes that the proper function of a controversy is the identification and evaluation of potential problems, as an informal method of technology assessment.

Controversies bring their share of problems. In delaying the implementation of a technology, they may deny to society important benefits, at least for awhile, as in the case of the lightning rod. It is difficult to judge whether we take a net gain or loss from such delays. Our children have more cavities than they would have if the whole nation had been fluoridated in the 1950s, but most of us do not regret that the SST was stopped, and no doubt England and France wish that someone had stopped theirs as well. Research on recombinant DNA has probably not lost much ground, in the long view, because of a pause for evaluation. Perhaps we will suffer from the slowdown in nuclear power-plant production, though one might speculate that if the costs of a slowdown become large, the building program will speed up in spite of some objections.

Another problem with technical controversies is that they are chaotic, and therefore we fail to achieve the potential benefits which are available to us. In a controversy each side is motivated to make the strongest case it can while at the same time finding flaws in the other side's position. When these arguments are juxtaposed . . . , flaws and polemic tricks are disposed of, and we see the beauty of first-class reasoning in support of one's own ends. Unfortunately, in most controversies the adversaries never confront one another. . . . They address different audiences, or the same audience at different times or in formats which allow them to speak past one another. We become confused when one expert seems to contradict the other, and we cannot locate the reason for their disagreement. . . . If we could bring some order to the chaos, so that we might better understand the bases for these disagreements, then the value of such arguments would be greatly enhanced.

A final problem with technical controversies is that it is too difficult to start one. I have a number of students deeply concerned about the problems of a technological society, energy issues being at the top of their lists. As they examine the various energy options, they see problems and benefits associated with each one. Yet there is only one energy option which they can do much about, and that is nuclear power, which has readily available channels for opposition right on campus. They have no potent way to express their concerns about the explosive potential of liquified natural gas, about the pollution of coal, or about the hazards of any energy source except nuclear power. It is the only protest in town; you either join it or go home and study.

The power to start a serious and credible technical controversy lies in few hands, notably the environmental and consumer groups, and some prestigious scientists who have good access to the nation's mass media.[17] Controversies which come from less orthodox sources have trouble gaining credibility in higher circles. The fluoridation controversy came from the grassroots and was never taken seriously in the academic community. I suggest that if most scientists looked today at the evidence that was available on the safety of fluoridation back in 1950, when the drive for national implementation began, they would agree that the decision was premature if not foolhardy. Yet antifluoridationists were lumped together as cranks and kooks, and social scientists called their arguments "antiscience." Nonetheless, many communities rejected fluoridation, putting the proponents on the defensive. They had to strengthen their case for the presence of benefit with minimal risk, and I believe that an impressive body of evidence to that effect now exists.

Every day my family drinks fluoridated water and uses fluoridated toothpaste without worry. About once a year I accept our dentist's judgment that the children's teeth should have a topical application of fluoride. It seems an innocuous procedure, and I suppose my parents felt the same way when a physician suggested that my chronically infected tonsils be irradiated. I do not remember that physician, but I doubt that he was any less cautious than our dentist. Both the working dentist and the working physician treat their patients according to the fashion in their profession at that time. Sometimes the fashions are unwise, and if so, it is best that we learn of it sooner rather than later, even at the expense of some controversy.

NOTES

1. A. Kantrowitz, "A Technologist Looks at Anti-Technology," Messenger Lectures, Cornell University, 1978; and R. Nisbet, "The Rape of Progress," Public Opinion, Vol. 2 (1979), pp. 2–6, 55.
2. T. Hughes, ed., Changing Attitudes Toward American Technology (New York: Harper and Row, 1975).
3. T. LaPorte and D. Metlay, "Technology Observed: Attitudes of a Wary Public," Science, Vol. 188 (1975), pp. 121–27; and E. Marshall, "Public Attitudes to Technological Progress," Science, Vol. 205 (1979), pp. 281–85.
4. A. McFarland, Public Interest Lobbies (Washington, D.C.: American Enterprise Institute, 1976).
5. D. Broder, "PAC Power Curbed," Syracuse Post-Standard, October 27, 1979, p. 4.
6. S. Verba and N. Nie, Participation in America (New York: Harper and Row, 1972).
7. McFarland, op. cit.
8. L. Winner, Autonomous Technology: Technics-out-of-control as a Theme in Political Thought (Cambridge, Mass.: MIT Press, 1977).
9. B. Casper, "Technology Policy and Democracy," Science, Vol. 194 (1976), pp. 29–35.
10. N. deNevers, "Enforcing the Clean Air Act of 1970," Scientific American, Vol. 228 (June 1973), pp. 14–21.
11. W. Thomas, "Judicial Treatment of Scientific Uncertainty in the Reserve Mining Case," Proceedings of the Fourth Symposium on Statistics and the Environment (Washington, D.C.: American Statistical Association, 1977), pp. 1–13.
12. D. Bazelon, "Risk and Responsibility," Science, Vol. 205 (1979), pp. 277–80.

13. S. Krimsky, "A Citizen Court in the Recombinant DNA Debate," *Bulletin of the Atomic Scientists* (October 1978), pp. 37–43.
14. D. Nelkin and M. Pollak, "The Politics of Participation and the Nuclear Debate in Sweden, the Netherlands and Austria," *Public Policy*, Vol. 25 (1977), pp. 333–57.
15. P. Boffey, *The Brain Bank of America* (New York: McGraw-Hill, 1975); and S. Schiefelbein, "The Invisible Threat," *Saturday Review*, September 15, 1979, pp. 16–20.
16. N. Rasmussen, *et al.*, *Reactor Safety Study* (Washington, D.C.: Nuclear Regulatory Commission, 1975), document number WASH-1400 (NUREG-75/014).
17. A. Mazur, "Media Coverage and Public Opinion on Scientific Controversies," *Journal of Communication* (1981).

Part III
RESHAPING TECHNOLOGY

Some critics and observers of the social dimensions of technology are persuaded that it is not enough to assess and control existing technological systems. What is needed, they believe, is a broad-based reshaping of technology. Those who approach technology from this perspective are not interested in abstract critiques. Some are concerned with the practical, long-term viability of main-stream industrial technology in the face of limits on investment capital, resource constraints, and the potential for human actions to cause long-term, possibly irreversible damage to the global environment. Others focus on the equity implications of technological systems—for example, the effects of such systems on women and ethnic or racial minorities. Still others see the need for democratic governance of technology as the central issue. All would alter the directions of technological change to better accommodate these perspectives.

Central to the first two readings in this section of *Technology and the Future* is the notion of *alternative* or *appropriate* technology. This concept, which in the United States was associated with the counterculture of the 1960s and 1970s, is a way of looking at technology itself rather than at a specific type of hardware. It is a set of design criteria that stress simplicity, individual self-worth and self-reliance, labor-intensiveness rather than capital intensiveness, minimum energy use, consistency with environmental quality, and decentralization rather than centralization.

Among the authors of the readings in this section are two of the best known and most creative thinkers on the subject of technological alternatives: E. F. Schumacher and Paul Goodman. Schumacher's name is practically synonymous with appropriate technology, especially as it applies to Third World development. His essay, "Buddhist Economics," though more than twenty years old, is still as thought-provoking as it is original. Paul Goodman's paper, "Can

Technology Be Humane?" which actually predates Schumacher's by several years, provides an idea of what the philosophical underpinnings of a really different style of technology might be like.

The next four chapters discuss the redirection of technology from somewhat different angles. Richard Sclove's recent essay, "Technological Politics As If Democracy Really Mattered," updates some of Goodman's ideas and suggests ways in which technological decision making can be made more democratic and technology can be used to create a more democratic society. Donald Norman speculates on how the power of modern technology, particularly advanced information processing, can be harnessed to serve the needs of humans, contrasting the potential results with the traditional approach in which people are forced to conform to the needs of machines.

Judy Wajcman critiques mainstream technology from a feminist standpoint, suggesting along the way how a technology based on women's values might look. Robert Johnson draws on an African-American perspective in his article, "Science, Technology and Black Community Development," originally published in *The Black Scholar*. He points out that technology has had an enormous effect on the lives of African-Americans and warns that they must take a more active role in shaping technology's future development if they want it to serve their needs and interests. In a provocative essay at the end of Part III, Langdon Winner suggests the ways in which technological "artifact/ideas" have inherent political dimensions. He lays out a series of "guiding maxims" to help direct discussions of technological choice toward the goal of increasing democratic control.

14. Buddhist Economics

E. F. SCHUMACHER

More than any other single individual, E. F. Schumacher is responsible for popularizing the notion of appropriate technology. It may seem strange that a man who was once chief economist of Britain's National Coal Board emerged as intellectual parent of a movement that seeks the radical restructuring of the whole essence of economics, but Schumacher was also a longtime advocate of organic farming, a student of Gandhi, and an activist for political decentralization.

Schumacher was born in Germany in 1911, trained in economics, and came to England as a Rhodes scholar. Like many Germans living in Britain, he was interned for a time during World War II. Later he was released to do farm work, an experience that strongly influenced his later work. While pursuing a career as a government economist, he became involved in organic farming, became president of the Soil Association, and in 1966 founded the Intermediate Technology Development Group, an organization that promotes small-scale technology tailored to the needs of specific developing countries. Schumacher died in 1977, not long after a visit to the United States in which he was accorded the recognition of a meeting with President Jimmy Carter to discuss his ideas.

His book Small Is Beautiful, *from which the brilliant essay "Buddhist Economics" is taken, became an underground classic soon after its publication in the early 1970s. As effective an introduction as can be found to the ideas of appropriate technology, "Buddhist Economics" provides Schumacher's answer to the question Leo Marx asks in the opening chapter of this book, "Does Improved Technology Mean Progress?"*

Source: Selection from *Small Is Beautiful: Economics As If People Mattered* by E. F. Schumacher. Copyright ©1973 by E. F. Schumacher. Reprinted by permission of HarperCollins Publishers, Inc. and Blond & Briggs, London WC1N 3HZ, England.

"Right Livelihood" is one of the requirements of the Buddha's Noble Eightfold Path. It is clear, therefore, that there must be such a thing as Buddhist economics.

Buddhist countries have often stated that they wish to remain faithful to their heritage. So Burma: "The New Burma sees no conflict between religious values and economic progress. Spiritual health and material well-being are not enemies: they are natural allies."[1] Or: "We can blend successfully the religious and spiritual values of our heritage with the benefits of modern technology."[2] Or: "We Burmans have a sacred duty to conform both our dreams and our acts to our faith. This we shall ever do."[3]

All the same, such countries invariably assume that they can model their economic development plans in accordance with modern economics, and they call upon modern economists from so-called advanced countries to advise them, to formulate the policies to be pursued, and to construct the grand design for development, the Five-Year Plan or whatever it may be called. No one seems to think that a Buddhist way of life would call for Buddhist economics, just as the modern materialist way of life has brought forth modern economics.

Economists themselves, like most specialists, normally suffer from a kind of metaphysical blindness, assuming that theirs is a science of absolute and invariable truths, without any presuppositions. Some go as far as to claim that economic laws are as free from "metaphysics" or "values" as the law of gravitation. We need not, however, get involved in arguments of methodology. Instead, let us take some fundamentals and see what they look like when viewed by a modern economist and a Buddhist economist.

There is universal agreement that a fundamental source of wealth is human labor. Now, the modern economist has been brought up to consider "labor" or work as little more than a necessary evil. From the point of view of the employer, it is in any case simply an item of cost, to be reduced to a minimum if it cannot be eliminated altogether, say, by automation. From the point of view of the workman, it is a "disutility"; to work is to make a sacrifice of one's leisure and comfort, and wages are a kind of compensation for the sacrifice. Hence the ideal from the point of view of the employer is to have output without employees, and the ideal from the point of view of the employee is to have income without employment.

The consequences of these attitudes both in theory and in practice are, of course, extremely far-reaching. If the ideal with regard to work is to get rid of it, every method that "reduces the work load" is a good thing. The most potent method, short of automation, is the so-called "division of labor," and the classical example is the pin factory eulogized in Adam Smith's *Wealth of Nations*. Here it is not a matter of ordinary specialization, which mankind has practiced from time immemorial, but of dividing up every complete process of production into minute parts, so that the final product can be produced at great speed without anyone having had to contribute more than a totally insignificant and, in most cases, unskilled movement of his limbs.

The Buddhist point of view takes the function of work to be at least threefold: to give a man a chance to utilize and develop his faculties; to enable him to overcome his egocenteredness by joining with other people in a common task; and to bring forth the goods and services needed for a becoming existence. Again, the consequences that flow from this view are endless. To organize work in such a manner that it becomes meaningless, boring, stultifying, or nerve-racking for the worker would be little short of criminal; it would indicate a greater concern with goods than with people, an evil lack of compassion and a soul-destroying degree of attachment to the most primitive side of this worldly existence. Equally, to strive for leisure as an alternative to work would be considered a complete misunderstanding of one of the basic truths of human existence, namely that work and leisure are complementary parts of the same living process and cannot be separated without destroying the joy of work and the bliss of leisure.

From the Buddhist point of view, there are therefore two types of mechanization which must be clearly distinguished: one that enhances a man's skill and power and one that turns the work of man over to a mechanical slave, leaving man in a position of having to serve the slave. How to tell the one from the other? "The craftsman himself," says Ananda Coomaraswamy, a man equally competent to talk about the modern West as the ancient East, "can always, if allowed to, draw the delicate distinction between the machine and the tool. The carpet loom is a tool, a contrivance for holding warp threads at a stretch for the pile to be woven round them by the craftsmen's fingers; but the power loom is a machine, and its significance as a destroyer of culture lies in the fact that it does the essentially human part of the work."[4] It is clear, therefore, that Buddhist economics must be very different from the economics of modern materialism, since the Buddhist sees the essence of civilization not in a multiplication of wants but in the purification of human character. Character, at the same time, is formed primarily by a man's work. And work, properly conducted in conditions of human dignity and freedom, blesses those who do it and equally their products. The Indian philosopher and economist J. C. Kumarappa sums the matter up as follows:

> If the nature of the work is properly appreciated and applied, it will stand in the same relation to the higher faculties as food is to the physical body. It nourishes and enlivens the higher man and urges him to produce the best he is capable of. It directs his free will along the proper course and disciplines the animal in him into progressive channels. It furnishes an excellent background for man to display his scale of values and develop his personality.[5]

If a man has no chance of obtaining work he is in a desperate position, not simply because he lacks an income but because he lacks this nourishing and enlivening factor of disciplined work which nothing can replace. A modern economist may engage in highly sophisticated calculations on whether full employment "pays" or whether it might be more "economic" to run an economy at less than full employment so as to ensure a greater mobility of labor, a better stability of wages, and so forth. His fundamental criterion of success is simply the total quantity of goods produced during a given period of time. "If the marginal urgency of goods is low," says Professor Galbraith in *The Affluent Society*, "then so is the urgency of employing the last man or the last million men in the labor force."[6] And again: "If . . . we can afford some unemployment in the interest of stability—a proposition, incidentally, of impeccably conservative antecedents—then we can afford to give those who are unemployed the goods that enable them to sustain their accustomed standard of living."

From a Buddhist point of view, this is standing the truth on its head by considering goods as more important than people and consumption as more important than creative activity. It means shifting the emphasis from the worker to the product of work, that is, from the human to the subhuman, a surrender to the forces of evil. The very start of Buddhist economic planning would be a planning for full employment, and the primary purpose of this would in fact be employment for everyone who needs an "outside" job: it would not be the maximization of employment nor the maximization of production. Women, on the whole, do not need an "outside" job, and the large-scale employment of women in offices or factories would be considered a sign of serious economic failure. In particular, to let mothers of young children work in factories while the children run wild would be as uneconomic in the eyes of a Buddhist economist as the employment of a skilled worker as a soldier in the eyes of a modern economist.

While the materialist is mainly interested in goods, the Buddhist is mainly interested in liberation. But Buddhism is "The Middle Way" and therefore in no way antagonistic to physical well-being. It is not wealth that stands in the way of liberation but the attachment to wealth; not the enjoyment of pleasurable things but the craving for them. The keynote of Buddhist economics, therefore, is simplicity and nonviolence. From an economist's point of view, the marvel of the Buddhist way of life is the utter rationality of its pattern—amazingly small means leading to extraordinarily satisfactory results.

For the modern economist this is very difficult to understand. He is used to measuring the "standard of living" by the amount of annual consumption, assuming all the time that a man who consumes more is "better off" than a man who consumes less. A Buddhist economist would consider this approach excessively irrational: since consumption is merely a means to human well-being, the aim should be to obtain the maximum of well-being with the minimum of consumption. Thus, if the purpose of clothing is a certain amount of temperature comfort and an attractive appearance, the task is to attain this purpose with the smallest possible effort; that is, with the smallest annual destruction of cloth and with the help of designs that involve the smallest possible input of toil. The less toil there is, the more time and strength is left for artistic creativity. It would be highly uneconomic, for instance, to go in for complicated tailoring, like the modern West, when a much more beautiful effect can be achieved by the skillful draping of uncut material. It would be the height of folly to make material so that it should wear out quickly and the height of barbarity to make anything ugly, shabby or mean. What has just been said about clothing applies equally to all other human requirements. The ownership and the consumption of goods is a means to an end, and Buddhist economics is the systematic study of how to attain given ends with the minimum means.

Modern economics, on the other hand, considers consumption to be the sole end and purpose of all economic activity, taking the factors of production—land, labor, and capital—as the means. The former, in short, tries to maximize human satisfactions by the optimal pattern of consumption, while the latter tries to maximize consumption by the optimal pattern of productive effort. It is easy to see that the effort needed to sustain a way of life which seeks to attain the optimal pattern of consumption is likely to be much smaller than the effort needed to sustain a drive for maximum consumption. We need not be surprised, therefore, that the pressure and strain of living is very much less in, say, Burma than it is in the United States, in spite of the fact that the amount of labor-saving machinery used in the former country is only a minute fraction of the amount used in the latter.

Simplicity and nonviolence are obviously closely related. The optimal pattern of consumption, producing a high degree of human satisfaction by means of a relatively low rate of consumption, allows people to live without great pressure and strain and to fulfill the primary injunction of Buddhist teaching: "Cease to do evil; try to do good." As physical resources are everywhere limited, people satisfying their needs by means of a modest use of resources are obviously less likely to be at each other's throats than people depending upon a high rate of use. Equally, people who live in highly self-sufficient local communities are less likely to get involved in large-scale violence than people whose existence depends on worldwide systems of trade.

From the point of view of Buddhist economics, therefore, production from local resources for local needs is the most rational way of economic life, while dependence on imports from afar and the consequent need to produce for export to unknown and distant peoples is highly uneconomic and justifiable only in exceptional cases and on a small scale. Just as the modern economist would admit that a high rate of consumption of transport services between a man's home and his place of work signifies a misfortune and not a high standard of life, so the Buddhist economist would hold that to satisfy human wants from faraway sources rather than from sources nearby signifies failure rather than success. The former tends to take statistics showing an increase in the number of tons/miles per head of the population carried by a country's transport system as proof of economic progress, while to the latter—the Buddhist economist—the same statistics would indicate a highly undesirable deterioration in the *pattern* of consumption.

Another striking difference between modern economics and Buddhist economics arises over the use of natural resources. Bertrand de Jouvenel, the eminent French political philosopher, has characterized "Western man" in words which may be taken as a fair description of the modern economist:

He tends to count nothing as an expenditure, other than human effort; he does not seem to mind how much mineral matter he wastes and, far worse, how much living matter he destroys. He does not seem to realize at all that human life is a dependent part of an ecosystem of many different forms of life. As the world is ruled from towns where men are cut off from any form of life other than human, the feeling of belonging to an ecosystem is not revived. This results in a harsh and improvident treatment of things upon which we ultimately depend, such as water and trees.[7]

The teaching of the Buddha, on the other hand, enjoins a reverent and nonviolent attitude not only to all sentient beings but also, with great emphasis, to trees. Every follower of the Buddha ought to plant a tree every few years and look after it until it is safely established, and the Buddhist economist can demonstrate without difficulty that the universal observation of this rule would result in a high rate of genuine economic development independent of any foreign aid. Much of the economic decay of southeast Asia (as of many other parts of the world) is undoubtedly due to a heedless and shameful neglect of trees.

Modern economics does not distinguish between renewable and non-renewable materials, as its very method is to equalize and quantify everything by means of a money price. Thus, taking various alternative fuels, like coal, oil, wood, or water-power: the only difference between them recognized by modern economics is relative cost per equivalent unit. The cheapest is automatically the one to be preferred, as to do otherwise would be irrational and "uneconomic." From a Buddhist point of view, of course, this will not do; the essential difference between nonrenewable fuels like coal and oil on the one hand and renewable fuels like wood and water-power on the other cannot simply be overlooked. Nonrenewable goods must be used only if they are indispensable, and then only with the greatest care and the most meticulous concern for conservation. To use them heedlessly or extravagantly is an act of violence, and while complete nonviolence may not be attainable on this earth, there is nonetheless an ineluctable duty on man to aim at the ideal of nonviolence in all he does.

Just as a modern European economist would not consider it a great economic achievement if all European art treasures were sold to America at attractive prices, so the Buddhist economist would insist that a population basing its economic life on nonrenewable fuels is living parasitically, on capital instead of income. Such a way of life could have no permanence and could therefore be justified only as a purely temporary expedient. As the world's resources of nonrenewable fuels—coal, oil, and natural gas—are exceedingly unevenly distributed over the globe and undoubtedly limited in quantity, it is clear that their exploitation at an ever-increasing rate is an act of violence against nature which must almost inevitably lead to violence between men.

This fact alone might give food for thought even to those people in Buddhist countries who care nothing for the religious and spiritual values of their heritage and ardently desire to embrace the materialism of modern economics at the fastest possible speed. Before they dismiss Buddhist economics as nothing better than a nostalgic dream, they might wish to consider whether the path of economic development outlined by modern economics is likely to lead them to places where they really want to be. Towards the end of his courageous book *The Challenge of Man's Future*, Professor Harrison Brown of the California Institute of Technology gives the following appraisal:

> Thus we see that, just as industrial society is fundamentally unstable and subject to reversion to agrarian existence, so within it the conditions which offer individual freedom are unstable in their ability to avoid the conditions which impose rigid organization and totalitarian control. Indeed, when we examine all of the foreseeable difficulties which threaten the survival of industrial civilization, it is difficult to see how the achievement of stability and the maintenance of individual liberty can be made compatible.[8]

Even if this were dismissed as a long-term view there is the immediate question of whether "modernization," as currently practiced without regard to religious and spiritual values, is actually producing agreeable results. As far as the masses are concerned, the results appear to be disastrous—a collapse of the rural economy, a rising tide of unemployment in town and country, and the growth of a city proletariat without nourishment for either body or soul.

It is in the light of both immediate experience and long-term prospects that the study of Buddhist economics could be recommended even to those who believe that economic growth is more important than any spiritual or religious values. For it is not a question of choosing between "modern growth" and "traditional stagnation." It is a question of finding the right path of development, the Middle Way between materialist heedlessness and traditionalist immobility, in short, of finding "Right Livelihood."

NOTES

1. *The New Burma* (Economic and Social Board, Government of the Union of Burma, 1954).
2. *Ibid.*
3. *Ibid.*
4. Ananda K. Coomaraswamy, *Art and Swadeshi* (Madras: Ganesh & Co.).
5. J. C. Kumarappa, *Economy of Permanence* (Sarva-Seva Sangh Publication, Rajghat, Kashi, 4th ed., 1958).
6. John Kenneth Galbraith, *The Affluent Society* (London: Penguin Books Ltd., 1962).
7. Richard B. Gregg, *A Philosophy of Indian Economic Development* (Ahmedabad: Navajivan Publishing House, 1958).
8. Harrison Brown, *The Challenge of Man's Future* (New York: The Viking Press, 1954).

15. Can Technology Be Humane?

PAUL GOODMAN

In his essay "Technology: The Opiate of the Intellectuals" (which appears in Part I of this book), John McDermott despairs of the possibility of creating a humane technology within our present system. In a footnote to a section not included in this book, he writes, "Any discussion of the reorganization of technology to serve human needs seems, at this point, so Utopian that it robs one of the conviction necessary to shape a believable vision." Paul Goodman, unwilling to accept such a hopeless view, asks in the title of his selection, "Can Technology Be Humane?"

In developing his response to this question, Goodman admits there is no certainty that technology will become humane. Yet, in the classic style of a prophet—partly predictive and partly prescriptive—he asserts that our society is "on the eve of a new protestant Reformation, and no institution or status will go unaffected." In this selection, he offers a number of suggestions for channeling the energies of this Reformation into directions that he sees as critical to its success: prudence in the application of technology, an ecological viewpoint, and decentralization. Although the term appropriate technology *was barely known when Goodman first published this essay in the late 1960s, it is easy to see its roots in his prescription for reshaping technology.*

Paul Goodman (1911–1972) was a philosopher and humanist whose book Growing Up Absurd *established him as "the philosopher of the New Left" in the 1960s. His work ranged widely, from* Communitas *(1947), a classic of community planning written with his brother, Percival Goodman, to* Gestalt Therapy *(1951), written with F. S. Perls and Ralph Hefferline. Goodman was an anarchist and a pacifist, very active in the antiwar movement of the sixties. He wrote theoretical and practical treatises on politics, education, lan-*

guage, and literature, but his own judgment was that his literary work—novels, stories, poems, and plays—was his best. He was born in Manhattan, attended City College of New York, and was trained in philosophy at the University of Chicago.

On March 4, 1969, there was a "work stoppage" and teach-in initiated by dissenting professors at the Massachusetts Institute of Technology, and followed at thirty other major universities and technical schools across the country, against misdirected scientific research and the abuse of scientific technology. Here I want to consider this event in a broader context than the professors did, indeed as part of a religious crisis. For an attack on the American scientific establishment is an attack on the worldwide system of belief. I think we are on the eve of a new protestant Reformation, and no institution or status will go unaffected.

March 4 was, of course, only [one] of a series of protests in the [over] twenty-five years since the Manhattan Project to build the atom bomb, during which time the central funding of research and innovation has grown so enormously and its purposes have become so unpalatable. In 1940 the federal budget for research and development was less than 100 million dollars, in 1967, 17 billion. Hitler's war was a watershed of modern times. We are accustomed, as H. R. Trevor-Roper has pointed out, to write Hitler off as an aberration, of little political significance. But, in fact, the military emergency that he and his Japanese allies created confirmed the worst tendencies of the giant states, till now they are probably irreversible by ordinary political means.

After Hiroshima, there was the conscience-stricken movement of the Atomic Scientists and the founding of their Bulletin. The American Association for the Advancement of Science pledged itself to keep the public informed about the dangerous bearings of new developments. There was the Oppenheimer incident. Ads of the East Coast scientists successfully stopped the bomb shelters, warned about the fallout, and helped produce the test ban. There was a scandal about the bombardment of the Van Allen belt. Scientists and technologists formed a powerful (and misguided) ad hoc group for Johnson in the 1964 election. In some universities, sometimes with bitter struggle, classified contracts have been excluded. There is a Society for Social Responsibility in Science. Rachel Carson's book on the pesticides caused a stir, until the Department of Agriculture rescued the manufacturers and plantation-owners. Ralph Nader has been on his rampage. Thanks to spectacular abuses like smog, strip-mining, asphalting, pesticides, and oil pollution, even ecologists and conservationists have been getting a hearing. Protest against the boom has slowed up the development of the supersonic transport [particularly in the United States]. Most recent has been the concerned outcry against the antiballistic missiles.

The target of protest has become broader and the grounds of complaint deeper. The target is now not merely the military, but the universities, commercial corporations, and government. It is said that money is being given by the wrong sponsors to the wrong people for the wrong purposes. In some of the great schools, such funding is the main support, e.g., at MIT, 90 percent of the research budget is from the government, and 65 percent of that is military.

Inevitably, such funding channels the brainpower of most of the brightest science students, who go where the action is, and this predetermines the course of American science and technology for the foreseeable future. At present nearly 200,000 American engineers and scientists spend all their time making weapons, which is a comment on, and perhaps explanation for, the usual statement that more scientists are now alive than since Adam and Eve. And the style of such research and development is not good. It is dominated by producing hardware, figuring logistics, and devising salable novelties. Often there is secrecy, always nationalism. Since the grants go overwhelmingly through a very few corporations and universities, they favor a limited number of scientific attitudes and preconceptions, with incestuous staffing. There is a premium on "positive results"; surprising "failures" cannot be pursued, so that science ceases to be a wandering dialogue with the unknown.

The policy is economically wasteful. A vast amount of brains and money is spent on crash programs to solve often essentially petty problems, and the claim that there is a spin-off of useful discoveries is derisory, if we consider the sums involved. The claim that research is neutral, and it doesn't matter what one works on, is shabby, if we consider the heavy funding in certain directions. Social priorities are scandalous: money is spent on overkill, supersonic planes, brand name identical drugs, annual model changes of cars, new detergents, and color television, whereas water, air space, food, health, and foreign aid are neglected. And much research is morally so repugnant, e.g., chemical and biological weapons, that one dares not humanly continue it.

The state of the behavioral sciences is, if anything, worse. Their claim to moral and political neutrality becomes, in effect, a means of diverting attention from glaring social evils, and they are in fact used—

or would be if they worked—for warfare and social engineering, manipulation of people for the political and economic purposes of the
powers that be. This is an especially sad betrayal since, in the not-too-
distant past, the objective social sciences were developed largely to
dissolve orthodoxy, irrational authority, and taboo. They were hereti-
cal and intellectually revolutionary, as the physical sciences had been
in their own Heroic Age, and they weren't getting government grants.

This is a grim indictment. Even so, I do not think the dissenting scientists understand how deep their trouble is. They still take themselves too
much for granted. Indeed, a repeated theme of the March 4 [1969] complaints was that the science budget was being cut back, especially in basic research. The assumption was that though the sciences are abused,
Science would rightly maintain and increase its expensive preeminence
among social institutions. Only Science could find the answers.

But underlying the growing dissent there is a historical crisis. There
has been a profound change in popular feeling, more than among the
professors. Put it this way: Modern societies have been operating as if
religion were a minor and moribund part of the scheme of things. But
this is unlikely. Men do not do without a system of "meanings" that
everybody believes and puts his hope in even if, or especially if, he
doesn't know anything about it; what Freud called a "shared psychosis,"
meaningful because shared, and with the power that resides in dream
and longing. In fact, in advanced countries it is science and technology
themselves that have gradually and finally triumphantly become the
system of mass faith, not disputed by various political ideologies and
nationalism that have also been mass religions. Marxism called itself
"scientific socialism" as against moral and utopian socialisms; and movements of national liberation have especially promised to open the
benefits of industrialization and technological progress when once they
have gotten rid of the imperialists.

For three hundred years, science and scientific technology had an
unblemished and justified reputation as a wonderful adventure, pour-
ing out practical benefits, and liberating the spirit from the errors of
superstition and traditional faith. During this century they have finally
been the only generally credited system of explanation and problem-
solving. Yet in our generation they have come to seem to many, and to
very many of the best of the young, as essentially inhuman, abstract,
regimenting, hand-in-glove with Power, and even diabolical. Young
people say that science is antilife, it is a Calvinist obsession, it has been
a weapon of white Europe to subjugate colored races, and manifestly—

in view of recent scientific technology—people who think that way become insane. With science, the other professions are discredited; and the academic "disciplines" are discredited.

The immediate reasons for this shattering reversal of values are fairly obvious. Hitler's ovens and his other experiments in eugenics, the first atom bombs and their frenzied subsequent developments, the deterioration of the physical environment and the destruction of the biosphere, the catastrophes impending over the cities because of technological failures and psychological stress, the prospect of a brainwashed and drugged 1984. Innovations yield diminishing returns in enhancing life. And instead of rejoicing, there is now widespread conviction that beautiful advances in genetics, surgery, computers, rocketry, or atomic energy will surely only increase human woe.

In such a crisis, in my opinion, it will not be sufficient to ban the military from the universities; and it will not even be sufficient, as liberal statesmen and many of the big corporations envisage, to beat the swords into ploughshares and turn to solving problems of transportation, desalinization, urban renewal, garbage disposal, and cleaning up the air and water. If the present difficulty is religious and historical, it is necessary to alter the entire relationship of science, technology, and social needs both in men's minds and in fact. This involves changes in the organization of science, in scientific education, and in the kinds of men who make scientific decisions.

In spite of the fantasies of hippies, we are certainly going to continue to live in a technological world. The question is a different one: is that workable?

PRUDENCE

Whether or not it draws on new scientific research, technology is a branch of moral philosophy, not of science. It aims at prudent goods for the commonweal and to provide efficient means for these goods. At present, however, "scientific technology" occupies a bastard position in the universities, in funding, and in the public mind. It is half tied to the theoretical sciences and half treated as mere know-how for political and commercial purposes. It has no principles of its own. To remedy this—so Karl Jaspers in Europe and Robert Hutchins in America have urged—technology must have its proper place on the faculty as a learned profession important in modern society, along with medicine, law, the humanities, and natural philosophy, learning from them and having something to teach them. As a moral philosopher, a technician should be able to criticize the programs given him to implement. As a professional in a community of learned professionals, a technologist must have a different kind of training and develop a different character than we see at present among technicians and engineers. He should know something of the social sciences, law, the fine arts, and medicine, as well as relevant natural sciences.

Prudence is foresight, caution, utility. Thus it is up to the technologists, not to regulatory agencies of the government, to provide for safety and to think about remote effects. This is what Ralph Nader is saying and Rachel Carson used to ask. An important aspect of caution is flexibility, to avoid the pyramiding catastrophe that occurs when something goes wrong in interlocking technologies, as in urban power failures. Naturally, to take responsibility for such things often requires standing up to the front office and urban politicians, and technologists must organize themselves in order to have power to do it.

Often it is clear that a technology has been oversold, like the cars. Then even though the public, seduced by advertising, wants more, technologists must balk, as any professional does when his client wants what isn't good for him. We are now repeating the same self-defeating congestion with the planes and airports: the more the technology is oversold, the less immediate utility it provides, the greater the costs, and the more damaging the remote effects. As this becomes evident, it is time for technologists to confer with sociologists and economists and ask deeper questions. Is so much travel necessary? Are there ways to diminish it? Instead, the recent history of technology has consisted largely of a desperate effort to remedy situations caused by previous overapplication of technology.

Technologists should certainly have a say about simple waste, for even in an affluent society there are priorities—consider the supersonic transport, which has little to recommend it. But the moon shot has presented the more usual dilemma of authentic conflicting claims. I myself believe that space exploration is a great human adventure, with immense aesthetic and moral benefits, whatever the scientific or utilitarian uses. Yet it is amazing to me that the scientists and technologists involved have not spoken more insistently for international cooperation instead of a puerile race. But I have heard some say that except for this chauvinist competition, Congress would not vote any money at all.

Currently, perhaps the chief moral criterion of a philosophic technology is modesty, having a sense of the whole and not obtruding more than a particular function warrants. Immodesty is always a danger of free enterprise, but when the same disposition is financed by big corporations, technologists rush into production with neat solutions that swamp the environment. This applies to packaging products and disposing of garbage, to freeways that bulldoze neighborhoods, high-rises that destroy landscape, wiping out a species for a passing fashion, strip mining, scrapping an expensive machine rather than making a minor repair, draining a watershed for irrigation because (as in Southern California) the cultivable land has been covered by asphalt. Given this disposition, it is not surprising that we defoliate a forest in order to expose a guerrilla and spray tear gas from a helicopter on a crowded campus.

Since we are technologically overcommitted, a good general maxim in advanced countries at present is to innovate in order to simplify the technical system, but otherwise to innovate as sparingly as possible. Every advanced country is overtechnologized; past a certain point, the quality of life diminishes with new "improvements." Yet no country is rightly technologized, making efficient use of available techniques. There are ingenious devices for unimportant functions, stressful mazes for essential functions, and drastic dislocation when anything goes wrong, which happens with increasing frequency. To add to the complexity, the mass of people tend to become incompetent and dependent on repairmen—indeed, unrepairability except by experts has become a desideratum of industrial design.

When I speak of slowing down or cutting back, the issue is not whether research and making working models should be encouraged or not. They should be, in every direction, and given a blank check. The point is to resist the temptation to apply every new device without a second thought. But the big corporate organization of research and development makes prudence and modesty very difficult; it is necessary to get big contracts and rush into production in order to pay the salaries of the big team. Like other bureaucracies, technological organizations are run to maintain themselves but they are more dangerous because, in capitalist countries, they are in a competitive arena.

I mean simplification quite strictly, to simplify the *technical* system. I am unimpressed by the argument that what is technically more complicated is really economically or politically simpler, e.g., by complicating the packaging we improve the supermarkets; by throwing away the machine rather than repairing it, we give cheaper and faster service all around; or even by expanding the economy with trivial innovations, we increase employment, allay discontent, save on welfare. Such ideas may be profitable for private companies or political parties, but for society they have proved to be an accelerating rat race. The technical structure of the environment is too important to be a political or economic pawn; the effect on the quality of life is too disastrous; and the hidden social costs are not calculated, the auto graveyards, the torn-up streets, the longer miles of commuting, the advertising, the inflation, etc. As I pointed out in *People or Personnel,* a country with a fourth of our per capita income, like Ireland, is not necessarily less well off; in some respects it is much richer, in some respects a little poorer. If possible, it is better to solve political problems by political means. For instance, if teaching machines and audiovisual aids are indeed educative, well and good; but if they are used just to save money on teachers, then not good at all—nor do they save money.

Of course, the goals of right technology must come to terms with other values of society. I am not a technocrat. But the advantage of raising technology to be a responsible learned profession with its own principles is that it can have a voice in the debate and argue for its proper contribution to the community. Consider the important case of modular sizes in building, or prefabrication of a unit bathroom: these conflict with the short-run interests of manufacturers and craft-unions, yet to deny them is technically an abomination. The usual recourse is for a government agency to set standards; such agencies accommodate to interests that have a strong voice, and at present technologists have no voice.

The crucial need for technological simplification, however, is not in the advanced countries—which can afford their clutter and probably deserve it—but in underdeveloped countries which must rapidly innovate in order to diminish disease, drudgery, and deepening starvation. They cannot afford to make mistakes. It is now widely conceded that the technological aid we have given to such areas according to our own high style—a style usually demanded by the native ruling groups—has done more harm than good. Even when, as frequently if not usually, aid has been benevolent, without strings attached, not military, and not dumping, it has nevertheless disrupted ways of life, fomented tribal wars, accelerated urbanization, decreased the food supply, gone wasted for lack of skills to use it, developed a do-nothing élite.

By contrast, a group of international scientists called Intermediate Technology argue that what is needed is techniques that use only native labor, resources, traditional customs, and teachable know-how, with the simple aim of remedying drudgery, disease, and hunger, so that people can then develop further in their own style. This avoids cultural imperialism. Such intermediate techniques may be quite primitive, on a level unknown among us for a couple of centuries, and yet they may pose extremely subtle problems, requiring exquisite scientific research and political and human understanding, to devise a very simple technology. Here is a reported case (which I trust I remember accurately): In Botswana, a very poor country, pasture was overgrazed, but the economy could be salvaged if the land were fenced. There was no local material for fencing, and imported fencing was prohibitively expensive. The solution was to find the formula and technique to make posts out of mud, and a pedagogic method to teach people how to do it.

In *The Two Cultures*, C. P. Snow berated the humanists for their irrelevance when two-thirds of mankind are starving and what is needed is science and technology. They have perhaps been irrelevant; but unless technology is itself more humanistic and philosophical, it is of no use. There is only one culture.

Finally, let me make a remark about amenity as a technical criterion. It is discouraging to see the concern about beautifying a highway and banning billboards, and about the cosmetic appearance of the cars, when there is no regard for the ugliness of bumper-to-bumper traffic and the suffering of the drivers. Or the concern for preserving a historical landmark while the neighborhood is torn up and the city has no shape. Without moral philosophy, people have nothing but sentiments.

ECOLOGY

The complement to prudent technology is the ecological approach to science. To simplify the technical system and modestly pinpoint our artificial intervention in the environment makes it possible for the environment to survive in its complexity evolved for a billion years, whereas the overwhelming instant intervention of tightly interlocked and bulldozing technology has already disrupted many of the delicate sequences and balances. The calculable consequences are already frightening, but of course we don't know enough, and won't in the foreseeable future, to predict the remote effects of much of what we have done. The only possible conclusion is to be prudent; when there is serious doubt, to do nothing.

Cyberneticists—I am thinking of Gregory Bateson—come to the same cautious conclusion. The use of computers has enabled us to carry out crashingly inept programs on the bases of willful analyses. But we have also become increasingly alert to the fact that things respond, systematically, continually, cumulatively; they cannot simply be manipulated or pushed around. Whether bacteria or weeds or bugs or the technologically unemployed or unpleasant thoughts, they cannot be eliminated and forgotten; repressed, the nuisances return in new forms. A complicated system works most efficiently if its parts readjust themselves decentrally, with a minimum of central intervention or control, except in case of breakdown. Usually there is an advantage in a central clearing house of information about the gross total situation, but decision and execution require more minute local information. The fantastically simulated moon landing hung on a last split-second correction on the spot. In social organization, deciding in headquarters means relying on information that is cumulatively abstract and irrelevant, and chain-of-command execution applies standards that cumulatively do not fit the concrete situation. By and large it is better, given a sense of the whole picture, for those in the field to decide what to do and do it.

But with organisms too, this has long been the bias of psychosomatic medicine, the Wisdom of the Body, as Cannon called it. To cite a classical experiment of Ralph Hefferline of Columbia: a subject is wired to suffer an annoying regular buzz, which can be delayed and finally eliminated if he makes a precise but unlikely gesture, say by twisting his ankle in a certain way; then it is found that he adjusts quicker if he is *not* told the method and it is left to his spontaneous twitching than if he is told and tries deliberately to help himself. He adjusts better without conscious control, his own or the experimenter's.

Technological modesty, fittingness, is not negative. It is the ecological wisdom of cooperating with Nature rather than trying to master her. (The personification of "Nature" is linguistic wisdom.) A well-known example is the long-run superiority of partial pest-control in farming by using biological deterrents rather than chemical ones. The living defenders work harder, at the right moment, and with more pinpointed targets. But let me give another example because it is so lovely—though I have forgotten the name of my informant: A tribe in Yucatan educates its children to identify and pull up all weeds in the region; then what is left is a garden of useful plants that have chosen to be there and now thrive.

In the life sciences there is at present a suggestive bifurcation in methodology. The rule is still to increase experimental intervention, but there is also a considerable revival of old-fashioned naturalism, mainly watching and thinking, with very modest intervention. Thus, in medicine, there is new diagnostic machinery, new drugs, spectacular surgery; but there is also a new respect for family practice with psychosomatic background, and a strong push, among young doctors and students, for a social-psychological and sociological approach, aimed at preventing disease and building up resistance. In psychology, the operant conditioners multiply and refine their machinery to give maximum control of the organism and the environment (I have not heard of any dramatic discoveries, but perhaps they have escaped me). On the other hand, the most interesting psychology in recent years has certainly come from animal naturalists, e.g., pecking order, territoriality, learning to control aggression, language of the bees, overcrowding among rats, trying to talk to dolphins.

On a fair judgment, both contrasting approaches give positive results. The logical scientific problem that arises is, What is there in the nature of things that makes a certain method, or even moral attitude, work well or poorly in a given case? This question is not much studied. Every scientist seems to know what "the" scientific method is.

Another contrast of style, extremely relevant at present, is that between Big Science and old-fashioned shoestring science. There is plenty of research, with corresponding technology, that can be done only by Big Science; yet much, and perhaps most, of science will always be shoestring science, for which it is absurd to use the fancy and expensive equipment that has gotten to be the fashion.

Consider urban medicine. The problem, given a shortage of doctors and facilities, is how to improve the level of mass health, the vital statistics, and yet to practice medicine, which aims at the maximum possible health for each person. Perhaps the most efficient use of Big Science technology for the general health would be compulsory biennial checkups, as we inspect cars, for early diagnosis and to forestall chronic conditions with accumulating costs. Then an excellent machine would be a total diagnostic bus to visit the neighborhoods, as we do chest X-rays. On the other hand, for actual treatment and especially for convalescence, the evidence seems to be that small personalized hospitals are best. And to revive family practice, maybe the right idea is to offer a doctor a splendid suite in a public housing project.

Our contemporary practice makes little sense. We have expensive technology stored in specialists' offices and big hospitals, really unavailable for mass use in the neighborhoods; yet every individual, even if he is quite rich, finds it almost impossible to get attention to himself as an individual whole organism in his setting. He is sent from specialist to specialist and exists as a bag of symptoms and a file of test scores.

In automating there is an analogous dilemma of how to cope with masses of people and get economies of scale, without losing the individual at great consequent human and economic cost. A question of immense importance for the immediate future is, Which functions should be automated or organized to use business machines, and which should not? This question also is not getting asked, and the present disposition is that the sky is the limit for extraction, refining, manufacturing, processing, packaging, transportation, clerical work, ticketing, transactions, information retrieval, recruitment, middle management, evaluation, diagnosis, instruction, and even research and invention. Whether the machines can do all these kinds of jobs and more is partly an empirical question, but it also partly depends on what is meant by doing a job. Very often, e.g., in college admissions, machines are acquired for putative economies (which do not eventuate); but the true reason is that an overgrown and overcentralized organization cannot be administered without them. The technology conceals the essential trouble, e.g., that there is no community of scholars and students are treated like things. The function is badly performed, and finally the system breaks down anyway. I doubt that enterprises in which interpersonal relations are important are suited to much programming.

But worse, what can happen is that the real function of the enterprise is subtly altered so that it is suitable for the mechanical system. (E.g., "information retrieval" is taken as an adequate replacement for critical scholarship.) Incommensurable factors, individual differences, the local context, the weighting of evidence are quietly overlooked though they may be of the essence. The system, with its subtly transformed purposes, seems to run very smoothly; it is productive, and it is more and more out of line with the nature of things and the real problems. Meantime it is geared in with other enterprises of society, e.g., major public policy may depend on welfare or unemployment statistics which, as they are tabulated, are blind to the actual lives of poor families. In such a case, the particular system may not break down, the whole society may explode.

I need hardly point out that American society is peculiarly liable to the corruption of inauthenticity, busily producing phony products. It lives by public relations, abstract ideals, front politics, show-business communications, mandarin credentials. It is preeminently over-technologized. And computer technologists especially suffer for the eu-phoria of being in a new and rapidly expanding field. It is so astonishing that the robot can do the job at all or seem to do it, that it is easy to blink at the fact that he is doing it badly or isn't really doing quite that job.

DECENTRALIZATION

The current political assumption is that scientists and inventors, and even social scientists, are "value-neutral," but their discoveries are "applied" by those who make decisions for the nation. Counter to this, I have been insinuating a kind of Jeffersonian democracy or guild socialism, that scientists and inventors and other workmen are responsi-ble for the uses of the work they do, and ought to be competent to judge these uses and have a say in deciding them. They usually are competent. To give a striking example, Ford assembly-line workers, according to Harvey Swados, who worked with them, are accurately critical of the glut of cars, but they have no way to vent their dissatisfac-tions with their useless occupation except to leave nuts and bolts to rattle in the body.

My bias is also pluralistic. Instead of the few national goals of a few decision-makers, I propose that there are many goods of many activities of life, and many professions and other interest groups each with its own criteria and goals that must be taken into account. A society that distributes power widely is superficially conflictful but fundamentally stable.

Research and development ought to be widely decentralized, the national fund for them being distributed through thousands of centers of initiative and decision. This would not be chaotic. We seem to have forgotten that for four hundred years Western science majestically pro-gressed with no central direction whatever, yet with exquisite interna-tional coordination, little duplication, almost nothing getting lost, in constant communication despite slow facilities. The reason was simply that all scientists wanted to get on with the same enterprise of testing the boundaries of knowledge, and they relied on one another.

What is noteworthy is that something similar holds also in inven-tion and innovation, even in recent decades when there has been such a concentration of funding and apparent concentration of oppor-tunity. The majority of big advances have still come from independents, partnerships, and tiny companies. (Evidence published by the Senate Subcommittee on Antitrust and Monopoly, May 1965.) To name a few, jet engines, xerography, automatic transmission, cellophane, air-conditioning, quick freeze, antibiotics, and tranquilizers. The big technological teams must have disadvantages that outweigh their advantages, like lack of singlemindedness, poor communications, awkward scheduling. Naturally, big corporations have taken over the innovations, but the Senate evidence is that 90 percent of the government subsidy has gone for last-stage development for production, which they ought to have paid out of their own pockets.

We now have a theory that we have learned to learn, and that we can program technical progress, directed by a central planning board. But this doesn't make it so. The essence of the new still seems to be that nobody has thought of it, and the ones who get ideas are those in direct contact with the work. *Too precise* a preconception of what is wanted discourages creativity more than it channels it; and bureaucratic memoranda from distant directors don't help. This is especially true when, as at present, so much of the preconception of what is wanted comes from desperate political anxiety in emergencies. Solutions that emerge from such an attitude rarely strike out on new paths, but rather repeat traditional thinking with new gimmicks; they tend to compound the problem. A priceless advantage of widespread decentralization is that it engages more minds, and more mind, instead of a few panicky (or greedy) corporate minds.

A homespun advantage of small groups, according to the Senate testimony, is that co-workers can talk to one another, without schedules, reports, clock-watching, and face-saving.

An important hope from decentralizing science is to develop knowledgeable citizens, and provide not only a bigger pool of scientists and inventors but also a public better able to protect itself and know how to judge the enormous budgets asked for. The safety of the environment is too important to be left to scientists, even ecologists. During the last decades of the nineteenth century and the first decade of the twentieth, the heyday of public faith in the beneficent religion of science and invention, say from Pasteur and Huxley to Edison and the Wright Brothers, philosophers of science had a vision of a "scientific way of life," one in which people would be objective, respectful of evidence, accurate, free of superstition and taboo, immune to irrational authority, experimental. All would be well, is the impression one gets from Thomas Huxley, if everybody knew the splendid Ninth Edition of the

Encyclopaedia Britannica with its articles by Darwin and Clerk Maxwell. Veblen put his faith in the modesty and matter-of-factness of engineers to govern. Sullivan and Frank Lloyd Wright spoke for an austere functionalism and respect for the nature of materials and industrial processes. Patrick Geddes thought that new technology would finally get us out of the horrors of the Industrial Revolution and produce good communities. John Dewey devised a system of education to rear pragmatic and experimental citizens to be at home in the new technological world rather than estranged from it. Now fifty years later, we are in the swamp of a scientific and technological environment and there are more scientists alive, etc., etc. But the mention of the "scientific way of life" seems like black humor.

Many of those who have grown up since 1945 and have never seen any other state of science and technology assume that rationalism itself is totally evil and dehumanizing. It is probably more significant than we like to think that they go in for astrology and the Book of Changes, as well as inducing psychedelic dreams by technological means. Jacques Ellul, a more philosophic critic, tries to show that technology is necessarily overcontrolling, standardizing, and voraciously inclusive, so that there is no place for freedom. But I doubt that any of this is intrinsic to science and technology. The crude history has been, rather, that they have fallen willingly under the dominion of money and power. Like Christianity or communism, the scientific way of life has never been tried.

THE NEW REFORMATION

To satisfy the March 4 dissenters, to break the military-industrial corporations and alter the priorities of the budget, would be to restructure the American economy almost to a revolutionary extent. But to meet the historical crisis of science at present, for science and technology to become prudent, ecological, and decentralized requires a change that is even more profound, a kind of religious transformation. Yet there is nothing untraditional in what I have proposed; prudence, ecology, and decentralization are indeed the high tradition of science and technology. Thus the closest analogy I can think of is the Protestant Reformation, a change of moral allegiance, liberation from the Whore of Babylon, return to the pure faith.

Science has long been the chief orthodoxy of modern times and has certainly been badly corrupted, but the deepest flaw of the affluent societies that has alienated the young is not, finally, their imperialism, economic injustice, or racism, bad as these are, but their nauseating phoniness, triviality, and wastefulness, the cultural and moral scandal that Luther found when he went to Rome in 1510. And precisely science, which should have been the wind of truth to clear the air, has polluted the air, helped to brainwash, and provided weapons for war. I doubt that most young people today have even heard of the ideal of the dedicated researcher, truculent and incorruptible, and unrewarded, for instance the "German scientist" that Sinclair Lewis described in *Arrowsmith*. Such a figure is no longer believable. I don't mean, of course, that he doesn't exist; there must be thousands of him, just as there were good priests in 1510.

The analogy to the Reformation is even more exact if we consider the school system, from educational toys and Head Start up through the universities. This system is manned by the biggest horde of monks since the time of Henry VIII. It is the biggest industry in the country. I have heard the estimate that 40 percent of the national product is in the Knowledge Business. It is mostly hocus-pocus. Yet the belief of parents in this institution is quite delusional and school diplomas are in fact the only entry to licensing and hiring in every kind of job. The abbots of this system are the chiefs of science, e.g., the National Science Foundation, who talk about reform but work to expand the school budgets, step up the curriculum, and inspire the endless catechism of tests.

These abuses are international, as the faith is. For instance, there is no essential difference between the military-industrial or the school system, of the Soviet Union and the United States. There are important differences in way of life and standard of living, but the abuses of technology are very similar: pollution, excessive urbanization, destruction of the biosphere, weaponry, and disastrous foreign aid. Our protesters naturally single out our own country, and the United States is the most powerful country, but the corruption we are speaking of is not specifically American nor even capitalist; it is a disease of modern times.

But the analogy is to the Reformation, it is not to primitive Christianity or some other primitivism, the abandonment of technological civilization. There is indeed much talk about the doom of Western civilization, and a few Adamites actually do retire into the hills; but for the great mass of mankind, and myself, that's not where it's at. There is not the slightest interruption to the universalizing of Western civilization, including most of its delusions, into the so-called Third World. (If the atom bombs go off, however?)

Naturally the exquisitely interesting question is whether or not this Reformation will occur, how to make it occur, against the entrenched worldwide system of corrupt power that is continually aggrandizing itself. I don't know. In my analogy I have deliberately been choosing the date 1510, Luther in Rome, rather than 1517 when, in the popular story, he nailed his Theses on the cathedral door. There are everywhere contradictory signs and dilemmas. The new professional and technological class is more and more entangled in the work, statuses, and rewards of the system, and yet this same class, often the very same people, are more and more protestant. On the other hand, the dissident young, who are unequivocally for radical change, are so alienated from occupation, function, knowledge, or even concern, that they often seem to be simply irrelevant to the underlying issues of modern times. The monks keep "improving" the schools and getting bigger budgets to do so, yet it is clear that high schools will be burned down, twelve-year-olds will play truant in droves, and the taxpayers are already asking what goes on and voting down the bonds.

The interlocking of technologies and all other institutions makes it almost impossible to reform policy in any part; yet this very interlocking that renders people powerless, including the decision-makers, creates a remarkable resonance and chain-reaction if any determined group, or even determined individual, exerts force. In the face of overwhelmingly collective operations like the space exploration, the average man must feel that local or grassroots efforts are worthless, there is no science but Big Science, and no administration but the State. And yet there is a powerful surge of localism, populism, and community action, as if people were determined to be free even if it makes no sense. A mighty empire is stood off by a band of peasants, and *neither* can win—this is even more remarkable than if David beats Goliath; it means that neither principle is historically adequate. In my opinion, these dilemmas and impasses show that we are on the eve of a transformation of conscience.

16. Technological Politics As If Democracy Really Mattered

RICHARD SCLOVE

"Of all the social impacts of technology," writes Richard Sclove in the following selection, "perhaps the most worrisome are the adverse effects on democracy." Technologies as diverse as microwave ovens, air conditioning, and urban sewage systems all have aspects that can prove detrimental to human communities and to democracy. Sclove has no desire to reject all technology outright, however. Rather, he would like us "to become more discriminating in how we design, choose, and use technologies"—a course that might force us to give democracy priority over short-run economic goals.

How would democratic technologies look? Sclove proposes a set of design criteria. He gives examples, including several from Scandinavian nations, of technologies that meet these criteria. And he suggests some of the ways in which our political system and the nation's R&D enterprise might contribute to the development and promotion of democratic technologies. Sclove's essay comes from the "progressive left" political tradition. Some might regard it as hopelessly idealistic, particularly in view of current political trends in the U.S. that seem to run in the opposite direction. Nevertheless, it is a provocative piece that should give readers from all parts of the political spectrum much food for thought.

Richard Sclove, executive director of the Loka Institute in Amherst, Massachusetts, is the author of Democracy and Technology *(Guilford Press, 1995). He is also the founder of FASTnet (the Federation of Activists on Science and Technology Network). Sclove, whose education combines a B.A. in environmental studies with a graduate degree in*

Source: Selection from *Technology for the Common Good*, Michael Shuman and Julia Sweig, eds. "Technological Politics As If Democracy Really Mattered: Choices Confronting Progressives" by Richard Sclove. Copyright ©1993 by Richard Sclove. Reprinted by permission of the author.

223

nuclear engineering and a Ph.D. in political science from MIT, founded the Loka Institute in 1987 as a vehicle to carry on his work, which is dedicated to making science and technology more responsive to democratically decided social and environmental concerns. He is a popular lecturer and serves as a consultant to a variety of organizations.

A century and a half ago Alexis de Tocqueville described a politically exuberant United States in which steaming locomotives could not restrain citizens' enthusiasm to involve themselves in politics and community life:

> In some countries the inhabitants seem unwilling to avail themselves of the political privileges which the law gives them; it would seem that they set too high a value upon their time to spend it on the interests of the community; and they shut themselves up in a narrow selfishness. . . . But if an American were condemned to confine his activity to his own affairs, he would be robbed of one half of his existence; he would feel an immense void in the life which he is accustomed to lead, and his wretchedness would be unbearable.[1]

That is not today's United States, in which a bare majority of eligible voters participate in presidential elections while usually even fewer engage in local politics.[2] The causes of Americans' political disengagement are complex, but one culprit, more significant and intricate than commonly believed, is technology. Consider an instructive story from across the Atlantic.

During the early 1970s running water was installed in the houses of Ibieca, a small village in northeast Spain. With pipes running directly to their homes, Ibiecans no longer had to fetch water from the village fountain. Families gradually purchased washing machines, and women stopped gathering to scrub laundry by hand at the village washbasin. Arduous tasks were rendered technologically superfluous, but village social life was unexpectedly altered. The public fountain and washbasin, once scenes of vigorous social interaction, became nearly deserted. Men began losing their sense of familiarity with the children and donkeys that once helped them haul water. Women stopped gathering at the washbasin to intermix scrubbing with politically empowering gossip about men and village life. In hindsight the installation of running water helped break down the Ibiecans' strong bonds—with one another, with their animals, and with the land—that had knit them together as a community.[3] Painful in itself, such loss of community carries a specific political cost as well: as social ties weaken, so does a people's capacity to mobilize for political action.[4]

Is this a parable for our time? Like Ibiecans, we acquiesce in seemingly benign or innocuous technological changes. Ibiecans opted for technological innovations promising convenience, productivity, and economic growth. But they did not anticipate the hidden costs: greater inequality, social alienation, and steps toward community disintegration and political disempowerment. Does technological change invariably embody a Faustian trade-off between economic reward and sociopolitical malaise? No, not invariably. But the best hope for escaping such trade-offs is to develop a full-blown democratic politics of technology—something that even political progressives have not begun to conceive.

* * *

TECHNOLOGY AND DEMOCRACY

The approach to technology policy proposed here is grounded morally in the belief that people should be able to shape the basic social circumstances of their lives. It is aimed at organizing society along relatively equal and participatory lines, at achieving a system of egalitarian decentralization and confederation that Rutgers political scientist Benjamin Barber calls "strong democracy."[5] Historic examples of strong democracy include New England town meetings, the confederation of self-governing Swiss villages and cantons, and the tradition of trial by a jury of peers. Strong democracy also is apparent in the methods or aspirations of various social movements, such as the late nineteenth century American Farmers Alliance, the 1960s civil rights movement, and the 1980s uprising of Solidarity in Poland.[6] In each of these cases ordinary people claimed the rights and responsibilities of active citizenship.

If citizens ought to be empowered to participate in determining their society's basic structure and if technologies *are* an important part of that structure, it follows that technological design and practice should be democratized. Substantively, technologies must be compatible with our fundamental interest in strong democracy. And procedurally, people from all walks of life must have expanding opportunities to shape the evolving technological order.

DESIGN CRITERIA FOR
DEMOCRATIC TECHNOLOGIES

Table 1 presents some criteria for distinguishing among technologies based on their compatibility with democracy. The criteria are labeled "provisional" because they are neither complete nor definitive. Rather, they are intended to provoke political debate that can lead to an improved set of criteria.

Each criterion is intended to fulfill the institutional requirements for strong democracy: democratic community, democratic work, or democratic politics.[7] Technological decisions should attend initially and foremost to strengthening democracy, because democracy provides the necessary circumstances for deciding freely and fairly what other considerations must be taken into account in technological (and nontechnological) decision making. Until we do this, technologies will continue to hinder the advancement of other social objectives in subtle yet significant ways.

A series of examples can help explain these criteria and the feasibility of designing technologies that can satisfy them. Before proceeding, however, one clarifying note is in order. Each of the following examples illustrates a worthy social and democratic goal in its own right. However, isolated technological changes of this sort cannot be expected to represent a significant improvement in the overall democratization of society. The latter result will require multiple democratic design criteria, applied simultaneously to diverse technologies by citizens who employ broadly democratized processes of technological decision making. In other words, all the elements of a complete democratic politics of technology should converge at one time.

Criterion A: Technology and Democratic Community

Egalitarian community life is important to strong democracy because it enhances citizens' mutual respect, shared understanding, political equality, and social commitment. It empowers individuals within collectivities to challenge unjust concentrations of power. Unfortunately, diverse technological developments have contributed to the decline of community. The noise and danger of automobile traffic, detached single-family homes, air conditioning, and television all have isolated families away from one another and undermined a

TABLE 1.   A Provisional System of Design Criteria for
Democratic Technologies

Toward DEMOCRATIC COMMUNITY:
A.   Seek a balance among communitarian/cooperative, individualized, and intercommunity technologies. Avoid technologies that establish authoritarian social relationships.

Toward DEMOCRATIC WORK:
B.   Seek a diverse array of flexibly schedulable, self-actualizing technological practices. Avoid meaningless, debilitating, or otherwise autonomy-impairing technological practices.

Toward DEMOCRATIC POLITICS:
C.   Seek technologies that can enable disadvantaged individuals and groups to participate fully in social and political life. Avoid technologies that support illegitimately hierarchical power relations between groups, organizations, or polities.

To Secure DEMOCRATIC SELF-GOVERNANCE:
D.   Keep the potentially adverse consequences (e.g., environmental or social harms) of technologies within the boundaries of local political jurisdictions.
E.   Seek local economic self-reliance. Avoid technologies that promote dependency and loss of local autonomy.
F.   Seek technologies (including an architecture of public space) compatible with globally-aware, egalitarian political decentralization and federation.

To Perpetuate DEMOCRATIC SOCIAL STRUCTURES:
G.   Avoid technologies that are ecologically destructive of human health, survival, and the perpetuation of democratic institutions.

---

sense of collective purpose. This has been exacerbated by the loss of public spaces (with, for instance, town commons being supplanted by shopping malls).[8]

Are there plausible alternatives? Zurich, Switzerland, has promoted a partial antidote by providing neighborhoods with legal advice and architectural assistance aimed at increasing community interaction.[9] Thanks to the program, neighbors have begun to remove backyard fences, to build new walkways, gardens, and other community facilities, and generally to refashion a system of purely private yards into a well-balanced blend of private, semipublic, and public spaces.[10]

A housing movement born in Denmark in the mid-1960s seeks, more ambitiously, to integrate desirable aspects of traditional village life with such contemporary realities as urbanization, smaller families, single-parent or working-parent households, and greater sexual equality. The result is "co-housing"—resident-planned communities ranging today from six to 80 households. More than 100 such communities now exist in Denmark and the Netherlands, and they are spreading to the United States and elsewhere.

The Trudeslund co-housing community, located near Copenhagen, comprises thirty-three families. Homes for each family cluster along two garden-lined pedestrian streets and are surrounded by ample open space and forested areas. Each home has its own kitchen, living room, and bedrooms, though these rooms have been somewhat downsized so that the savings can be used to construct and maintain common facilities. The latter include picnic tables, sandboxes, a parking lot, and, most importantly, a "common house" with a large kitchen and dining room, playrooms, a darkroom, a workshop room, a laundry room, and a community store. Each night residents have the option of eating in the common dining room; cooking responsibilities rotate among all adults in the community (which means everyone cooks one evening a month). Because the community is designed to have residents walk past the common house on the way from the parking lot to any house, the common house becomes a natural gathering spot.

Trudeslund is successful by many measures. The common facilities save time and money, day care and baby-sitting flow naturally from the pattern of community life, social interaction flourishes without sacrificing privacy, and safety and conviviality both prosper by banishing cars to the outskirts of the community. Over time cooperation has grown, with resident families choosing to purchase and share collectively tools, a car, a sailboat, and a vacation home. Rather than becoming insular, residents are actively involved in social and political life outside Trudeslund, with the common house serving as an organizing base for other activities.[11]

Insofar as mutual respect and equality are fundamental democratic values, an egalitarian community represents a democratic gain in its own right. Moreover, if one could envision creating an interacting network of such communities, one could expect to see greater respect, tolerance, and commonality emerging *between* communities, with beneficial implications for democratization on a broader scale.[12]

Criterion B: Democratic Work

Social scientists have hypothesized that the quality of our work life influences our moral development and our readiness to function as engaged citizens—that is, as active participants in a strong democracy.[13] Technology, in turn, plays a critical role in shaping our work experiences. Some years ago sociologist and one-time union organizer Robert Schrank discussed alternative work arrangements with a group of union representatives at the General Motors Corporation. After describing several experiments in Scandinavian factories that permitted more interesting work routines and greater worker involvement in the day-to-day decision making, Schrank asked the men to imagine how they might redesign their own factories if given a chance. Their response was skeptical and unenthusiastic. Later Schrank reflected, "[T]he frame of reference of these workers was the linear assembly line as they experienced it. Even to think beyond that seemed difficult. . . ."[14]

Linear assembly lines not only tend to restrict possibilities for worker self-management, conviviality, and meaningful work but also to impair the ability of workers to envision technological alternatives. Schrank, however, was eventually able to show the GM workers more democratic automobile manufacturing technologies that have been in use for some years. For example, an innovative Volvo factory in Kalmar, Sweden, uses independently movable electronic dollies—each carrying an individual auto chassis—in place of a traditional assembly line. The dollies enable small teams of workers to plan and vary their daily routines for assembling automobile subsystems.[15]

A more creative and self-managed workplace is democratically desirable in itself. But it also can help workers develop the moral commitment, skills, and confidence to participate politically beyond the workplace.

Criterion C: Technology and Power

While political equality is essential for strong democracy, all contemporary political systems encompass groups whose opportunities for participating in social and political life are circumscribed. Today's technologies help reproduce this inequitable constellation of power. For instance, the technologies and architecture with which women must cope every day often help exclude them from the corridors of power. "Labor-saving" appliances "liberate" many wives to do housework that was once performed by other family members. (During the bygone era of open hearth cooking, for example, men chopped wood and children hauled water, thus contributing more equally to household maintenance.)[16] Likewise, modern neighborhood designs often isolate women socially, heighten their risk of physical abuse, and limit their opportunities to organize child care. The typical suburb lacks sidewalks, common gathering spaces, or the opportunity to work within a short distance of home.[17] Most public-transit systems have been designed without regard to women's typical social responsibilities. How is a mother supposed to get a baby carriage up onto a traditional bus or down the steps of a New York City subway station?[18] Many workplaces have jobs stereotyped as female that carry special risks of isolation, domination, stress, or harm. Secretaries and key-punch operators, who are preponderantly female, suffer unusually high levels of stress-related emotional and physical disorders. The marketing techniques of the mass media often degrade women and erect punishingly unattainable beauty standards.[19]

All of these consequences of technology limit women's opportunities to participate on equal terms in social and political life—including technological decision making. To explain these results, one need not invoke theories of misogyny or conspiracy (although the temptation may be strong). Generally it seems more plausible to blame the indifference and insensitivity of male-dominated institutions and design professions, in which women's evaluations of their own needs rarely qualify as even a discussion topic.

Criterion D: Translocal Harms

Local self-governance is a key building block for strong democracy. The average citizen can exert much more influence locally than nationally, and local political equality and autonomy provide crucial opportunities for citizens to influence translocal politics.

Technologies can affect a community's ability to govern itself in several ways. For example, a technology that harms people in neighboring communities can provoke intercommunity conflict, which in turn can precipitate intervention by higher political authorities that subverts local self-governance. In the late 19th and early 20th century, American cities imported clean water, or filtered and treated incoming water, while discharging raw sewage into rivers and lakes. Various methods of sewage treatment were known or under development, but few cities adopted them (unless the raw sewage caused local harm). As the buildup of sewage increased illness and death in downstream communities, state governments passed preemptive laws protecting water quality, established state boards of health to help administer the laws, and created new regional governmental authorities ("special districts") charged with integrating and managing the systems of water supply and sewage treatment. The result of this state intervention was that water quality and public health dramatically improved—but local autonomy dramatically declined. Indeed, regional water management set a precedent that influenced the development of institutions governing transportation, electrification, and telephone communication. The failure of municipal governments to assume technological responsibility toward neighboring communities wound up subverting their own autonomy and the tradition of local self-governance.[20]

Today a related pattern continues to play out as large corporations repeatedly use cross-border pollution as a rationale to justify environmental regulation at ever higher levels of political aggregation (shifting, that is, from local to state, national, and ultimately international authorities). When "successful," this reallocation of power has transposed environmental decision making to arenas relatively inaccessible to grassroots participation, where corporations have secured weak environmental standards that preempt stronger standards favored at the local level. This logic helps to explain industry support for the 1970 U.S. Clean Air Act and for the 1990 amendments to the Montreal Protocol (a treaty that regulates emissions of industrial chemicals hazardous to the earth's atmospheric ozone shield).[21] The erosion of local authority to control pollution has thus led to a violation of Criterion G in Table 1—"seek ecological sustainability."

Criterion E: Local Economic Self-Reliance

How can citizens meaningfully decide the fate of their community if economic survival depends on institutions or forces utterly beyond their control? Just as a measure of local political autonomy is essential for strong democracy, so, in turn, is a measure of local and regional economic self-reliance essential for political autonomy. Many modern technologies, however, can subvert self-reliance.

A century ago London differed from other leading world cities in eschewing reliance on a single major electric company, large generating plants, or even a city-wide electric grid. Instead there were dozens of small electric companies scattered throughout London—some privately owned, some public—deploying a diverse array of small-scale electrical generating technologies. Was London just backwards and irrational, as engineers from elsewhere commonly supposed? Not obviously. London's electric companies operated at a profit and provided reliable and affordable power adapted to local needs. London's borough governments, perceiving their own political significance and autonomy as inextricable from the infrastructures upon which they depended economically, consistently opposed Parliamentary efforts to consolidate the grid. The boroughs favored a highly decentralized electrical system that each could control more easily.[22]

For analogous reasons, a number of American cities, towns, and neighborhoods have begun to develop locally owned businesses oriented toward production for nearby markets. The Rocky Mountain Institute has developed an analytical process, along with supporting instructional materials, which a growing number of towns in economic difficulty are using to reduce their consumption of imported energy, water, and food (rather than to depend on distant supplies) and to reinvest local capital (rather than to put it in the hands of bankers thousands of miles away).[23] Once these communities are more secure against distant market forces or multinational corporate decisions, they are more empowered to conceive and undertake local democratic initiatives.[24]

This strategy of self-reliance contrasts strikingly with the prevalent strategy used by communities—self-defeating when it is not futile—of using concessionary tax breaks, waivers on environmental standards, or the promise of low wages to try to entice geographically fickle corporations. While generally decrying these corporate inducements, most proposed progressive technological strategies nonetheless remain preoccupied with advancing U.S. international competitiveness in ways that will assuredly *erode* local self-reliance.[25]

DEMOCRATIC DESIGN VS. PROGRESSIVE PROPOSALS

Several of the preceding examples suggest an important deficiency in the familiar progressive call to "rebuild America's crumbling technological infrastructure."[26] If our infrastructure needs repair or modernization, shouldn't we rebuild it in ways amenable to local democratic governance? For instance, there are technologies for managing industrial and municipal waste and conserving energy that can be deployed and administered with extensive local involvement.[27] Facilities to store solar energy for heating homes and buildings, as pioneered in Scandinavia, comprise another example of neighborhood-scale technology. Indeed, unless localities regain more control of their own infrastructures, there will be diminished incentive for the kind of grassroots political involvement essential to strong democracy. People only will participate in local politics when they have the power to affect important local decisions.

Another significant feature of all the preceding criteria is that they are designed to work together as a complementary system applied to an entire technological order. This too can improve the technology policies advocated by progressives. For instance, advocates of workplace democracy aim admirably to fulfill Criteria A and B, but generally fail to inquire whether the resulting goods and services are socially benign.[28] If we competitively and democratically produce democratically dubious technologies—say, chemical weapons, or certain consumer electronics like Walkmen that erode social interaction and solidarity—are we really making progress?[29]

Similarly, in an era of growing popular concern over acid rain, atmospheric ozone depletion, and global warming, few doubt the necessity of devising more ecologically sustainable technologies (Criterion G). Yet environmentalists sometimes imagine that sustainability alone is a sufficient basis for technological design.[30] To see the incompleteness here, recall the old sewage system configurations that both protected public health *and* subverted local self-governance, or consider Singapore's relatively stringent environmental policies that are coupled with a starkly authoritarian political regime.[31] In evaluating technology, we must learn to take into account all technologies, all their focal and nonfocal effects, and all the manifold ways in which technologies influence political relations.

TOWARD A DEMOCRATIC POLITICS OF TECHNOLOGY

Democratic design criteria are essential to a democratic politics of technology, but only if coupled with institutions for greater popular involvement in all domains of technological decision making. This suggests the need to establish new opportunities for popular participation to contest and apply the design criteria (in communities, workplaces, and other social realms), to set research and development (R&D) priorities, and to govern important technological systems. For instance, perhaps corporate R&D tax credits could be scaled up if a business introduces a democratic process or uses democratic design criteria to guide its R&D.

Our basic goal must be to open, democratize, and partly decentralize pertinent government agencies, to create avenues for worker and community involvement in corporate R&D and strategic planning, and to strengthen the capabilities of public institutions to monitor and, as needed, guide the political and social consequences of technology. We also need political strategies to accomplish these objectives, preferably built on popular movements and technological initiatives that already exist.[32]

One pertinent example of how to democratize technological decision making began in the early 1970s, when natural gas was found beneath the frigid and remote northwest corner of Canada. Energy companies soon proposed building a high-pressure, chilled pipeline across thousands of miles of wilderness, the traditional home of the Inuit (Eskimos) and various Indian tribes. At that point a government ministry, anticipating significant environmental and social repercussions, initiated a public inquiry under the supervision of a respected Supreme Court justice, Thomas R. Berger.

The MacKenzie Valley Pipeline Inquiry opened its preliminary hearings to any Canadian who felt remotely affected by the proposal. Berger and his staff developed a novel format to encourage a thorough, open, and accessible inquiry. Formal, quasi-judicial hearings were held that combined conventional expert testimony with cross-examination. Berger also conducted a series of informal "community hearings." Traveling 17,000 miles to 35 remote villages and settlements, the MacKenzie Inquiry took testimony from nearly 1,000 native witnesses. And it provided funding to disadvantaged groups to support travel and legal counsel for more competent participation. The Canadian Broadcasting Company carried daily radio summaries of all the hearings in English and in six native languages.

One of the MacKenzie Inquiry's important lessons was that laypeople can produce useful social *and* technical information. According to one technical adviser:

Input from nontechnical people played a key role in the Inquiry's deliberations over even the most highly technical and specialized scientific and engineering subjects . . . [The final report] discusses the biological vulnerability of the Beaufort Sea based not only on the evidence of the highly trained biological experts who testified at the formal hearings but also on the views of the Inuit hunters who spoke at the community hearings. . . . [Moreover,] when discussion turned to . . . complex socioeconomic issues of social and cultural impact, [native] land claims, and local business involvement—it became apparent that the people who live their lives with the issues are in every sense the experts. . . . Their perceptions provided precisely the kind of information necessary to make an impact assessment.[33]

Quoting generously from expert and citizen witnesses, Berger's final report became a national best-seller. Within months the original pipeline proposal was rejected, and the Canadian Parliament instead approved an alternate route paralleling the existing Alaska Highway.[34] One can fault the MacKenzie Inquiry for depending so much on the democratic sensibilities and good faith of one man—Judge Berger—rather than empowering the affected native groups to play a role in formulating the conclusions. But the process was nevertheless vastly more open and egalitarian than comparable decisionmaking efforts in other industrial societies.[35]

There are many other prototypes for institutions or processes that enable greater popular involvement in technological decisions. For instance, the nascent Community Health Decisions (CHD) movement has developed grassroots procedures to forge popular consensus on ethical principles governing medical policy and technology. One of the accomplishments of the movement has been to organize dozens of community meetings throughout Oregon to debate proposed reforms in the state's health care system.[36] The CHD movement could provide a model for future forums in which citizens debate more general democratic design criteria for technology.

In several states, including Maine, Washington, and California, coalitions of peace activists, labor unions, business leaders, community groups, and government officials have created democratic processes to wean regional economies away from their dependence on military production. For example, responding to grassroots pressure, the state of Washington has established a citizen advisory group that monitors military spending in the state, assesses post–Cold War economic needs and opportunities, and promulgates action plans to help defense-dependent communities diversify their productive base.[37] This shows how a region can use social criteria to evaluate and redirect an entire technological order.

During the past twenty years the Dutch have developed a network of street corner "science shops," supported by nearby university staff and students, where citizen groups receive free assistance to address social issues with technical components. One science shop helped a local environmental group document the contamination of heavy metals in vegetables, which pressured the Dutch government to sponsor a major cleanup in metalworking plants. The science shops have empowered citizens to participate in technological decision making so successfully that they have prompted similar efforts throughout much of Western Europe.[38]

Traditional Amish communities, often misperceived as technologically naive or backwards, have pioneered popular deliberative processes for screening technologies based on their cumulative social impacts, in effect attending to many of the criteria listed in Table 1. One of their methods is to place the adoption of a new technology on probation for one year to discover what the social effects might be. For instance, Amish dairy communities in east-central Illinois ran a one-year trial with diesel-powered bulk milk tanks before judging them socially acceptable; other Amish communities used social trials to prohibit once-probationary household telephones or personal computers.[39]

The preceding examples are, of course, atypical. Most technological choices are made by experts, bureaucratic machination, or unregulated market interactions. But the exceptions provide crucial evidence that, given the right institutional circumstances, lay citizens can make reasonable technological decisions reflecting their own priorities. Even federal agencies, such as the Office of Technology Assessment, the National Science Foundation, and the National Institutes of Health (NIH), have occasionally supported or incorporated citizen participation in their decisionmaking procedures.[40] For instance, the NIH has used both expert and lay advisory panels to evaluate research proposals. The experts judge the scientific merits, while laypeople help weigh the social value, political import, or ethical propriety. One can envision a wide range of private or public institutions using such models to develop a new system of democratic procedures for choosing among technological alternatives.

PARTICIPATORY RESEARCH, DEVELOPMENT, AND DESIGN

Democratic processes for technological choice and oversight are vital, yet hardly worth the effort unless participants have a broad range of alternative technologies from which to choose. Hence it is essential to weave democracy into the fabric of technological research, development, and design (RD&D).[41] Consider four examples of how this has been done.

Democratic Design of Workplace Technology

In Scandinavia during the early 1980s unionized newspaper-graphics workers—in collaboration with sympathetic university researchers, a Swedish government laboratory, and a state-owned publishing company—succeeded in inventing a form of computer software unique in its day. Instead of following trends toward routinized or mechanized newspaper layout, this software contained some of the capabilities later embodied in desktop publishing programs that enable printers and graphic artists to exercise considerable creativity in page design.[42] Known as UTOPIA, this project demonstrated how broadened participation in the RD&D process could lead to a design innovation that, in turn, supported one condition of democracy—creative work (Criterion B).

UTOPIA is less ambitious than several other attempts at participatory design within the workplace. For example, in the 1970s workers at Britain's Lucas Aerospace Corporation sought not only to democratize their own work processes but also to produce more socially useful products.[43] But UTOPIA demonstrated that workers could go beyond just developing prototypes. It also should be noted that this instance of collaboration between workers and technical experts—initially limited to a single technology within a single industry—occurred under unusually favorable social and political conditions. Sweden's workforce is 85 percent unionized, and the nation's pro-labor Social Democratic party has held power during most of the past 50 years.

Participatory Architecture

Compared with the relatively few examples of participatory design of machinery, appliances, or technical infrastructure, there is a rich history of citizen participation in architectural design. One example is the "Zone Sociale" at the Catholic University of Louvain Medical School in Brussels. In 1969 students insisted that new university housing miti-gate the alienating architecture of the adjacent hospital. Architect Lucien Kroll established an open-ended, participatory design process that elicited intricate organic forms (e.g., support pillars shaped like gnarled tree trunks), richly diverse patterns of social interaction (e.g., a nursery school situated near administrative offices and a bar), and a dense network of pedestrian paths, gardens, and public spaces. Walls and floors of dwellings were movable, so that students could design their own living spaces. Construction workers were given design princi-ples and constraints rather than finalized blueprints, and they were encouraged to create and display their own sculptures. Initially baffled by the level of spontaneity and playfulness, the project's structural engineers gradually adapted themselves to the diversity of competent participants.

Everything proceeded splendidly for some years until the university administration became alarmed at the extent to which they could not control the process. When the students were away on vacation, they fired Kroll and halted further construction.[44]

Feminist Design

What would happen if women played a greater part in RD&D? One answer comes from feminists who have long been critical of housing designs and urban layouts that reinforce the social isolation and the low, unpaid status of women as housewives.[45] If women were more actively engaged in community design, they might set up more shared neighborhood facilities for day care, laundry, or food preparation, or they might locate homes, workplaces, stores, and public facilities more closely together. Realized examples of feminist design exist in London, Stockholm, and Providence, Rhode Island.[46]

Another approach has been pioneered by an artist and former over-worked mother named Frances GABe, who devoted several decades to inventing a self-cleaning house. "In GABe's house," according to au-thor Jan Zimmerman, "dishes are washed in the cupboard, clothes are cleaned in the closets, and the rest of the house sparkles after a humid misting and blow dry!"[47]

Other feminists have established women's computer networks and designed alternatives to dreary female office work and to transportation networks insensitive to women's needs.[48] An explicit feminist complaint against current reproductive technologies—such as infertility treatments, surrogate mothering, hysterectomy, and abortion—is that women have played a negligible role in guiding the medical RD&D agendas which have imposed on women agonizing moral dilemmas that might otherwise be averted or structured differently.[49]

Barrier-Free Design

During the past two decades there has been substantial innovation in the design of "barrier-free" equipment, buildings, and public spaces responsive to the needs of people with physical disabilities. Much of the impetus came from disabled citizens who organized themselves to assert their needs or helped invent design solutions. For example, prototypes of the Kurzweil Reading Machine, which uses computer voice-synthesis to read typed text aloud, were tested by over a hundred and fifty blind users. In an eighteen-month period these users made over a hundred recommendations, many of which were incorporated into later versions of the device.[50]

Technology by the People

All these examples of participatory design demonstrate that it is possible to have a much wider range of people participating in technological research, development, and design.[51] Moreover, participatory design broadens the menu of technological choices. But many participatory design exercises also have encountered fierce opposition from powerful institutions—opposition engendered, not because the exercises were failing, but because they were succeeding.

Still, some advocates of participatory RD&D have elected to state their case entirely in terms of the material interests of the nonparticipants. Others have noted the contribution that participation can make to improved productivity or to better design solutions. These are all fair and reasonable arguments. What is rarely articulated is the specific moral argument that the opportunity to participate in RD&D should be a matter of right, because it is essential to individual moral autonomy, to human dignity, and to democratic self-governance.

A powerful moral case for participatory design has been made by people with disabilities who have demanded barrier-free design. The movement's achievements are now apparent in the profusion of ramps and modified rest rooms in public places (responsive to Criterion C), and they will soon become even more apparent with the promulgation of new regulations under the Americans with Disabilities Act of 1990. The movement not only opposes antidemocratic design but also has a constructive, hopeful thrust. Nonmarket, democratic design criteria—often formulated and applied by disabled laypeople—are now being used to define individual and collective needs, including access to public spaces. Moreover, participants do not evaluate just one technology at a time—the norm in conventional technology assessment—but entire technological and architectural environments. When the range of democratic criteria broadens and when the participants expand beyond the disabled population, we will be well on our way toward ensuring that our technology is compatible with democracy.

CONCLUSION

Current technological orders are generally short on communitarian or cooperative activities, and long on isolation and authoritarianism (violating Criterion A). Work is frequently stultifying and tends to impair moral growth and political efficacy (violating Criterion B). Illegitimate power asymmetries are reproduced through technological means (violating Criterion C).

The opportunity to engage in a vibrant civic life is often preempted by shopping malls, suburban subdivisions, unconstrained automobilization, and an explosive proliferation in home entertainment devices. Thus we have diminished access to local mediating institutions or to public spaces that could support democratic empowerment within the broader society (violating Criteria A and F). The need to manage translocal harms, coupled with widespread dependence on centrally managed technological systems and with the growing integration of the global economy, has helped render local governments relatively powerless, thereby reducing anyone's incentive to participate (violating Criteria D, E, and F). Meanwhile, there is little compensating incentive to engage directly in national politics, which television reduces to a passive spectator sport, where powerful corporations exert disproportionate influence, where deep questions of social structure are slighted, and where the average citizen has negligible effect.

While it is not always easy to establish causal connections running from structural deficiencies to other social ills, it hardly seems conceivable that weak community ties, atrophied local political capabilities, and authoritarian and degraded work processes have had no influence upon illiteracy, stress, illness, divorce rates, teen pregnancy, crime, drug abuse, psychological disorders, and so on. Perhaps, as de Tocqueville foresaw, many of us *do* sometimes feel shut up in a narrow selfishness, robbed of one half of our existence, left with an immense void in our lives.

Progressive technological strategists face a dilemma. We can couch our nostrums in terms of prevailing economic goals like competitiveness and try to win short-run victories. Or we can strive for a world worthy of our ideals. But we can no longer pretend that the progressive policies so far proposed for improving national economic performance, any more than conservative policies, are going to avoid exacerbating the United States' most profound social and political maladies. Has not the time arrived to mobilize for a democratic politics of technology?

NOTES

1. Alexis de Tocqueville, *Democracy in America*, ed. Phillips Bradley, rev. ed. (1848; reprint, New York: Vintage Books, 1945), Vol. 1, p. 260.
2. John J. Kushma, "Participation and the Democratic Agenda: Theory and Praxis," in Marc V. Levine *et al.*, *The State and Democracy: Revitalizing America's Government* (New York: Routledge, 1988), pp. 14–48. According to the Harwood Group, many Americans would like to be more involved in public affairs, but feel locked out of the current system. See *Citizens and Politics: A View from Main Street America* (Dayton, OH: The Kettering Foundation, 1991).
3. Susan Friend Harding, *Remaking Ibieca: Rural Life in Aragon under Franco* (Chapel Hill: University of North Carolina Press, 1984).
4. Samuel Bowles and Herbert Gintis, *Democracy and Capitalism: Property, Community, and the Contradictions of Modern Social Thought* (New York: Basic Books, 1986). The converse causal tie between community strength and political empowerment is suggested, for instance, by solidaristic Amish communities' success in resisting mandatory public schooling, military conscription, and participation in the federal social security system. See Donald B. Kraybill, *The Riddle of Amish Culture* (Baltimore: Johns Hopkins University Press, 1989).
5. Benjamin Barber, *Strong Democracy: Participatory Politics for a New Age* (Berkeley: University of California Press, 1984).
6. See, for example, Sara M. Evans and Harry C. Boyte, *Free Spaces: The Sources of Democratic Change in America* (New York: Harper and Row, 1986).

7. For the complete derivation and justification for these and additional democratic design criteria, see Richard E. Sclove, *Democracy and Technology* (New York: Guilford Press, 1995).

8. See, for example, Kenneth T. Jackson, *Crabgrass Frontier: The Suburbanization of the United States* (New York: Oxford University Press, 1985); and Barber, *Strong Democracy*, pp. 267–73, 305–06.

9. I define technology broadly as material artifacts and the practices or beliefs that accompany their creation or use. Hence I regard architecture and community planning as a sub-domain of technology.

10. Dolores Hayden, *Redesigning the American Dream: The Future of Housing, Work, and Family Life* (New York: W. W. Norton, 1984), pp. 189–91.

11. Kathryn McCamant and Charles Durrett, *Cohousing: A Contemporary Approach to Housing Ourselves* (Berkeley, CA: Habitat Press, 1988).

12. Some supporting evidence can be found in the emerging international network of "sister cities." See, for example, Michael Shuman, "From Charity to Justice," *Bulletin of Municipal Foreign Policy*, 2:4 (Autumn 1988), pp. 50–59.

13. Edward S. Greenberg, *Workplace Democracy: The Political Effects of Participation* (Ithaca: Cornell University Press, 1986); William M. Lafferty, "Work as a Source of Political Learning Among Wage-Laborers and Lower-Level Employees," in *Political Learning in Adulthood: A Sourcebook of Theory and Research*, ed. Roberta S. Sigel (Chicago: University of Chicago Press, 1989), pp. 102–42; Melvin L. Kohn et al., "Position in the Class Structure and Psychological Functioning in the United States, Japan, and Poland," *American Journal of Sociology*, 95:4 (January 1990), pp. 964–1008.

14. Robert Schrank, *Ten Thousand Working Days* (Cambridge, MA: MIT Press, 1978), p. 226.

15. *Ibid.*, 221–27. On remaining democratic shortcomings in the Volvo factories, see Stephen Hill, *Competition and Control at Work: The New Industrial Sociology* (Cambridge, MA: MIT Press, 1981), pp. 49 and 104–05. For further recent examples of both democratic and nondemocratic workplace technology, see Shoshana Zuboff, *In the Age of the Smart Machine: The Future of Work and Power* (New York: Basic Books, 1988).

16. See Ruth Schwartz Cowan, *More Work for Mother: The Ironies of Household Technology from the Open Hearth to the Microwave* (New York: Basic Books, 1983).

17. Hayden, *Redesigning the American Dream*; Ray Oldenburg, *The Great Good Place: Cafes, Coffee Shops, Community Centers, Beauty Parlors, General Stores, Bars, Hangouts, and How They Get You through the Day* (New York: Paragon House, 1989).

18. See Women and Transport Forum, "Women on the Move: How Public Is Public Transport?" in *Technology and Women's Voices: Keeping in Touch*, ed. Cheris B. Kramarae (New York: Routledge & Kegan Paul, 1988), pp. 116–34.

19. Barbara Drygulski Wright, *Women, Work, and Technology: Transformations* (Ann Arbor: University of Michigan Press, 1987); Naomi Wolf, *The Beauty Myth: How Images of Beauty Are Used against Women* (New York: William Morrow, 1991).

20. See Joel A. Tarr, "Sewerage and the Development of the Networked City in the United States, 1850–1930," in *Technology and the Rise of the Networked City*, eds. Joel A. Tarr and Gabriel Dupuy (Philadelphia: Temple University Press, 1988), pp. 159–85; and Gerald E. Frug, "The City as a Legal Concept," *Harvard Law Review*, 93:6 (April 1980), pp. 1057–1154.

21. See Samuel P. Hays, *Beauty, Health, and Permanence: Environmental Politics in the United States, 1955–1985* (Cambridge, England: Cambridge University Press, 1987), pp. 443–5, 456–7; Gareth Porter and Janet Welsh Brown, *Global Environmental Politics* (Boulder, CO: Westview Press, 1991), pp. 64, 66; Wolfgang Sachs, "Environment and Development: The Story of a Dangerous Liaison," *The Ecologist*, 21:6 (November/December 1991), pp. 252–57. Local ability to pressure corporations to reduce pollution could be much advanced by supportive legislation such as the Environmental Bill of Rights proposed in Samuel Bowles, David M. Gordon, and Thomas E. Weisskopf, *Beyond the Wasteland: A Democratic Alternative to Economic Decline* (Garden City: Anchor Press/Doubleday, 1983), pp. 344–6.

22. Thomas Parke Hughes, *Networks of Power: Electrification in Western Society, 1880–1930* (Baltimore: Johns Hopkins University Press, 1983), Chapter 9.

23. See Robert Gilman, "Four Steps to Self-Reliance: The Story Behind Rocky Mountain Institute's Economic Renewal Project," *In Context*, 14 (Autumn 1986), pp. 41–6; Barbara A. Cole, *Business Opportunities Casebook* (Snowmass, CO: Rocky Mountain Institute, 1988); and David Morris, "Self-Reliant Cities: The Rise of the New City-States," in *Resettling America: Energy, Ecology and Community*, ed. Gary J. Coates (Andover, MA: Brick House Publishing Co., 1981), pp. 240–62.

24. See John Gaventa, *Power and Powerlessness: Quiescence and Rebellion in an Appalachian Valley* (Urbana: University of Illinois Press, 1980), Chapter 8; and Frug, "The City as a Legal Concept."

25. See, for example, Stephen S. Cohen and John Zysman, *Manufacturing Matters: The Myth of the Post-Industrial Economy* (New York: Basic Books, 1987); Michael L. Dertouzos, Richard K. Lester, Robert M. Solow, and the MIT Commission on Industrial Competitiveness, *Made in America: Regaining the Productive Edge* (Cambridge, MA: MIT Press, 1989); Lester C. Thurow, *The Zero-Sum Solution: Building a World-Class American Economy* (New York: Simon & Schuster, Inc 1985); and Joel S. Yudken and Michael Black, "Targeting National Needs: A New Direction for Science and Technology Policy," *World Policy Journal*, 7:2 (Spring 1990), pp. 282–3. While sharply critical of economic nationalism, Robert B. Reich's *The Work of Nations* (New York: Vintage Books, 1992) assumes increased integration into an ever more intensively globalized economy. On the importance of granting greater *local* power over national self-reliance, see Ann J. Tickner, *Self-Reliance versus Power Politics: The American and Indian Experiences in Building Nation States* (New York: Columbia University Press, 1987).

26. See, for example, Bowles *et al.*, *Beyond the Wasteland*; Yudken and Black, "Targeting National Needs", and Robert B. Reich, "The Real Economy," *Atlantic Monthly*, 267:2, February 1991, pp. 35–52.

27. Amory B. Lovins, *Soft Energy Paths: Toward a Durable Peace* (Cambridge,

MA: Ballinger, 1977); National Center for Appropriate Technology, *Wastes to Resources: Appropriate Technologies for Sewage Treatment and Conversion,* DOE/CE/15095–2 (Washington, DC: U.S. Government Printing Office, July 1983); Ken Darrow and Mike Saxenian, *Appropriate Technology Sourcebook: A Guide to Practical Books for Village and Small Community Technology* (Stanford, CA: Volunteers in Asia, 1986); Valjean McLenighan, *Sustainable Manufacturing: Saving Jobs, Saving the Environment* (Chicago: Center for Neighborhood Technology, 1990).
28. See, for example, Michael J. Piore and Charles F. Sabel, *The Second Industrial Divide: Possibilities for Prosperity* (New York: Basic Books, 1984); Cohen and Zysman, *Manufacturing Matters.*
29. David F. Noble's deservedly influential essay, "Social Choice in Machine Design: The Case of Automatically Controlled Machine Tools," lauds Norwegian factory worker involvement in technology choices that helped workers maintain autonomy and creativity. True enough, but the factory in question was a state-owned weapons production plant. In *Case Studies on the Labor Process,* ed. Andrew Zimbalist (New York: Monthly Review Press, 1979), pp. 18–50.
30. See, for example, John Todd and Nancy Jack Todd, *Bioshelters, Ocean Arks, City Farming: Ecology as the Basis of Design* (San Francisco: Sierra Club Books, 1984).
31. Stan Sesser, "A Reporter at Large: A Nation of Contradictions," *The New Yorker,* 13 January 1992, pp. 37–68.
32. See Richard E. Sclove, "The Nuts and Bolts of Democracy: Toward a Democratic Politics of Technological Design," in *Critical Perspectives on Non-Academic Science and Engineering,* ed. Paul T. Durbin (Bethlehem, PA: Lehigh University Press, 1991), pp. 239–62; and Sclove, *Technology and Freedom.*
33. D.J. Gamble, "The Berger Inquiry: An Impact Assessment Process," *Science,* 199:4332 (3 March 1978), pp. 950–51.
34. *Ibid.,* pp. 946–52; Thomas R. Berger, *Northern Frontier, Northern Homeland: The Report of the MacKenzie Valley Pipeline Inquiry,* 2 vols. (Ottawa: Minister of Supply and Services, Canada, 1977); Organisation for Economic Cooperation and Development (OECD), *Technology on Trial: Public Participation in Decision-Making Related to Science and Technology* (Paris: OECD, 1979).
35. See, for example, Barry M. Casper and Paul David Wellstone, *Powerline: The First Battle of America's Energy War* (Amherst: University of Massachusetts Press, 1981); David Dickson, *The New Politics of Science* (Chicago: University of Chicago Press, 1988).
36. Bruce Jennings et al., "Grassroots Bioethics Revisited: Health Care Priorities and Community Values," *Hastings Center Report,* 20:5 (September/October 1990), pp. 16–23.
37. Kevin J. Cassidy, "Defense Conversion: Economic Planning and Democratic Participation," *Science, Technology, and Human Values,* 17:3 (Summer 1992), pp. 334–48.
38. Seth Shulman, "Mr. Wizard's Wetenschapswinkel," *Technology Review,* 91:5 (July 1988), pp. 8–9.

39. On Amish technological decision making see, for example, Marc A. Olshan, "Modernity, the Folk Society, and the Old Order Amish: An Alternative Interpretation," *Rural Sociology*, 46:2 (Summer 1981), pp. 297–309; Victor Stoltzfus, "Amish Agriculture: Adaptive Strategies for Economic Survival of Community Life," *Rural Sociology*, 38:2 (Fall 1973), pp. 196–206; Kraybill, *The Riddle of Amish Culture*.

40. See, for example, U.S. Congress, Office of Technology Assessment, *Coastal Effects of Offshore Energy Systems* (Washington, DC: U.S. Government Printing Office, 1976); Rachelle Hollander, "Institutionalizing Public Service Science: Its Perils and Promise," in *Citizen Participation in Science Policy*, ed. James C. Petersen (Amherst: University of Massachusetts Press, 1984), pp. 75–95.

41. For additional arguments in support of participatory design, see Richard E. Sclove, "The Nuts and Bolts of Democracy: Democratic Theory and Technological Design," in *Democracy in a Technological Society*, ed. Langdon Winner (Dordrecht: Kluwer Academic Publisher, 1992), pp. 139–157.

42. Andrew Martin, "Unions, the Quality of Work, and Technological Change in Sweden," in *Worker Participation and the Politics of Reform*, ed. Carmen Sirianni (Philadelphia: Temple University Press, 1987), pp. 95–139.

43. Hilary Wainwright and Dave Elliott, *The Lucas Plan: A New Trade Unionism in the Making?* (London: Allison and Busby, 1982).

44. Lucien Kroll, "Anarchitecture," in *The Scope of Social Architecture*, ed. Richard C. Hatch (New York: Van Nostrand Reinhold, 1984), pp. 166–85.

45. Hayden, *Redesigning the American Dream*, Chapter 4.

46. *Ibid.*, pp. 163–70.

47. Jan Zimmerman, *Once Upon the Future: A Woman's Guide to Tomorrow's Technology* (New York: Pandora, 1986), pp. 36–7.

48. Wright, *Women, Work, and Technology*; Kramarae, *Technology and Women's Voices*.

49. See Sarah Franklin and Maureen McNeil, "Reproductive Futures: Recent Literature and Current Feminist Debates on Reproductive Technologies," *Feminist Studies*, 14:3 (Fall 1988), pp. 545–60.

50. See Michael Hingson, "The Consumer Testing Project for the Kurzweil Reading Machine for the Blind," pp. 89–90, and Raymond Kurzweil, "The Development of the Kurzweil Reading Machine," pp. 94–96, in Virginia W. Stern and Martha Ross Redden, eds., *Technology for Independent Living: Proceedings of the 1980 Workshops on Science and Technology for the Handicapped* (Washington, DC: American Association for the Advancement of Science, 1982).

51. For a number of additional examples, see Sclove, "The Nuts and Bolts of Democracy: Toward a Democratic Politics of Technological Design" or contact the Loka Institute, P.O. Box 355, Amherst, MA 01004; e-mail: Loka@amherst.edu.

17. Toward Human-Centered Design

DONALD A. NORMAN

Every so often, one reads an essay or a book that makes one look at familiar things in an entirely new way. Again and again, one nods in agreement, saying to oneself, "that makes sense" or "of course," having had inklings of these thoughts or ideas but never having quite articulated them. The following selection, Donald Norman's "Toward Human-Centered Design," is such an essay. Norman asks the simple question, "What would technologies be like if they were made to conform to human needs instead of forcing humans to conform to theirs?"

Thinking along these lines, he sees machines and technological systems—ranging from commercial aircraft to the allocation of television channels to manufacturing production lines to telephones—that seem to have been designed to suit their own convenience or that of the designers rather than the convenience of the people who have to operate or work with them. How many "human errors," Norman asks, are caused by systems that automate tasks that should be done by people while failing to provide automated assistance when it is most needed? For example, when automated aircraft systems suddenly stop working, crew members, who may be lulled into inattention, are "thrust into the middle of the problem and required to immediately figure out what has gone wrong and what should be done. There is not always enough time." Today's information technology has the potential to complement and augment human abilities, leading to human-centered technologies that meet our needs

Source: Excerpted from *Things That Make Us Smart* by Donald A. Norman. Copyright © 1993 Donald A. Norman. Reprinted by permission of Addison-Wesley Publishing Company, Inc.

much more effectively than existing systems. It is an approach well worth pursuing.

Donald Norman is a fellow at Apple Computer in Cupertino, California. This selection was adapted from his book, Things That Make Us Smart *(Addison-Wesley, 1993).*

Consider the free throw in basketball. A player stands a fixed distance from the hoop and is given one or two chances to toss the ball through it, unimpeded by the other players.

What is it about the free throw that is hard for the player? Throwing the ball with the required accuracy. Becoming good at the free throw requires continual practice and concentration. Amateurs will miss frequently. Even professionals will occasionally miss. What is easy about the free throw? Many things, including seeing the hoop: it would never occur to anyone to spend time practicing seeing the hoop.

If a machine were to attempt the same task, what would be hard? Seeing the hoop. What would be easy for the machine? Throwing the ball. If the machine could distinguish the hoop from other objects, tossing the ball into it would be trivial, a simple matter of computing the appropriate trajectory and applying the required forces. The mathematics would be easy, the perception would be hard.

Now consider a second situation. Helene is showing slides but the projector beam is too low. She looks at the corner of the room and says to her friend, "Over there." Her friend goes to the table in the corner of the room, gets a book, and brings it to Helene, who puts it under the front of the projector and proceeds with the slide show.

What is easy for the person? Realizing the nature of the problem, finding an unconventional use of an existing object, asking someone to help. The friend found it easy to understand what was needed, and fetching the book was simple—so simple that normally we wouldn't even talk about it.

What would be hard for a machine, even an intelligent robot? First of all, noticing the problem and having some empathy for Helene and her situation. Empathy is a hard trait to build into machines, even the most artificially intelligent of them. And, if the robot were acting as Helene's assistant, there is no way that the look followed by the words "over there" would be understood. Where is the verb? What is the command? Suppose the command were precise: "Please go over to the red book with the three glasses of wine on it that is lying on its side on the small wooden table in the southeast corner of the room. Bring the book here and give it to me." Then what? The hard part for the robot would be moving to the proper table without knocking something over along the way, managing to pick up the book properly (after dealing with the three glasses of wine), and so on.

But there is an easy solution for a robot, if it ever managed to understand the problem with the slide projector. It could just lift the appropriate side to the proper height and hold the projector for the next hour or two during the show. It would be hard to imagine a person doing that.

The things people are good at are the things natural to humankind: Creativity. Invention. Adapting to changing circumstances. Thinking of the problem in the first place. Perceiving the world. Feeling emotions such as joy, love, hope, and excitement. Enjoying humor and experiencing wonder.

Every one of these and similar things is hard for a machine. We are capable of building machines that can perform flawlessly many of the things we are bad at but it is very difficult to build machines that can do the things we are good at.

"Why, that's wonderful," you should be saying. "Between us and our machines, we could accomplish anything. People are good at the creative side and at interpreting ambiguous situations. Machines are good at precise and reliable operation."

Unfortunately, this is not the approach engineers have followed in reacting to advances in technology. Instead, they've adopted a machine-centered view of life: machines have certain needs, humans are adaptable. Give the machines priority, technologists' thinking goes, and tailor human operations to fulfill the requirements of machines.

It's not that technologists don't care about people. They do—after all, they are people, too. But they also focus predominantly on the issue of machine performance. When there is an industrial accident, review teams pore over the site, looking for signs of equipment failure. If none is found, humans are blamed. Thus, in 75 percent of commercial aviation accidents, pilots are supposedly at fault—"human error." People, as we know, are distractible and imprecise. They make errors in remembering things, in doing actions they shouldn't have done, or in failing to do things they should have. As soon as we take the machine point of view, everything automatically leads to a focus upon human weaknesses rather than strengths.

TO AUTOMATE OR INFORMATE?

Airplanes today are flown more and more by automated controls. But did the designers do a careful analysis of the cockpit to decide which tasks were best done by people and which were in need of some machine assistance? Of course not. Instead, the parts that could be automated *were* automated, and the leftovers given to the humans. Earl Wiener has studied the use of automation for NASA's aviation safety program in the most advanced, so-called "glass cockpits" (the name derives from the fact that so many of the old-fashioned mechanical instruments have been replaced by "glass" display screens).

Wiener finds that automation works best when conditions are normal. When conditions become difficult—say an engine, a radio, or an electrical generator fails—the automation is also more likely to fail. In other words, automation isn't available to help when it is needed most. When an automated system suddenly stops working, often with no advance warning, the crew is thrust into the middle of the problem and required to immediately figure out what has gone wrong and what should be done. There is not always enough time.

The same type of machine-centered approach causes problems in manufacturing. Back when people walked the production lines, they could tell what was happening by the sounds, the vibrations, even the smells. When computers took over, the machines had to be located in air-conditioned, air-filtered rooms, away from the heat, noise, and vibration of the factory. The people who tended the production lines were relocated to be with the automation controls, thus changing the way they interacted with the operation. Whereas before they were physically able to keep an eye on things, often catching problems before they arose, now they are connected to the real world by second- or third-order representations: graphs, trend lines, flashing lights. The problem is that the representations people receive are most often those used by the machines themselves: numbers. But while the machines may use numbers internally, human operators should receive information in the format most appropriate to the task they must perform.

In her influential book, *The Age of the Smart Machine* [see the excerpt in chapter 25], Shoshana Zuboff distinguishes between systems that automate and those that "informate"—systems that provide people with access to a rich variety of information that would not be otherwise available. In an informated system, technology is used to inform, not to take over. Workers are encouraged to make use of on-line databases and real-time measurements of critical variables to get immediate answers to questions about any aspect of the task.

In an automated system, workers are relegated to the role of meter watchers, staring at automatic displays, waiting for an alarm bell that calls them to action. In an informed system, people are always active, analyzing patterns, continually able to find out the state of whatever aspect of the job is relevant at the moment. This involvement empowers the worker not only to become more interested in the job, but also to make more intelligent decisions. When the equipment fails in an informed system, workers are much more knowledgeable about the source of the problems and about possible remedies.

HUMANE COMMUNICATIONS

Consider television or radio channels, which are organized by channel number or frequency, not by content. This organization is convenient for the broadcasters and the newspaper listings but it does not always serve the consumer's purposes very well.

In my hometown, the newspaper lists 48 television channels: the shows for one evening (from 6 p.m. to midnight) occupy most of a page. You should watch *Nova*, someone says to me, but the TV guide doesn't make the task easy. If 6 hours of television listings require almost a full newspaper page, an entire week of 24 hours and 7 days might take as much as 28 pages. Must I search every channel for every day until I find the show? The last time I was in a strange city, I failed to find what I wanted. Among other problems, the same show is carried by differently numbered stations in different cities (or even by differently numbered stations in the same city if they have different cable services).

Today's organization of television channels is somewhat as if a bookstore were to organize its books by publisher instead of by topic, content, or author. Worse, as if the publishers were each given identification numbers instead of names, with each bookstore numbering the publishers differently.

And the problem may get even more unwieldy, with systems that offer 500 or more channels of information. But, on the other hand, we could combine the power of digital transmission with computer technology to produce a human-centered design, where shows can be selected by name, content, or actors. We could even point to a desired program and say, in effect, "show me that," and have the program appear on the screen whenever it is most convenient for us.

The traditional telephone is another example of machine-centered technology. Its initial design did not allow a caller to follow the normal rules of social courtesy. Even today, after decades of adaptation, there is still no way the caller can know the activities of the recipient, no way to tell whether the person is busy, unwilling to be interrupted, or anxiously awaiting the call. Instead, the ringing telephone simply informs us that someone is on the line.

People have used add-ons such as phone machines and cordless extensions to address some of the telephone's shortcomings. But the technology is still designed around the needs of the system: the goal is to make proper connections as efficiently as possible to minimize the demands on, and the cost of, equipment at central switching offices.

But it is possible to develop a more humane technology for the telephone—one that addresses the needs of both callers and call recipients. For one thing, callers do not always wish to contact recipients— it is sometimes more efficient to leave a message without conversing or explaining to various intermediaries who might answer the call. Wouldn't it be nice if the caller could specify whether the call is intended for a message center or for the actual recipient? Similarly, from the recipient's point of view, wouldn't it be nice to know who is on the phone, and maybe even why, before deciding whether to answer? The service that telephone companies now offer known as Caller ID is not the answer. Caller ID identifies the telephone number, not the person, and raises serious privacy concerns.

The answering machine helps by allowing the recipient to screen a call and intercept it before the caller has hung up. But the technical capabilities of the phone itself are currently limited by the low bandwidth of the wires that connect the home to the central switching office, and by the standard home telephone's limited keypad and lack of visual displays. Fortunately, these limitations are evaporating.

Suppose instead that upon placing a call, the caller could also give a three-second message that would be delivered along with the ring. The recipient would hear "ring . . . *this is Julie for Don* . . . *ring*" or perhaps "*ring* . . . *Petersen's Auto Repair with a question about your car* . . . *ring.*" Better yet, just as in normal conversation, suppose the ring were interpreted as a request to schedule a conversation—not, as it is now, a demand for instant talk. Let the recipient choose whether to answer or to signal something like "in five minutes, please." The caller could decide whether to call back then (the telephone could do that automatically) or to deliver a message to the message center. This would be a true, people-centered technology: caller and recipient alike could use it in whatever way best fits their needs.

APPROPRIATE TRANSFORMATIONS

Technology has provided many benefits to society. We would not want to go without them. But at the same time, technology has too often trapped us into a machine-centered mode of life, dominated by the needs of technology itself. This wasn't deliberate. It came about naturally as an unintended side-effect of the rapid expansion of machines into human activities. But there is an alternative approach.

Today's technology is increasingly centered around information. Because information is essentially invisible and information-processing machines have no moving parts, they have no natural way of showing their operation to us. We are entirely dependent on the skills of the designers to present us with an intelligible, humane means of interacting with and understanding the information that the systems provide.

The good news is that these machines can take on whatever appearance and mode of operation are best suited for their users. For the first time in history, we are truly free to make machines that fit human needs, independent of mechanical constraints.

New information technologies can enhance the power of human thinking, for machine plus person can do more than either alone, but only if the technology complements human abilities. A calculator is a good example of a complementary skill: calculators work very differently from the human mind, which is why they are so valuable. They can perform arithmetic operations rapidly and efficiently without error. Humans are good at inventing the problems: calculators are good tools for solving them. Calculators are well matched to our needs, expanding our mental capabilities.

We need more information-processing tools that complement our thinking, reasoning, and memory skills as comfortably as the calculator enhances our arithmetic skills.

The best of these technologies provide us with rich information about a topic of interest and leave control of the process—and what to do with the results—in our hands. The daily newspaper is a fine example of a large database that allows ready access to information about events in the local region and throughout the world. The technology is nonobtrusive, perhaps because it has no choice: it is a passive, surface artifact that requires all the major work to be done by the user. This does not stop it from being very effective.

But newspapers, like dictionaries and encyclopedias, are restricted to a linear, sequential structure. Dictionaries and encyclopedias are further restricted by the convention of alphabetical order. But today's technology can free us from the limits of yesterday's technology. Dictionaries and encyclopedias no longer have to have any single, fixed organization. Instead, they can have any organizational structure that the user needs at the moment. Merge the dictionary, encyclopedia, and thesaurus so that information can be selected by name, description, or content. Insert a good spelling program so that even misspelled entries can get the desired information to the user.

We can transform the hard technology of computers and information processing into soft technology suitable for people if we start with the needs of the human users, not with the requirements of the technology. With some thought, it is possible to transform even the most inhumane of systems.

18. Feminist Perspectives on Technology

JUDY WAJCMAN

The study of the relationship between gender and technology has received increasing attention in recent years as feminists (both scholars and activists) have become concerned with the impact of new technology on women's lives and work, while "social constructivist" researchers on technology (see the headnote for chapter 4) have begun to examine how women's roles have influenced the evolution of technologies. This selection, "Feminist Perspectives on Technology" by Judy Wajcman, provides an overview of this area of study.

In order to understand the relationship between women and technology, it is important, first of all, to distinguish between science and technology and to be sensitive to the different layers of meaning of "technology." Women have contributed to the development of technology not just in terms of their conventional inventive activity but in terms of other kinds of activities in which they have engaged, but which may not have been recognized as "technological" in a gender-stereotyped view of technology. The emerging sociology of technology lacks a gender dimension, which Wajcman aims to provide, both by the conceptual analysis presented here (and in her book, Feminism Confronts Technology, *from which this essay is taken), and by her own sociological research.*

Judy Wajcman is a professor of sociology at the University of New South Wales in Sydney, Australia. She has been a research fellow at the Industrial Relations Research Unit at the University of Warwick in England, where she has studied the differing experiences of men and women who are senior managers. She has been active in the women's movement in both Britain and Australia. She is coeditor, with Donald MacKenzie, of The Social Shaping of Technology *(1985).*

FROM SCIENCE TO TECHNOLOGY

While there has been a growing interest in the relationship of science to society over the last decade, there has been an even greater preoccupation with the relationship between technology and social change. Debate has raged over whether the "white heat of technology" is radically transforming society and delivering us into a postindustrial age. A major concern of feminists has been the impact of new technology on women's lives, particularly on women's work. The introduction of word processors into the office provided the focus for much early research. The recognition that housework was also work, albeit unpaid, led to studies on how the increasing use of domestic technology in the home affected the time spent on housework. The exploitation of Third World women as a source of cheap labor for the manufacture of computer components has also been scrutinized. Most recently there has been a vigorous debate over developments in reproductive technology and the implications for women's control over their fertility.

Throughout these debates there has been a tension between the view that technology would liberate women—from unwanted pregnancy, from housework and from routine paid work—and the obverse view that most new technologies are destructive and oppressive to women. For example, in the early seventies, Shulamith Firestone (1970) elaborated the view that developments in birth technology held the key to women's liberation through removing from them the burden of biological motherhood. Nowadays there is much more concern with the negative implications of the new technologies, ironically most clearly reflected in the highly charged debate over the new reproductive technologies.

A key issue here is whether the problem lies in men's domination of technology, or whether the technology is in some sense inherently patriarchal. If women were in control, would they apply technology to more benign ends? In the following discussion on gender and technology, I will explore these and related questions.

An initial difficulty in considering the feminist commentary on technology arises from its failure to distinguish between science and technology. Feminist writing on science has often construed science purely as a form of knowledge, and this assumption has been carried over into much of the feminist writing on technology. However just as science includes practices and institutions, as well as knowledge, so too does technology. Indeed, it is even more clearly the case with technology because technology is primarily about the creation of artifacts. This points to the need for a different theoretical approach to the analysis of the gender relations of technology, from that being developed around science.

Perhaps this conflation of technology with science is not surprising given that the sociology of scientific knowledge over the last ten years has contested the idea of a noncontroversial distinction between science and technology. John Staudenmaier (1985, pp. 83–120) comments that although the relationship between science and technology has been a major theme in science and technology studies, the discussion has been plagued by a welter of conflicting definitions of the two basic terms. The only consensus to have emerged is that the way in which the boundaries between science and technology are demarcated, and how they are related to each other, change from one historical period to another.

In recent years, however, there has been a major reorientation of thinking about the form of the relationship between science and technology. The model of the science–technology relationship which enjoyed widespread acceptance over a long period was the traditional hierarchical model which treats technology as applied science. This view that science discovers and technology applies this knowledge in a routine uncreative way is now in steep decline. "One thing which practically any modern study of technological innovation suffices to show is that far from applying, and hence depending upon, the culture of natural science, technologists possess their own distinct cultural resources, which provide the principal basis for their innovative activity" (Barnes and Edge, 1982, p. 149). Technologists build on, modify and extend existing technology but they do this by a creative and imaginative process. And part of the received culture technologists inherit in the course of solving their practical problems is nonverbal; nor can it be conveyed adequately by the written word. Instead it is the individual practitioner who transfers practical knowledge and competence to another. In short, the current model of the science–technology relationship characterizes science and technology as distinguishable subcultures in an interactive symmetrical relationship.

Leaving aside the relationship between technology and science, it is most important to recognize that the word "technology" has at least three different layers of meaning. Firstly, "technology" is a form of knowledge, as Staudenmaier emphasizes.[1] Technological "things" are meaningless without the "know-how" to use them, repair them, design them and make them. That know-how often cannot be captured in words. It is visual, even tactile, rather than simply verbal or mathematical. But it can also be systematized and taught, as in the various disciplines of engineering.

Few authors however would be content with this definition of technology as a form of knowledge. "Technology" also refers to what people do as well as what they know. An object such as a car or a vacuum cleaner is a technology, rather than an arbitrary lump of matter, because it forms part of a set of human activities. A computer without programs and programmers is simply a useless collection of bits of metal, plastic and silicon. "Steelmaking," say, is a technology; but this implies that the technology includes what steelworkers do, as well as the furnaces they use. So "technology" refers to human activities and practices. And finally, at the most basic level, there is the "hardware" definition of technology, in which it refers to sets of physical objects, for example, cars, lathes, vacuum cleaners and computers.

In practice the technologies dealt with here cover all three aspects, and often it is not useful to separate them further. My purpose is not to attempt to refine a definition. These different layers of meaning of "technology" are worth bearing in mind in what follows.

The rest of this [essay] will review the theoretical literature on gender and technology, which in many cases mirrors the debates about science outlined above. However, feminist perspectives on technology are more recent and much less theoretically developed than those which have been articulated in relation to science. One clear indication of this is the preponderance of edited collections which have been published in this area.[2] As with many such collections, the articles do not share a consistent approach or cover the field in a comprehensive fashion. Therefore I will be drawing out strands of argument from this literature rather than presenting the material as coherent positions in a debate.

HIDDEN FROM HISTORY

To start with, feminists have pointed out the dearth of material on women and technology, especially given the burgeoning scholarship in the field of technology studies. Even the most perceptive and humanistic works on the relationship between technology, culture and society rarely mention gender. Women's contributions have by and large been left out of technological history. Contributions to *Technology and Culture*, the leading journal of the history of technology, provide one accurate barometer of this. Joan Rothschild's (1983, pp. xii–xiv) survey of the journal for articles on the subject of women found only four in twenty-four years of publishing. In a more recent book about the journal, Staudenmaier (ibid., p. 180) also notes the extraordinary bias in the journal towards male figures and the striking absence of a women's perspective. The history of technology represents the prototype inventor as male. So, as in the history of science, an initial task of feminists has been to uncover and recover the women hidden from history who have contributed to technological developments.

There is now evidence that during the industrial era, women invented or contributed to the invention of such crucial machines as the cotton gin, the sewing machine, the small electric motor, the McCormick reaper, and the Jacquard loom (Stanley, 1992). This sort of historical scholarship often relies heavily on patent records to recover women's forgotten inventions. It has been noted that many women's inventions have been credited to their husbands because they actually appear in patent records in their husbands' name. This is explained in terms of women's limited property rights, as well as the general ridicule afforded women inventors at that time (Pursell, 1981; Amram, 1984; Griffiths, 1985). Interestingly, it may be that even the recovery of women inventors from patent records seriously underestimates their contribution to technological development. In a recent article on the role of patents, Christine MacLeod (1987) observes that prior to 1700 patents were not primarily about the recording of the actual inventor, but were instead sought in the name of financial backers.[3] Given this, it is even less surprising that so few women's names are to be found in patent records.

For all but a few exceptional women, creativity alone was not sufficient. In order to participate in the inventive activity of the Industrial Revolution, capital as well as ideas were necessary. It was only in 1882 that the Married Women's Property Act gave English women legal possession and control of any personal property independently of their husbands. Dot Griffiths (1985) argues that the effect of this was to virtually exclude women from participation in the world of the inventor–entrepreneur. At the same time women were being denied access to education and specifically to the theoretical grounding in mathematics and mechanics upon which so many of the inventions and innovation of the period were based. As business activities expanded and were moved out of the home, middle-class women were increasingly left to a life of enforced leisure. Soon the appropriate education for girls became "accomplishments" such as embroidery and music—accomplishments hardly conducive to participation in the world of the inventor–entrepreneur. In the current period, there has been considerable interest in the possible contributions which Ada Lady Lovelace, Grace Hopper and other women may have made to the development of computing. Recent histories of computer programming provide substantial evidence for the view that women played a major part.[4]

To fully comprehend women's contributions to technological development, however, a more radical approach may be necessary. For a start, the traditional conception of technology too readily defines technology in terms of male activities. As I have pointed out above, the concept of technology is itself subject to historical change, and different epochs and cultures had different names for what we now think of as technology. A greater emphasis on women's activities immediately suggests that females, and in particular black women, were among the first technologists. After all, women were the main gatherers, processors and storers of plant food from earliest human times onward. It was therefore logical that they should be the ones to have invented the tools and methods involved in this work such as the digging stick, the carrying sling, the reaping knife and sickle, pestles and pounders. In this vein, Autumn Stanley (1992) illustrates women's early achievements in horticulture and agriculture, such as the hoe, the scratch plow, grafting, hand pollination, and early irrigation.

If it were not for the male bias in most technology research, the significance of these inventions would be acknowledged. As Ruth Schwartz Cowan notes:

> The indices to the standard histories of technology . . . do not contain a single reference, for example, to such a significant cultural artifact as the baby bottle. Here is a simple implement . . . which has transformed a fundamental human experience for vast numbers of infants and mothers, and been one of the more controversial exports of Western technology to underdeveloped countries—yet it finds no place in our histories of technology. (1979, p. 52)

There is important work to be done not only in identifying women inventors, but also in discovering the origins and paths of development of "women's sphere" technologies that seem often to have been considered beneath notice.

A TECHNOLOGY BASED ON WOMEN'S VALUES?

During the eighties, feminists have begun to focus on the gendered character of technology itself. Rather than asking how women could be more equitably treated within and by a neutral technology, many feminists now argue that Western technology itself embodies patriarchal values. This parallels the way in which the feminist critique of science evolved from asking the "woman question" in science to asking the more radical "science question" in feminism. Technology, like science, is seen as deeply implicated in the masculine project of the domination and control of women and nature.[5] Just as many feminists have argued for a science based on women's values, so too has there been a call for a technology based on women's values. In Joan Rothschild's (1983) preface to a collection on feminist perspectives on technology, she says that: "Feminist analysis has sought to show how the subjective, intuitive, and irrational can and do play a key role in our science and technology." Interestingly, she cites an important male figure in the field, Lewis Mumford, to support her case. Mumford's linking of subjective impulses, life-generating forces and a female principle is consistent with such a feminist analysis, as is his endorsement of a more holistic view of culture and technological developments.

Other male authors have also advocated a technology based on women's values. Mike Cooley is a well-known critic of the current design of technological systems and he has done much to popularize the idea of human-centered technologies. In *Architect or Bee?* (1980, p. 43) he argues that technological change has "male values" built into it: "the values of the White Male Warrior, admired for his strength and speed in eliminating the weak, conquering competitors and ruling over vast armies of men who obey his every instruction . . . Technological change is starved of the so-called female values such as intuition, subjectivity, tenacity and compassion." Cooley sees it as imperative that more women become involved in science and technology to challenge and counteract the built-in male values: that we cease placing the objective above the subjective, the rational above the tacit, and the digital above analogical representation. In *The Culture of Technology*, Arnold Pacey (1983) devotes an entire chapter to "Women and Wider Values." He outlines three contrasting sets of values involved in the practice of technology—firstly, those stressing virtuosity, secondly, economic values and thirdly, user or need-oriented values. Women exemplify this third "responsible" orientation, according to Pacey, as they work with nature in contrast to the male interest in construction and the conquest of nature.

Ironically the approach of these male authors is in some respects rather similar to the eco-feminism that became popular among feminists in the eighties. This marriage of ecology and feminism rests on the "female principle," the notion that women are closer to nature than men and that the technologies men have created are based on the domination of nature in the same way that they seek to dominate women. Eco-feminists concentrated on military technology and the ecological effects of other modern technologies. According to them, these technologies are products of a patriarchal culture that "speaks violence at every level" (Rothschild, 1983, p. 126). An early slogan of the feminist antimilitarist movement, "Take the Toys from the Boys," drew attention to the phallic symbolism in the shape of missiles. However, an inevitable corollary of this stance seemed to be the representation of women as inherently nurturing and pacifist. The problems with this position have been outlined above in relation to science based on women's essential values. We need to ask how women became associated with these values. The answer involves examining the way in which the traditional division of labor between women and men has generally restricted women to a narrow range of experience concerned primarily with the private world of the home and family.

Nevertheless, the strength of these arguments is that they go beyond the usual conception of the problem as being women's exclusion from the processes of innovation and from the acquisition of technical skills. Feminists have pointed to all sorts of barriers—in social attitudes, girls' education and the employment policies of firms—to account for the imbalance in the number of women in engineering. But rarely has the problem been identified as the way engineering has been conceived and taught. In particular, the failure of liberal and equal opportunity policies has led authors such as Cynthia Cockburn (1985) to ask whether women actively resist entering technology. Why have the women's training initiatives designed to break men's monopoly of the building trades, engineering and information technology not been more successful? Although schemes to channel women into technical trades have been small-scale, it is hard to escape the conclusion that women's response has been tentative and perhaps ambivalent.

I share Cockburn's view that this reluctance "to enter" is to do with the sex-stereotyped definition of technology as an activity appropriate for men. As with science, the very language of technology, its symbolism, is masculine. It is not simply a question of acquiring skills, because these skills are embedded in a culture of masculinity that is largely coterminous with the culture of technology. Both at school and in the workplace this culture is incompatible with femininity. Therefore, to enter this world, to learn its language, women have first to forsake their femininity.

TECHNOLOGY AND THE DIVISION OF LABOR

I will now turn to a more historical and sociological approach to the analysis of gender and technology. This approach has built on some theoretical foundations provided by contributors to the labor process debate of the 1970s. Just as the radical science movement had sought to expose the class character of science, these writers attempted to extend the class analysis to technology. In doing so, they were countering the theory of "technological determinism" that remains so widespread.

According to this account, changes in technology are the most important cause of social change. Technologies themselves are neutral and impinge on society from the outside; the scientists and technicians who produce new technologies are seen to be independent of their social location and above sectional interests. Labor process analysts were especially critical of a technicist version of Marxism in which the development of technology and productivity is seen as the motor force of history. This interpretation represented technology itself as beyond class struggle.

With the publication of Harry Braverman's *Labor and Monopoly Capital* (1974), there was a revival of interest in Marx's contribution to the study of technology, particularly in relation to work. Braverman restored Marx's critique of technology and the division of labor to the center of his analysis of the process of capitalist development. The basic argument of the labor process literature which developed was that capitalist–worker relations are a major factor affecting the technology of production within capitalism. Historical case studies of the evolution and introduction of particular technologies documented the way in which they were deliberately designed to deskill and eliminate human labor.[6] Rather than technical inventions developing inexorably, machinery was used by the owners and managers of capital as an important weapon in the battle for control over production. So, like science, technology was understood to be the result of capitalist social relations.

This analysis provided a timely challenge to the notion of technological determinism and, in its focus on the capitalist division of labor, it paved the way for development of a more sophisticated analysis of gender relations and technology. However, the labor process approach was gender-blind because it interpreted the social relations of technology in exclusively class terms. Yet, as has been well established by the socialist feminist current in this debate, the relations of production are constructed as much out of gender divisions as class divisions. Recent writings (Cockburn, 1983, 1985; Faulkner and Arnold, 1985; McNeil, 1987) in this historical vein see women's exclusion from technology as a consequence of the gender division of labor and the male domination of skilled trades that developed under capitalism. In fact, some argue that prior to the Industrial Revolution women had more opportunities to acquire technical skills, and that capitalist technology has become more masculine than previous technologies.

I have already described how, in the early phases of industrialization, women were denied access to ownership of capital and access to education. Shifting the focus, these authors show that the rigid pattern of gender divisions which developed within the working class in the context of the new industries laid the foundation for the male dominance of technology. It was during this period that manufacturing moved into factories, and home became separated from paid work. The advent of powered machinery fundamentally challenged traditional craft skills because tools were literally taken out of the hands of workers and combined into machines. But as it had been men who on the whole had technical skills in the period before the Industrial Revolution, they were in a unique position to maintain a monopoly over the new skills created by the introduction of machines.

Male craft workers could not prevent employers from drawing women into the new spheres of production. So instead they organized to retain certain rights over technology by actively resisting the entry of women to their trades. Women who became industrial laborers found themselves working in what were considered to be unskilled jobs for the lowest pay. "It is the most damning indictment of skilled working-class men and their unions that they excluded women from membership and prevented them gaining competences that could have secured them a decent living" (Cockburn, 1985, p. 39). This gender division of labor within the factory meant that the machinery was designed by men with men in mind, either by the capitalist inventor or by skilled craftsmen. Industrial technology from its origins thus reflects male power as well as capitalist domination.

The masculine culture of technology is fundamental to the way in which the gender division of labor is still being reproduced today. By securing control of key technologies, men are denying women the practical experience upon which inventiveness depends. I noted earlier the degree to which technical knowledge involves tacit, intuitive knowledge and "learning by doing." New technology typically emerges not from sudden flashes of inspiration but from existing technology, by a process of gradual modification to, and new combinations of, that existing technology. Innovation is to some extent an imaginative process, but that imagination lies largely in seeing ways in which existing devices can be improved, and in extending the scope of techniques successful in one area into new areas. Therefore giving women access to formal technical knowledge alone does not provide the resources necessary for invention. Experience of existing technology is a precondition for the invention of new technology.

The nature of women's inventions, like that of men's, is a function of time, place and resources. Segregated at work and primarily confined to the private sphere of household, women's experience has been severely restricted and therefore so too has their inventiveness. An interesting illustration of this point lies in the fact that women who were employed in the munitions factories during the First World War are on record as having redesigned the weaponry they were making.[7] Thus, given the opportunity, women have demonstrated their inventive capacity in what now seems the most unlikely of contexts.

MISSING: THE GENDER DIMENSION IN THE SOCIOLOGY OF TECHNOLOGY

The historical approach is an advance over essentialist positions which seek to base a new technology on women's innate values. Women's profound alienation from technology is accounted for in terms of the historical and cultural construction of technology as masculine. I believe that women's exclusion from, and rejection of, technology is made more explicable by an analysis of technology as a culture that expresses and consolidates relations among men. If technical competence is an integral part of masculine gender identity, why should women be expected to aspire to it?

Such an account of technology and gender relations, however, is still at a general level.[8] There are few cases where feminists have really got inside the "black box" of technology to do detailed empirical research, as some of the most recent sociological literature has attempted. Over the last few years, a new sociology of technology has emerged which is studying the invention, development, stabilization and diffusion of specific artifacts.[9] It is evident from this research that technology is not simply the product of rational technical imperatives. Rather, political choices are embedded in the very design and selection of technology.

Technologies result from a series of specific decisions made by particular groups of people in particular places at particular times for their own purposes. As such, technologies bear the imprint of the people and social context in which they developed. David Noble (1984, p. xiii) expresses this point succinctly as follows: "Because of its very concreteness, people tend to confront technology as an irreducible brute fact, a given, a first cause, rather than as hardened history, frozen fragments of human and social endeavor." Technological change is a process subject to struggles for control by different groups. As such, the outcomes depend primarily on the distribution of power and resources within society.

There is now an extensive literature on the history of technology and the economics of technological innovation. Labor historians and sociologists have investigated the relationship between social change and the shaping of production processes in great detail and have also been concerned with the influence of technological form upon social relations. The sociological approach has moved away from studying the individual inventor and from the notion that technological innovation is a result of some inner technical logic. Rather, it attempts to show the effects of social relations on technology that range from fostering or inhibiting particular technologies, through influencing the choice between competing paths of technical development, to affecting the precise design characteristics of particular artifacts. Technological innovation now requires major investment and has become a collective, institutionalized process. The evolution of a technology is thus the function of a complex set of technical, social, economic, and political factors. An artifact may be looked on as the "congealed outcome of a set of negotiations, compromises, conflicts, controversies and deals that were put together between opponents in rooms filled with smoke, lathes or computer terminals" (Law, 1987, p. 406).

Because social groups have different interests and resources, the development process brings out conflicts between different views of the technical requirements of the device. Accordingly, the stability and form of artifacts depend on the capacity and resources that the salient social groups can mobilize in the course of the development process. Thus in the technology of production, economic and social class interests often lie behind the development and adoption of devices. In the case of military technology, the operation of bureaucratic and organizational interests of state decision making will be identifiable. Growing attention is now being given to the extent to which the state sponsorship of military technology shapes civilian technology.

So far, however, little attention has been paid to the way in which technological objects may be shaped by the operation of gender interests. This blindness to gender issues is also indicative of a general problem with the methodology adopted by the new sociology of technology. Using a conventional notion of technology, these writers study the social groups which actively seek to influence the form and direction of technological design. What they overlook is the fact that the absence of influence from certain groups may also be significant. For them, women's absence from observable conflict does not indicate that gender interests are being mobilized. For a social theory of gender, however, the almost complete exclusion of women from the technological community points to the need to take account of the underlying structure of gender relations. Preferences for different technologies are shaped by a set of social arrangements that reflect men's power in the wider society. The process of technological development is socially structured and culturally patterned by various social interests that lie outside the immediate context of technological innovation.

More than ever before technological change impinges on every aspect of our public and private lives, from the artificially cultivated food that we eat to the increasingly sophisticated forms of communication we use. Yet, in common with the labor process debate, the sociology of technology has concentrated almost exclusively on the relations of paid production, focusing in particular on the early stages of product development. In doing so [it has] ignored the spheres of reproduction, consumption and the unpaid production that takes place in the home.

By contrast, feminist analysis points us beyond the factory gates to see that technology is just as centrally involved in these spheres.

Inevitably perhaps, feminist work in this area has so far raised as many questions as it has answered. Is technology valued because it is associated with masculinity or is masculinity valued because of the association with technology? How do we avoid the tautology that "technology is masculine because men do it"? Why is women's work undervalued? Is there such a thing as women's knowledge? Is it different from "feminine intuition"? Can technology be reconstructed around women's interests? These are the questions that abstract analysis has so far failed to answer. The character of salient interests and social groups will differ depending on the particular empirical sites of technology being considered. Thus we need to look in more concrete and historical detail at how, in specific areas of work and personal life, gender relations influence the technological enterprise. . . . [In the book from which this excerpt is drawn I stress] that a gendered approach to technology cannot be reduced to a view which treats technology as a set of neutral artifacts manipulated by men in their own interests. While it is the case that men dominate the scientific and technical institutions, it is perfectly plausible that there will come a time when women are more fully represented in these institutions without transforming the direction of technological development. To cite just one instance, women are increasingly being recruited into the American space–defense program but we do not hear their voices protesting about its preoccupations. Nevertheless, gender relations are an integral constituent of the social organization of these institutions and their projects. It is impossible to divorce the gender relations which are expressed in, and shape technologies from, the wider social structures that create and maintain them. In developing a theory of the gendered character of technology, we are inevitably in danger of either adopting an essentialist position that sees technology as inherently patriarchal, or losing sight of the structure of gender relations through an overemphasis on the historical variability of the categories of "women" and "technology." [My work seeks] to chart another course.

NOTES

1. Staudenmaier (1985, pp. 103–20) outlines four characteristics of technological knowledge—scientific concepts, problematic data, engineering theory, and technological skill.

2. A good cross-section of this material can be found in Trescott (1979); Rothschild (1983); Faulkner and Arnold (1985); McNeil (1987); Kramarae (1988). McNeil's book is particularly useful as it contains a comprehensive bibliography which is organized thematically.
3. MacLeod (1987) suggests that although George Ravenscroft is credited in the patent records with being the "heroic" inventor of lead-crystal glass, he was rather the purchaser or financier of another's invention. This study alerts us to the danger of assuming that patent records have always represented the same thing.
4. For a biography of Lady Lovelace, which takes issue with the view of her as a major contributor to computer programming, see Stein (1985). However, both Kraft (1977) and more recently Giordano (1988) have documented the extensive participation of women in the development of computer programming.
5. Technology as the domination of nature is also a central theme in the work of critical theorists, such as Marcuse, for whom it is capitalist relations (rather than patriarchal relations) which are built into the very structure of technology. "Not only the application of technology but technology itself is domination (of nature and men)—methodical, scientific, calculated, calculating control. Specific purposes and interests of domination are not foisted upon technology 'subsequently' and from the outside; they enter the very construction of the technical apparatus" (Marcuse, 1968, pp. 223–4).
6. This point is elaborated in chapter 2 of *Feminism Confronts Technology*. See also Part Two of MacKenzie and Wajcman (1985) for a collection of these case studies.
7. Amram (1984) provides a selection of the patents granted to women during the First World War.
8. Cockburn's (1983, 1985) work is one important exception discussed at greater length in chapter 2 of *Feminism Confronts Technology*.
9. For an introduction to this literature, see MacKenzie and Wajcman (1985); Bijker, Hughes and Pinch (1987).

REFERENCES

Amram, F. 1984. The Innovative Woman. *New Scientist,* 24 May 1984, 10–12.
Barnes, Barry, and David Edge, eds. 1982. *Science in Context: Readings in the Sociology of Science.* Milton Keynes, England: Open University Press/Cambridge, MA: MIT Press.
Bijker, Wiebe E., Thomas P. Hughes, and Trevor J. Pinch, eds. 1987. *The Social Construction of Technological Systems.* Cambridge, MA: MIT Press.
Braverman, Harry. 1974. *Labor and Monopoly Capital: The Degradation of Work in the Twentieth Century.* New York: Monthly Review Press.
Cockburn, Cynthia. 1983. *Brothers: Male Dominance and Technological Change.* London: Pluto Press.
Cockburn, Cynthia. 1985. *Machinery of Dominance: Women, Men, and Technical Know-How.* London: Pluto Press.
Cooley, Mike. 1980. *Architect or Bee? The Human/Technology Relationship.* Slough, England: Langley Technical Services.

Cowan, Ruth S. 1979. From Virginia Dare to Virginia Slims: Women and Technology in American Life. *Technology and Culture* 20; 51–63.

Faulkner, Wendy, and Erik Arnold, eds. 1985. *Smothered by Invention: Technology in Women's Lives.* London: Pluto Press.

Firestone, Shulamith. 1970. *The Dialectic of Sex.* New York: William Morrow & Co.

Giordano, R. 1988. *The Social Context of Innovation: A Case History of the Development of COBOL Programming Language.* Columbia University Department of History.

Griffiths, Dot. 1985. The Exclusion of Women from Technology. In Faulkner and Arnold, op. cit.

Kraft, P. 1977. *Programmers and Managers: The Routinization of Computer Programming in the United States.* New York: Springer Verlag.

Kramarae, Chris, ed. 1988. *Technology and Women's Voices.* New York: Routledge & Kegan Paul.

Law, John. 1987. Review Article: The Structure of Sociotechnical Engineering—A Review of the New Sociology of Technology. *Sociological Review* 35; 404–425.

MacKenzie, Donald, and Judy Wajcman, eds. 1985. *The Social Shaping of Technology: How the Refrigerator Got Its Hum.* Milton Keynes, England: Open University Press.

MacLeod, Christine. 1987. Accident or Design? George Ravenscroft's Patent and the Invention of Lead-Crystal Glass. *Technology and Culture* 28, 4:776–803.

McNeil, Maureen, ed. 1987. *Gender and Expertise.* London: Free Association Books.

Marcuse, H. 1968. *Negations.* London: Allen Lane.

Noble, David. 1984. *Forces of Production: A Social History of Industrial Automation.* New York: Knopf.

Pacey, Arnold. 1983. *The Culture of Technology.* Oxford, England: Basil Blackwell/Cambridge, MA: MIT Press.

Pursell, C. 1981. Women Inventors in America. *Technology and Culture* 22, 3:545–549.

Rothschild, Joan, ed. 1983. *Machina ex Dea: Feminist Perspectives on Technology.* New York: Pergamon Press.

Stanley, Autumn. 1992. *Mothers and Daughters of Invention: Notes for a Revised History of Technology.* Metuchen, NJ: Scarecrow Press.

Staudenmaier, John M. 1985. *Technology Storytellers: Reweaving the Human Fabric.* Cambridge, MA: MIT Press.

Stein, D. 1985. *Ada: A Life and Legacy.* Cambridge, MA: MIT Press.

Trescott, Martha M., ed. 1979. *Dynamos and Virgins Revisited: Women and Technological Change in History.* Metuchen, NJ: Scarecrow.

19.   Science, Technology, and Black Community Development

ROBERT C. JOHNSON

Many studies of the impacts of specific technologies on U.S. society allude to the different ways in which these technologies affect African-Americans vis-à-vis white Americans. Few writers, however, have examined in a comprehensive manner the impacts of science and technology on African-Americans. Robert C. Johnson takes a first cut at such a comprehensive analysis in his selection, "Science, Technology, and Black Community Development."

Johnson, examining such pervasive technologies as automobiles, television, and industrial and agricultural mechanization, concludes that, taken together, these technologies have had a profound effect on the lives of African-Americans in the twentieth century. He looks at such seemingly diverse issues as the cultural impacts of electronics on African-American music and the potential influence of genetic engineering on the lives of African-Americans, and he argues that only by consciously taking a more active role in science and innovation can African-Americans reshape the course of technological development to address their own needs and interests. Central to this are education and a willingness to make science and technology integral parts of the African-American socialization process.

A scholar, researcher, and writer, Robert C. Johnson possesses a varied educational background that includes a master's degree in technology and human affairs and a Ph.D. in educational policy and research. He is currently professor and director of minority studies at St. Cloud State University of Minnesota, where he conducts math and science programs to increase the interest of women and people of color in scien-

Source: Selection from *The Black Scholar* (March/April 1984), pp. 32–44. By permission of *The Black Scholar*.

tific and technical careers. Johnson has held grants, fellowships, and awards from several federal and state agencies, as well as the Ford, Bush, and Danforth foundations.

Some may consider it odd to find included on the agenda of Black people a topic such as that being presented in this article. Some may ask what do science and technology have to do with Black survival and progress? What do they have to do with urban areas? It is my purpose [here] to explore the thesis that scientific breakthroughs and technological developments affect Black Americans in very profound, pervasive and substantial manners. So much so that I would argue that without examining and being aware of these various developments, Black Americans risk being subjugated to the vicissitudes of scientific and technological forces which are as oppressive, demeaning and domineering as are the socioeconomic and political forces of racism and exploitation.

I would further maintain that we should go beyond simply becoming aware of these occurrences and begin to make such concerns a top priority in our agendas and programs. Our political and economic advocacies and strategies should entail concerns of a technological and scientific nature, and these should have as much priority as economic development and civil rights. As a matter of fact, I hope to illustrate how these technological and scientific concerns are intricately interwoven with these other political, economic and social matters. I will, indeed, argue that our very existence, and continued survival, growth, and even prosperity in this society and in the world are very much dependent upon the various developments and emerging political debates occurring within the areas of science and technology.

One reason why many people may express reservations about the role of science and technology in Black community affairs is because of the popular view of these two phenomena. Science is generally viewed as some remote, isolated process that occurs by a handful of people in faraway laboratories, wearing white coats and working with test tubes and animals. Technology is viewed as that wonderful process by which marvelous inventions and gadgets are brought into existence. Furthermore, this vision of people in white coats performing miracles is further complicated by the fact that these persons wearing white coats also wear white faces in the popular image. We may have had occasion to hear some Black people use the expression, "W-W-M-S-T," which translates to mean "Wonders of the White Man's Science and Technology." Those five letters summarize quite succinctly the whole idea that science and technology are wonders and miracle processes that are somehow created magically and mysteriously by whites.

Such a view is both erroneous and dangerous. It is erroneous because the history of Black Americans and the history of Africans tells us that people of African descent have been in the forefront of scientific and technological breakthroughs and have laid the foundation for much of what we currently know as science and technology. It is further erroneous because it assumes that Blacks have been excluded from the scientific and technological process, that process of discovering, manipulating, and creating from natural phenomena goods, products, and devices that can be used for social and human purposes. The very fact that Black people have survived and have built and maintained cultures and civilizations demonstrates that they have made judicious use of scientific and technological principles, and have manipulated natural phenomena for their own collective benefit. This popular view of science and technology is also dangerous because it implies that only whites can do science and technology, that Blacks either cannot or should not be part of this process. Furthermore, in a very insidious, albeit indirect manner, it implies that Black people are not capable of producing in these domains. It is dangerous because it leaves the scientific and technological fate of Black people to others who may not and probably do not have their interests at heart. Such a view removes us from direct participation in and involvement with two aspects of our lives which are growing more and more important in governing and determining the quality of life that we will face on this planet.

Science is generally thought of, at least by the scientific community, as being the process of discovering, explaining and predicting natural phenomena, usually for the purpose of trying to acquire some type of control over these phenomena. We may simply define science as knowledge. (It is defined in Webster's as "the body of truth, information, and principles acquired by mankind.") Technology, on the other hand, is the application of knowledge. It is know-how applied to meet social and human needs and ends. While both of these definitions deal with natural forces of the physical environment it must be realized that both science and technology are the creations of man and that they derive their essence and significance from human and social contexts. Science and technology have their origins in humanity. And their development is linked to the development of humanity. If we as Black people define ourselves as part of humanity, then, by definition, we are, in part, scientific and technological beings.

THE NEED FOR BLACK AWARENESS OF SCIENCE AND TECHNOLOGY

Let us now look briefly at the need and importance of Blacks being involved with and aware of scientific and technological developments in our society. Science and technology in American society in the twentieth century are characterized by rapid occurrences and major breakthroughs. Perhaps one of the best accounts of the rapidity and effects of change of a technological nature on the American society has been presented by Alvin Toffler (1970) in his book entitled *Future Shock*. In this work, Toffler not only chronicles the major scientific and technological changes that have occurred in our society over the past few decades, but more importantly, describes "future shock," which is the "shattering stress and disorientation that we induce in individuals by subjecting them to too much change in too short a time." We can all identify elements of this change in our own personal lives as well as in the world around us—the new gadgets, the new modes of doing things, the new life-styles, all brought about by technology and science.

In addition to the psychological changes induced by scientific and technological developments, the cultural, economic, political and social institutions of this society have all been affected by science and technology as well. We will touch upon the topic of technology and culture later in this [article]. I will show how Black people have been disproportionately and more adversely affected by these developments than has been the general society at large. It is this heightened state of vulnerability of Blacks that makes technology and science two forces that should compete for our attention, knowledge and efforts along with some of the major issues facing Black people today.

In becoming aware of the fact that science and technology are powerful forces in our lives, we must also come to the realization that the scientific and technological developments are not separate nor removed from the other activities and processes of the society. We must realize that these scientific and technological forces are governed and influenced by the political processes and economic realities of our society. Public policy has as much to do with scientific and technological developments as do intellectual considerations, if not more so. With this awareness of the relationship between science and technology on the one hand and politics and economics on the other, we will be in a position to understand that we can influence technological and scientific developments as they affect the cultural, social, political and economic realities of our lives. Such an awareness will lead us to realize that scientific and technological decisions oftentimes are made in the public arena through the processes of public policy. Hence, our political and economic clout can be harnessed and directed toward these forces.

Earl Graves points out in an issue of *Black Enterprise* (October 1979, p. 9) that "politics and economics cannot be separated." He goes on to say "we have reached a stage where we need to develop structures, mobilizing our economic clout. A people who have an annual income of 101 billion dollars and who spend far in excess of that possess the potential for clout." He also notes that we have yet to realize and maximize either our political clout or our economic potential. Clearly then, if we lag behind in areas in which we at least recognize the need for maximizing potential, i.e., economics and politics, then we are sorely behind in the area of science and technology, because, as indicated before, we have failed to realize that we have a stake in and the ability to influence these domains. Thus, our potential is even more greatly untapped in these areas, which renders us more vulnerable to these forces.

Understanding the political and economic aspects and implications of science and technology clarifies the need for Black people to be in the position to thwart undesirable innovations and consequences arising from these forces. Such an understanding shows how imperative it is to have a say in policies governing science and technology. Furthermore, such understanding and awareness show the necessity for devising or creating answers or countermeasures to the real and potential adverse effects of these two areas.

TECHNOLOGY AND CULTURE

I would now like to turn my attention to reviewing in summary form some of the more salient effects of scientific innovations and technological inventions on the American society, and show how these have impacted the Black community. It is generally agreed that several technological inventions have affected our manner of living in major ways this century. Among these are the automobile, the television, the computer, and the telephone.

Observers of the phenomenon known as "automobility" have claimed that the automobile has had such industrial, sociological, economical, and environmental impact that it is responsible for the development of a new consumer goods–oriented society and economy that has persisted from the 1920s into the present; that it is responsible for the expansion of local tax systems; that it set the precedent during the 1920s for the great extension of consumer installment credit; that it was a major force unifying Americans during the period between the two World Wars; that the automobile was destructive of "the beneficial, as well as the repressive aspects of community" (Flink, 1975). And they even claim that automobility was also in large part responsible for the depressed conditions of agriculture in the 1920s. It is even pointed out that the automobile affected relationships between individuals and their communities. James Flink, in his book, *The Car Culture*, cites a study which says, "the opportunities afforded by the automobile provide a basis for new mobility for whites as well as Negroes based upon personal standards rather than community mores—upon what the individual wants to do rather than what the community does not want him to do" (pp. 156–157). Flink goes further in indicating the effects of the automobile on race relations in this society. He notes that the automobile accelerated the exodus of the middle class and businesses to suburbia, thus creating affluent suburbs and isolated ghettos in the central cities, paving way for the widespread riots during the mid-1960s.

While some of these claims about the pervasive influence of the automobile on American society may be open to question and skepticism, it is crystal clear that the automobile has had a substantial impact on individual mores as well as psychological states of citizens of this country in addition to its profound economic, technological and social impacts. One observer, in noting the central role of automobility, concluded "no one has or perhaps can reliably estimate the vast size of capital invested in reshaping society to fit the automobile" (quoted in Flink, p. 141). Flink notes: "that American urban life would conform to the needs of automobility rather than vice versa was obvious by the early 1920s." This phenomenon known as automobility has undoubtedly affected our ways of living, ranging from how and where we shop, where we live and work to the dating and mating habits of teenagers and adults. Among other things cars become indicators of social status and social acceptance. It has become a concrete symbol of the crass materialistic attitude that says, "I am somebody because I have something." Unfortunately, in the Black community this attitude appears to be quite rampant as we can tell by the steady stream of long, late-model gas guzzling "rides" that dot our community.

Beyond the attitudinal and social impacts of the automobile, we are becoming increasingly aware of the technological and environmental problems posed by the automobile. We have to face daily the smog, traffic congestion, parking problems created by the automobile culture. There is noise pollution and the problem of junk and abandoned cars. Motor vehicle safety is another major concern and by-product of the car culture in this society. According to Flink, in the last decade over 4 million Americans a year were injured in automobile accidents. There were 570 fatalities for every 10 billion miles of travel by car, versus only 14 for every 10 billion miles flown, 13 for every 10 billion traveled by bus, and 5 for every 10 billion miles traveled by train (p. 215). Flink has identified other disenchanting consequences of the car culture. He points out that the infrastructure built to support the car culture has had these effects: "the building of urban expressways has destroyed cohesive urban neighborhoods and city parks, further alienated racial minorities and contributed to the declining tax base of the central city" (pp. 214–215).

Television has had just as pervasive an impact on American society as the automobile even though it may not be as dramatically obvious. The rise of television has been meteoric. It was possible in 1945 for a national pollster to ask the country—"Do you know what a television is? Have you ever seen a television set in operation?" (quoted in Comstock, et al., 1978; p. 1). Today there are few households in the U.S. without at least one television set and many households are multiset homes. In a most comprehensive review of television and its effects on human behavior, a group of Rand researchers have delineated some of the principal characteristics of this technological medium (Comstock et al., 1978). It is estimated that there are about 4.7 million hours of programming shown annually in the U.S. On any given evening at the peak hour, television has an audience of more than 95 million Americans. While people complain about certain aspects of television, namely the commercials, sex-related content judged inappropriate for children, and violence, most people in this society consider it as the number one medium for receiving forms of entertainment, knowledge and information. Television is the principal source of news; it is considered to be the provider of the most complete coverage and also the most credible or believable source. Television is known to shape the behavior of those viewing it as well. In addition to consuming large amounts of hours of their time, the television also influences the way people act within their households. It has affected relationships within families and between family members. As the Rand study points out:

> The impact of television on the expenditure of leisure time has been sizable. Television has markedly increased the total amount of time spent with the mass media. Television viewing as a primary activity, excluding very disruptive viewing while doing something else, consumes more of the leisure time of Americans than any other activity. Among almost forty kinds of primary activities, exclusive categories into which the day can be divided, television viewing falls behind only sleep and time spent at work (p. 10).

This study goes on to say that:

> Television's absorption of leisure time naturally occurs at the expense of other activities. One of television's most marked effects appears to have been on time spent sleeping. It also appears to have reduced time spent in social gatherings away from home, in radio listening, in reading books, in miscellaneous leisure, in moviegoing, in religious activities and in household tasks (p. 10).

Television viewing also affects other aspects of life in this society. Commercials entice us to buy products that we ordinarily would not be aware of. Television may very well have influence on matters such as health practices, dietary preferences, or basic values (such as the desire to acquire material goods). Television has been considered an important factor in the political socialization of Americans since it dispenses information and because information is a major ingredient in political socialization.

The relationship between television viewing and aggressive or antisocial behavior is also important to consider. The Rand report states: "The viewing of television violence appears to increase the likelihood of subsequent aggressiveness" (p. 13).

The available evidence on the relationship between Blacks and the television shows that Blacks tend to watch more television, that Black children typically spend more time in front of televisions than do white children, and that Blacks are more likely to look upon television as a source of knowledge and information. Additionally, research shows that Blacks are usually portrayed on television either in very trite fashion, in lesser roles, or relatively infrequently. The implications of these various findings are profound and should be noted. If Blacks are among the heavy viewers of television and they are similarly treated in a very negative and stereotypic manner, but yet and still they put a great deal of faith in television as a source of information and knowledge, then the media tells them that they are nothing, that they are nobodies and they tend to believe it, and therefore act upon it. In this light, let me add parenthetically that we need to ask the question: given that Blacks are major viewers of television and television violence appears to increase the likelihood of subsequent aggressiveness, is there any relationship between television-viewing and the substantial increases that we have noted in Black-on-Black crime?

We can conclude that the net effect of these various situations with television is that they reinforce the position society imposes upon Blacks. Therefore, it can be argued that television is oppressive because it maintains and reinforces the status quo of this society which itself is oppressive to Blacks.

This society and our people within it have been affected by other technological innovations this century. The combined effects of industrialization, cybernation and agricultural mechanization serve to shape the employment patterns, or should I say the unemployment patterns, that the Black community faces today. We can even go so far as to say that they even shaped the very existence of the Black community. At the beginning of this century, the Black community in America was essentially a rural one. As agricultural mechanization was introduced on a large scale, massive rural exodus occurred, including that of Blacks. The industries of the North attracted many southern Black migrants to that region, leading to increased Black urbanization in the North and in the South as well. Some scholars (Frazier, Martin and Martin) have pointed to this process as a key factor in the loss of Black community cohesiveness and family spirit. Life in crowded cities increased the opportunities for crime, alcoholism, marital breakup and many other adverse social consequences that take place in Black communities.

An obvious key to the viability of the Black community is employment, or economic stability. Yet, technology has affected Black economic stability. The sociologist Sidney Wilhelm (1971) captures this point in a most poignant and succinct manner. He writes:

The Negro becomes a victim of neglect as he becomes useless to an emerging economy of automation. With the onset of automation the Negro moves out of his historical state of oppression into one of uselessness. Increasingly, he is not so much economically exploited as he is irrelevant. The tremendous historical change is taking place in these terms: he is not needed. He is not so much oppressed as unwanted; not so much unwanted as unnecessary; not so much abused as ignored. The dominant whites no longer need to exploit the black minority; as automation proceeds, it will be easier for the former to disregard the latter. In short, White America, by a more perfect application of mechanization and a vigorous reliance upon automation, disposes of the Negro; consequently, the Negro transforms from an exploited labor force into an outcast. The Negro's anguish does not rise only out of brutalities of past oppression; the anxiety stems, more than ever before, out of being discarded as a waste product of technological production (p. 210).

Thus, agricultural mechanization, one form of technology, drove us off the farms in the South and forced us to the industrialized cities of the South and the North. Industrialization, another form of technology, therefore served as a magnet to draw us to the urban areas.

But what has life been like in the urban areas? Again, we are affected by technological forces in this environment. In recent years, studies of cancer rates over the last 25 years show that the cancer incidence rate for Blacks is higher than that for whites. During this period of time the overall cancer incident rate for Blacks rose 8 percent, while for whites it dropped 3 percent. During this 25-year period, cancer mortality for Blacks increased 26 percent, while for whites it rose 5 percent. Why are these cancer rates so different for Blacks and whites? Those who have studied this situation point out that environmental and social factors are the principal causes more so than natural consequences. In a U.S. government document on the topic of Blacks and cancer, the following has been pointed out concerning the higher incidence of cancer among Black Americans:

. . . more Blacks have been exposed to environmental pollutants that have been linked to cancer, especially in the years since World War II. Also more Blacks are living in cities and the rates of all forms of cancer in all races are higher in more crowded industrialized areas than in rural areas. The increased risks in cities also may be related to lifestyle. People who live in cities generally smoke and drink more and the use of both tobacco and alcohol is related to certain forms of cancer. (U.S. Department of Health, Education, and Welfare, p. 4. See also Silverberg and Poindexter, 1979.)

Thus, the forces of technology, which forced us into the urban areas, have seen to it that Blacks run a greater risk of being exposed to industrial carcinogens. This risk, coupled with the principal factor of the lack of adequate medical care and facilities in the Black community, explains in large part our greater mortality and illness due to cancer.

Our increased urbanization makes us vulnerable to other problems which are basically technological in nature as well. We can look at the situation of food, food technology, agricultural science, nutrition and the related issues of Black health and life expectancy. Urban dwellers do not produce food. Farmers do. As we became more and more urbanites and less and less farmers, we produced less of our own food. This means, purely and simply, that we are now dependent upon others for the very substance of life—food. Various developments in food technology and agricultural practices have increased the stock of food and the production of food, but they have also increased the risks of food consumption. Because of a greater variety of additives, preservatives, artificial ingredients, and chemicals, we are beginning to note adverse effects upon the health of people. Americans are overfed but malnourished. We eat a lot, but a lot of the wrong things. It has been estimated that collectively, people in this country are one billion pounds overweight.

We do not eat proper diets nor in a very healthy manner. In fact, our eating habits may cause a deterioration in our health status and abort our longevity. It has been strongly suspected for a long time that our diet plays a role in the incidence of heart disease, cancer, hypertension, strokes, hyperactivity in children, and many other medical and health malfunctions. Recently, the Surgeon General of the United States issued a benchmark but highly controversial study on the prevention of disease. This report, which confirms many of the admonitions and recommendations of health food advocates, calls for Americans to eat less saturated fat and cholesterol, less salt, less sugar, more whole grains, cereals, fruits and vegetables, and relatively more fish, poultry, legumes (which are peas, beans, peanuts, etc.) and less red meat. On the other hand, the meat, dairy and egg industries are up in arms about this report, and can be expected to issue alternative views on the value of these products in our diet and [to] our health.

Clearly then, this is another arena in which we need to exercise some scientific astuteness and judgment in order to determine the validity of the various claims, and to be in a position to affect an area of our life which is too critical to our existence—food.

Obviously, the food problem can be considered very important given that the food is generally not grown in the cities, particularly the inner cities. There are several ways to deal with this problem. One is the home gardening program that has been implemented in many cities, where patches of unused land are utilized for agricultural purposes. Thus one grows one's own food and therefore knows what one is eating. Another possibility is what is known as the Farm to Market Project, a program implemented by a group of Blacks at Cornell University. This program has been described as a "strategy for controlling food costs for urban poor people in conjunction with support mechanisms for small farmer cooperatives in the rural southeast." More specifically,

> The Farm to Market Project is conceived as an approach to solving a series of acute problems confronting limited resource farmers in America, particularly in the South. And secondly, to increase the flow of affordable fresh and nutritious food to America's urban centers, particularly to inner-city areas where many poor people, who are in need of additional quantities of fresh food, live (Agricultural Teams, Inc., Food to Market Project handout).

There are many current situations which have scientific and technological aspects to them that can be noted. There is the energy crisis which is well known to all of us. Lerone Bennett, in an article in *Ebony* magazine (October 1979), comments on how this problem, which is both economic and technological in nature, is threatening to set Black people back to a pre-1960 state. This "energy siege," as Bennett and others call it, has already had deadly effect for the less fortunate of those in our race. So much so, in fact, that in St. Louis a predominantly white church in a wealthy section of this city bought wood-burning stoves to give to needy families, many of them Black. Here we are seeing Black people turning to a technological throwback of the beginning of the century in order to survive in present day America. Of the many technological and scientific issues facing this country, only on this one situation have Black people voiced some concern and opinion. Unfortunately, this commentary from the Black community has been too little, too late, and too divisive, rendering it at this point and time almost ineffectual. It is clear that we will have to "get our act together" in order to have some impact on the transition era from nonrenewable energy sources to other viable, productive and safe forms of energy. Technical, political, and economic influences will have to come to bear on this matter of vital importance to the continued survival of Black institutions and communities.

Another area where the same combination of influence and interest, i.e., technical, political, and economic, will have to be employed is housing. We are facing a housing shortage in this nation and the solution is in part technological in nature. In fact, the National Aeronautics and Space Administration (NASA) has built what it calls the NASA Technology Utilization House. Its billing claims "(the) house of the future is ready today." This agency predicts that this type of home, which has as its unique features solar energy use, energy and water conservation, safety, security, and cost, will be available for widespread dissemination in the very near future. To my knowledge, . . . this new development has [not] been assessed for its potential impact on Black communities, with a critical housing shortage facing them. . . . Clearly, such an important development should not occur unscrutinized, especially since it has widespread implications for the building, construction, and materials industry, which in turn has broad implications for Black employment.

As the economic and social position of Blacks in this society declines, many Black leaders are hinting that mass social protest and civil unrest may be in store as a response to our deteriorating status. In light of this, we must be mindful of the advances occurring in law enforcement technology. A vast armory of sophisticated, technological weaponry has been created for controlling "urban unrest" and where it is not already employed, it sits stockpiled, awaiting deployment in the case of "civil disobedience." The widespread use of technological and electronic surveillance of Black leaders and civil rights organizations is well known. As a matter of fact, Andy Young's downfall came in part because of electronic surveillance of his premises as well as those of the foreign governments and agents with whom he was dealing.

The influence of scientific developments and technological innovations is so pervasive that they have even entered the traditional realms of our culture. For good or bad, electronics has transformed Black music and has added various dimensions to it, and miniaturization has allowed us to have this music available to us in many different forms and many different places, i.e., in our homes, our cars, our offices, at our worksites.

Thus, to date it is no exaggeration to claim that science and technology have dramatically affected and even altered traditional Black life as we have known it up until this point in the twentieth century. So while the advent of television and the automobile, for example, have been major technological forces this far in the twentieth century, we can look to other technologies and scientific breakthroughs to be major influences on our lives in the near and long range future.

The computer is a coming force in the decades of the eighties and the nineties. For the twenty-first century, which is only a [few] years away, technological spin-offs from the space program and developments in biomedical fields will surely dominate in that century, as the computer, television, and the automobile have dominated the twentieth century. For the sake of brevity, I will only highlight some of the possible consequences and implications of these technological and scientific developments. Let us begin by looking at the application of the computer in the [remainder] of this century.

The computer will be the basis for many systems important in our lives—communications, education, medicine, finance, economics, transportation, law, and politics. The computer, coupled with other technologies, such as communications satellites, microwave relays, cable television, will bring about new systems. We will have *telemedicine*, where diagnoses and health care are dispensed remotely via various forms of technology. *Tele-education* will come into existence, where learning laboratories will be established in the homes, based on microcomputers, various input and output devices, software programs, and linked to databases all over the world. Various institutions such as museums, universities, businesses, and other settings will be hooked up via the computer and telecommunications systems into an *educational tele-communication technological network* beamed to the child in his own home or some other environment, rendering the concepts of schools as we know them today obsolete. We will have *tele-finance*, where it will be possible to conduct financial and economic matters without money. In essence, we will be converting to a cashless society.

Numerous other illustrations of the application of computers, satellites, and various other combinations of technologies can be cited. While these accounts may sound fascinating, exciting, promising, and glowing we need to be mindful that widespread use of the computer and the various other technologies increases the chances of violation of certain basic civil liberties, especially the invasion of privacy. Since the computer is expected to be a dominating force for the rest of this century, we need to be critically aware of its potencies.

Technological spin-offs from the space program have already entered our daily lives, and we can expect to see other direct and indirect applications of space age technology to our ways of living. It is not inconceivable that in the twenty-first century space technology will have advanced and have been applied in such a way that we will have people living in outer space and in enclaves known as space colonies. Space shuttles between these extraterrestrial colonies and the earth will be as much a part of our transportation system as airplanes, cars and buses are today. We have already seen the beginnings of such operations with moon shots, the building of space platforms, the orbiting of various planets and with many of the other feats accomplished in outer space by the Russians and the Americans. Research and plans are already underway and have been so for several years now to put factories, health care facilities, farms, and the like in outer space, on other planets, or on the moon. It doesn't take much imagination to envision large numbers of people fleeing a polluted, decaying, junk-filled, overcrowded earth for the roominess and solitude of outer space cities and farms; just as in the 1950s and 60s, large numbers of people fled the cities for the suburbs. Nor does it take much imagination to see who will be left behind in this decayed, cancer-producing environment, with fouled rivers and oceans, depleted resources and bankrupt institutions as the endangered species. After outer space has been colonized, generations later Black people and other people of color may find themselves in a situation of trying to prevent resettlement of the earth by those who have fled it, just as we are now faced with the situation of others who fled the cities now trying to reclaim them.

Of all the technological developments and scientific breakthroughs that I have noted I find most disturbing and most frightening those that are related to the biomedical fields. With these developments we see not only the possibility of our life-styles, our culture, our social and political institutions being changed, or our environment being altered, but we face the very real possibility of having ourselves, as a species of people, transformed, or if you will, deformed into something else. Developments in the biomedical fields are occurring rapidly, but disturbingly, very quietly. Persons familiar with the occurrences in this field make no bones about the profound effects of biological metamorphosis. Pamela Sargent, in her book entitled *Bio-Futures*, states: "Lest we deliberately turn our backs on biological innovation, or restrict research, biological advances will change our institutions and our attitudes" (p. xviii). She goes on to note: "Biological changes could in time affect our notions of what a human being *is*, we would become many different and divergent species. Each designed for different tasks" (p. xix).

Let us pause and ask what tasks would Black people be assigned? What variety of species will we be destined to become? The means for such biological changes are already available to us through a variety of techniques. Cloning, the use of artificial wombs, test tube fertilization, and hybridization of animal species and humans are new ways of reproducing human beings other than the traditional way that we have customarily used. Other possible biological innovations include somatic alternation, which involves making changes in a grown person's biological construction, and genetic manipulation. People in research labs are now deep freezing tissues and organs. This work may lead to the preservation of sick people until they can be treated by newly discovered methods. As Sargent notes, "a more speculative possibility is that people now alive might be frozen and revived in a future when immortality is a reality." She goes on to say, "immortality itself or a greatly extended lifespan may be realized in the future" (p. xix).

As noted earlier I found these developments to be very frightening. Others share this view. Howard and Rifkin (1977), in a book entitled *Who Should Play God?* find genetic engineering more dangerous than nuclear technology. They write:

> For many years social commentators have looked upon nuclear weaponry as the most powerful and dangerous tool at the disposal of humanity. With the development of human genetic engineering, a tool even more awesome is now available. It is true that nuclear weaponry poses the ever present threat of annihilation of human life on this planet. But with genetic engineering there is a threat of a very different kind: that by calculation and planning, not accident or the precipitous passion of the moment, some people will make conscious and deliberate decisions to irreversibly alter the biological structure of millions of other men and women and their descendants for all time. This is a form of annihilation, every bit as deadly as nuclear holocaust, and even more profound—whatever forms of future beings are developed will be forced to live the consequences of the biological designs that were molded for them (pp. 9–10).

Biogenetic manipulations, cloning, test-tube babies and the like bode ill for those of us who have been victims of forced sterilization, brutal castration, psychosurgery, aversion therapy, biased psychological testing, calculated syphilis experiments, persistent campaigns of birth and population control, and many other insidious and inhumane forms of so-called scientific research and human experimentation.

Blacks have always been the favorite sources of supply for medical and scientific research. Transplants of organs in dead or dying Black bodies provide life for white recipients. We are targets of many too many theories positing our alleged mental, biological, genetic, and cultural inferiority. It is only natural and logical that if this movement of genetic engineering grows, we will be likely targets and victims of it.

NEEDED BLACK ACTION

Earlier in this paper, the need for Black political awareness of and participation in the public debates surrounding these scientific and technological issues was pointed out and stressed. Many of these various developments are still underway. The public and scientific debates on them are just beginning. The Black community, which has so much to suffer from these occurrences, should be participating in these debates and should be trying to influence these developments at both the policy and technical levels. I dare say, however, in terms of policy and technical competence, we are thoroughly unprepared to deal with these complex, but yet vital issues facing us today.

National Black organizations need to identify and pull together groups of committed, concerned and capable Black scientists, technologists, engineers, policy analysts, social scientists, community workers, and activists, to review these situations and to devise plans, strategies, and tactics for coming to terms with them. Long-term planning and in-depth analysis must go into the solutions and proposals needed to address these issues. Simply convening a conference, meeting, or seminar or a series of them is not adequate to muster the type of analysis and action demanded by the gravity of these issues. Instead, support must be provided that allows these scholars, technicians, practitioners, and laypersons to devote a considerable amount of time, thought, and energy to these concerns. I argue that these matters should be ranked with our other priorities, mainly because they are inseparable from our concerns for decent and fair living conditions for Black Americans.

On another level, that of educational preparation, we must begin to address ourselves to becoming prepared to live in an increasingly technological society and being able to master a social environment that depends more and more upon technology and science. In order to have sufficient person-power, capable of fathoming the depth of scope of these issues, capable of proposing countermeasures and of informing policy, we must begin to prepare our young for training in the sciences, mathematics, and in the various fields of technology.

Blacks, in order to be competitive in the arenas of scientific accomplishment and technological innovation, will have to be exposed to the sciences at an early age. They will have to be inspired to enter these fields motivated to achieve and given the confidence that allows all the rest to happen. In short, science and technology should become a part of our collective life-style as much as sports, music, dancing, and religion have become. We should seek to reclaim this part of our cultural heritage and to integrate it into our current living patterns. *Science and technology should be integral parts of our socialization process.* They should permeate our personal and social development as Black people. In some of my other writings I have proposed how this may be done, utilizing various strategies and techniques as well as our cultural heritage. And from my current research I am also attempting to identify factors which inhibit or facilitate our participation in math-related fields as a first step, and hope to move beyond this to develop programs that will enable us to overcome the fears and ignorances that we have about these disciplines.

Without these educational and political strategies and tactics, we are in a very vulnerable position vis-à-vis scientific developments and technological innovations in this society. If we do not begin to address ourselves to these issues, our chances for survival, be it in a rural, urban or extraterrestrial environment, are almost nil.

REFERENCES

Bennett, Lerone Jr. "Black America and 'The Energy Siege,' " *Ebony,* October, 1979, pp. 31–32, 34, 36, 38, 42.
Comstock, George; Chaffee, Stephen; Katzman, Natan; McCombs, Maxwell; and Roberts, Daniel. *Television and Human Behavior.* New York: Columbia University Press, 1978.
Flink, James J. *The Car Culture.* Cambridge, Mass.: MIT Press, 1975.
Frazier, E. Franklin. *The Negro Family in the United States.* Chicago: University of Chicago Press, 1939.
Howard, Ted, and Rifkin, Jeremy. *Who Should Play God?* New York: Dell Publishing Company, Inc., 1977.

Martin, E., and Martin, J. *The Black Extended Family.* Chicago: University of Chicago Press, 1978.

Sargent, Pamela (ed.). *Bio-Futures.* New York: Random House Inc., 1976.

Silverberg, Edwin, and Poindexter, Syril E. "Cancer Facts and Figures for Black Americans—1979." New York: American Cancer Society, Inc.

Toffler, Alvin. *Future Shock.* New York: Bantam Books, Inc., 1970.

U.S. Department of Health, Education and Welfare, Public Health Service, National Institutes of Health. "What Black Americans Should Know About Cancer." DHEW Pub. No. (NIH) 78–1635.

Wilhelm, Sidney M. *Who Needs the Negro?* Garden City, New York: Doubleday and Company, Inc., 1971.

20. Artifact/Ideas and Political Culture

LANGDON WINNER

Langdon Winner's essay, "Artifact/Ideas and Political Culture," first appeared in the Winter 1991 issue of Whole Earth Review, *a special issue of the magazine devoted to "Questioning Technology." In the selection, Winner, a political scientist and longtime student of technology and society, outlines alternative perspectives on the prospects for technological change. Expanding on an earlier essay in which he developed the notion that "artifacts have politics," Winner explains the consequences of different technological forms for such aspects of society as the distribution of power and wealth, social mobility, and gender equity. Rather than focusing on the need to innovate or to improve productivity and competitiveness, Winner suggests that our discussions of technology should cultivate ways of democratizing technology policy-making—involving all groups and social interests likely to be affected by technological development in defining what that technology will be.*

A prolific writer whose papers have appeared in both scholarly and popular magazines and who served as a contributing editor to Rolling Stone *magazine from 1969 to 1971, Langdon Winner is professor of political science in the Department of Science and Technology Studies at Rensselaer Polytechnic Institute in Troy, New York. Prior to coming to RPI in 1985, he taught at MIT, the University of Leiden (in the Netherlands), and the University of California at Berkeley and Santa Cruz. Among Winner's books are* Autonomous Technology *(1977) and* The Whale and the Reactor *(1988). He spent the 1991–92 academic year as a visiting research scholar at the Center for Technology and Culture, University of Oslo, Norway.*

Source: Selection from *Whole Earth Review*, No. 73 (Winter 1991), pp. 18–24. Whole Earth Review, 27 Gate Five Road, Sausalito, CA 94965. Subscriptions available for $20.00 a year. Reprinted by permission.

This is a time of great excitement about the fruitful possibilities of new technology, but also a time of grave concern about what those possibilities mean for the future of our society. Horizons visible in microelectronics and photonics, biotechnology, composite materials, computing, and other fields hold out prospects of sweeping change in our way of life. How should we regard these prospects?

As individuals, groups and nations anticipate technological change nowadays, they usually focus upon three questions.

First: How will the technology be used? What are its functions and practical benefits?

Second: How will the technology change the economy? What will it contribute to the production, distribution and consumption of material wealth?

Third: How will the technology affect the environment? What will its consequences be for global climate change, pollution of the biosphere, and other environmental problems?

While these are important issues, another crucial question is seldom mentioned: What kind of world are we building here? As we develop new devices, techniques and technical systems, what qualities of social, moral and political life do we create in the process? Will this be a world friendly to human sociability or not?

These are questions about the relationship of technological change to the evolution of modern political culture. In what ways do the development, adoption and use of instrumental things affect our shared experience of freedom, power, authority, community and justice? How might we respond creatively to the role technology plays in contemporary political life?

In the titles of a great many books, articles, and conferences these days, the topic is often described as "technology and society" or "technology and culture" or "technology and politics." But if one takes a closer look, such distinctions no longer have much validity. In the late twentieth century technology and society, technology and culture, technology and politics are by no means separate. They are closely woven together in a multiplicity of settings in which many forms of human living are dependent upon and shaped by technological devices and systems of various kinds. Our useful artifacts reflect who we are, what we aspire to be. At the same time, we ourselves mirror the technologies which surround us; to an increasing extent social activities and human consciousness are technically mediated.

In this light, any attempt to understand the matter might well begin

from either of two basic starting points: (1) the technological world
seen from the point of view of human beings and (2) the same world
seen from the point of view of the artifacts. Although it may seem
perverse to do so, I shall begin with the second perspective.

Many of the things that we like to think of as mere tools or instruments now function as virtual members of our society. It makes sense to
ask: Which roles, responsibilities and possibilities for action have been
delegated to technological things? Which social features are associated
with a particular artifact? For example, does a computer in the
workplace function as a servant, slave, controller, guard, supervisor,
etc.?

The social roles delegated to the phone answering machine provide a
good illustration. It used to be that only executives in business and
government could afford to keep a full-time secretary answering the
phone, screening calls and taking messages. Now it is possible to buy a
small, inexpensive answering machine that does at least some of that
work. An alternative would be to answer the phone yourself, have
someone else do it for you or simply miss some calls. The machine
serves as a surrogate, a kind of nonhuman agent that has been given
certain kinds of work to do.

An interesting fact about these machines is that their initial use
often brings some embarrassment. In the little taped message that
precedes the beep, there is often something like an apology. "I'm sorry I
can't be here to answer your call . . . " or "I'm sorry you have to talk to
this machine, but . . . ." What one sees in cases like this is, I believe,
quite common in modern life: the uneasy feeling that accompanies the
renegotiation of social and moral boundaries around a technological
change. But what is sometimes at first a source of discomfort eventually
becomes a widely accepted pattern—"second nature," if you will.

It is clear that in decades to come a great many things like telephone
answering machines and automatic bank tellers will become, in effect,
members of our society. As their use spreads, the tone of embarrassment that surrounds their early introduction will gradually vanish. For
better or worse, the renegotiation of boundaries will be complete.
When I phoned a friend recently, I heard a recorded message that said
simply: "It's 1991. You know what to do!"

One can also consider technological innovations from the alternate
viewpoint—noticing the roles, responsibilities and possibilities for action delegated to human beings within and around technological systems of various kinds. Now one can ask: Is a person's guiding hand

required for the system to function? Does the human give orders or receive them? Is the person active or acted upon? What social qualities accompany the human presence?

I will offer some illustrations in a moment. But first I want to call attention to the fact that once one has entered the twofold perspective I've suggested, one has the beginning of a social and political vision of technology quite different from the one that economists, engineers, and technology policymakers usually employ. One recognizes, first and foremost, that technologies are not merely tools that one "picks up and uses." They can be seen as "forms of life" in which human and inanimate objects are linked in various kinds of relationships. The interesting question becomes: How can we describe and evaluate technologies seen as "forms of life"?

By comparison, in the conventional view of things, the story usually goes that people employ technologies as simple tools for rather specific instrumental purposes, attempting to wrest new advantages over nature and to gain various economic benefits. Once these instrumental advantages and economic benefits have been obtained, other things may happen. There are what are called secondary, tertiary, and other distant consequences of our action, often called the "impacts" or "unintended" consequences, the broader social, cultural, political, and environmental effects of technological applications of various kinds.

For some purposes, it is perfectly acceptable to view technological change in the conventional manner. However, if you take a longer view of history, an interesting fact soon emerges. In the fullness of time, the so-called "secondary" consequences or impacts of technological change are often far more significant than the results thought to be "primary" at the time. This is certainly true, for example, of the kinds of changes we associate with the Industrial Revolution of the eighteenth and nineteenth centuries. One could list the thousands upon thousands of instrumental advantages and economic benefits obtained during that period—techniques for making textiles, extracting coal, making locomotives run, etc. But that is not what is truly important about the Industrial Revolution. What matters is the fact that a whole new kind of society was created. The truly enduring part of that revolution, the truly significant aspect is the multiplicity of relationships between people and between humans and technology we call Industrial Society, results many of which arose largely as so-called "secondary" consequences of technological change.

If one looks carefully at contemporary technological innovations in their broader human context, one often finds emerging forms of political culture. Several years ago Maevon Garrett, a woman who had worked as a telephone operator in Baltimore for 18 years, was called into her supervisor's office and abruptly fired. She was informed that a computer had been installed to monitor the performance of telephone operators and that data gathered by the computer showed that she was less efficient than the average worker in processing phone calls. At that moment Maevon Garrett became the victim of norms of productivity and efficiency embodied in the workings of a new technological system.

What is interesting, however, is not only the fact of Ms. Garrett's firing, but her response to it. She pointed out that some portion of her time each day was spent talking with people who dial a telephone operator because they are lonely or in distress—elderly people who live alone, or "latchkey children," youngsters who come home after school to an empty house because their parents are still at work. Ms. Garrett argued she would not hang up on such people just to meet the phone company's hourly quota.

It is reasonable to conclude that she was behaving responsibly, serving a role in civic culture, but not a role recognized by the norms of efficiency and productivity in the system that employed her. This is a case in which conditions of technical rationality and cultural rationality meet in flagrant conflict.

The good news is that after a union protest Maevon Garrett's job was restored. The bad news, however, is that the systems design, the technopolitical regime that caused the problem, still exists and looms before us as a rapidly spreading form of life. A study released by the Office of Technology Assessment of the U.S. Congress several years ago noted that approximately seven million American workers now live under rapidly spreading systems of computerized surveillance, an unhappy spin-off of office automation. The title of that report is, appropriately, *The Electronic Supervisor*. To an increasing extent in today's workplaces, computers are delegated the role of supervising; human beings have been assigned roles that involve working faster and faster while engaging in less social conversation—all in the name of a system called "communications," but one that drastically limits people's ability to communicate in a human sense.

The term "regime" seems perfectly appropriate in such cases. For once they have been designed, built and put in operation, sociotechnical systems comprise regimes with features that can be described in a political way. It makes perfect sense to talk about freedom or its absence, equality or inequality, justice or injustice, authoritarianism or democracy, and the kinds of power relationships technological instruments and systems contain.

This is true of extremely simple as well as complex technologies. For example, if one visits the agricultural fields of the southwestern U.S.A., one finds workers using a hoe, "el cortito," a tool with a short handle. There's nothing political about the length of a wooden handle, is there? Well, that depends on the broader social relationships and activities in which it plays a part. To use "el cortito" you must bend over or get down on your knees. A casual observer might say: If you're digging in the ground, isn't it sometimes more comfortable to stand up?

Why, then, has the handle been shortened? The reason is, in large part, that the foremen who manage the work can look across a field, even at a great distance, and tell who is working and who is not. Those who are bending over are the ones working; those standing upright are not and the foreman can apply discipline accordingly. In that light, even the length of the handle of a hoe expresses a regime, a regime of power, authority and control.

Embodied in the tools and instruments of modern technology is a political world. I am suggesting that we use metaphors and rhetorical devices of political speech to unpack the meaning of various technologies for how we live.

Everyone understands that political ideas can be expressed in language. But ideas of this kind present themselves in material objects as well. In this form they might be called artifact/ideas. In their very silence, artifact/ideas have a great deal to say. They tell us who we are, where we are situated in the social order, what is normal, what is possible, what is excluded. The technological world is filled with artifact/ideas of great consequence for modern political culture. Things often speak louder than words. Among the many ideas present in the structure of contemporary technological devices and systems are the following:

- Power is centralized.
- The few talk and the many listen.
- There are barriers between social classes.
- The world is hierarchically structured.
- The good things are distributed unequally.
- Women and men have different kinds of competence.
- One's life is open to continual inspection.

As they are expressed in the shape of material objects, ideas of this kind are covert. They seldom become topics for discussion in the political sphere as it is usually understood. One reason that artifact/ideas tend to be covert is that most people buy the functional account of the meaning of material things. We are inclined to say: "This is a car which enables us to go from point A to point B." "This is a hoe which helps us to dig in the fields."

Another reason why ideologies in things tend to be covert is that they have been implanted there by those who do not wish those ideas to be known or widely discussed. The apparent solidity of useful things sometimes provides a mask for persons and groups who wish to exercise power while avoiding responsibility. Their alibi is usually something like: "This is the most effective way to do things" or "This is most efficient."

But whatever the source of specific beliefs and instrumental conditions, it is often true that ideas embodied in material things are painful or even dangerous to acknowledge. Artifact/ideas can involve astonishing contradictions. In particular, the mapping of the world encountered in the shape of things frequently contradicts the political ideology to which most people in Western societies claim to be committed.

In particular, many of the artifact/ideas prevalent in our time stand in flagrant contradiction to the ideology of modern democracy. That ideology holds that human beings flourish, achieving what is best in their potential, under conditions of freedom, equality, justice, and self-government. In that light, societies ought to create social conditions and political institutions that make it possible for each human being's potential to develop. Both victories and setbacks in this regard are clearly visible in the laws, constitutions, and political practices that prevail in each historical period.

From this vantage point a technological society is unique only in the sense that it presents new and seemingly unlikely domains—domains of instrumentality—in which the ends of democratic freedom, equality and justice must somehow be recognized and realized. I take it to be the fundamental failure of modern civilization to have ignored again and again how such questions present themselves in the guise of what appear to be "neutral" technologies. To a considerable extent the ideas embodied in the realm of material things stand in opposition to the central ideas that we believe describe and guide our political culture.

There is an important way in which freedom and justice depend in human communities upon the existence of suitable material environments—the creation and maintenance of arrangements in which the goal of becoming free, self-determining individuals is nurtured rather than destroyed. As we look at the kinds of sociotechnical innovations being introduced today, it is often beside the point to ask whether or not they are optimally efficient; by someone's definition they are usually very efficient indeed. Instead the crucial questions concern the kinds of cultural environments such technologies present to us. What one finds are far too many instances of developments of the following kind:

1. communications technologies employed in attempts to control people's thoughts, desires and behaviors;
2. computer technologies used to whittle away people's privacy and erode freedom;
3. information technologies that eliminate what were formerly places of community life;
4. energy systems that make people dependent upon, or even hostage to, sources of fuel over which they exercise no control;
5. systems of manufacturing that seek control by eliminating as much human initiative and creativity as possible.

The appropriate moment to examine and debate conditions such as these is the time during which they are designed and first introduced into the fabric of human activity. At present our society persists in designing a great many technical artifacts in ways that make people feel passive, superfluous, stupid, and incapable of initiating action. Such systems bear the cultural embryos of tomorrow's citizenry. For as we invent new technical systems, we also invent the kinds of people who will use them and be affected by them. The structures and textures of future social and political life can be seen in the blueprints of technologies now on the drawing board.

We often hear these days that the world is engaged in a "technology race" in which nations rise or fall according to their ability to use technologies to competitive advantage. Unfortunately, some of the design strategies that look fabulous from the point of view of efficiency, productivity and global competitiveness involve what amounts to an ingenious synthesis of oriental feudalism and capitalism. Many people in freedom-loving countries like the United States seem eager to embrace repressive models of social integration expressed in automation, electronic surveillance and pseudodemocratic "quality circles." But must we embrace these merging patterns of technofeudalism as "the wave of the future"? Would it not be a wiser approach to resist, choosing to explore ways of extending our ideas about freedom and a just society into the realm of technology itself?

In fact, one obvious path that may still be open to us is to cultivate ways of democratizing the process of technology policymaking and, indeed, the process of technological innovation. If this is to be done, both citizens and experts will need to become aware of the social, moral and political dimensions of choices made in technological policy and technological design. They will need to find ways to act directly and democratically within settings in which the important choices are made.

In that light I would offer three guiding maxims as a way to focus discussion about the relationship between technological choices and the future of political culture. These maxims can be raised at times in which unquestioned assumptions about "productivity," "competitiveness," "the need to innovate," or "technology transfer" seem to provide the only language for talking about the choices at hand.

1. *No innovation without representation.* This suggests that all the groups and social interests likely to be affected by a particular kind of technological change ought to be represented at a very early stage in defining what that technology will be. Yes, let us accept the idea that particular technologies are social creations that arise through a complex, multicentered process. But let us see to it that all the relevant parties are included rather than kept in the dark in this process. If we find that we do not have the kinds of social institutions that make this possible, then let's change our institutions to create such opportunities.

2. *No engineering without political deliberation.* Proposed technological projects should be closely examined to reveal the covert political conditions and artifact/ideas their making would entail. This ought to become an interpretive skill of people in all modern societies. It is especially important for engineers and technical professionals whose wonderful creativity is often accompanied by an appalling narrow-mindedness. The education of engineers ought to prepare them to evaluate the kinds of political contexts, political ideas, political arguments, and political consequences involved in their work. Skill in the arts of democratic citizenship ought to become part of the "tool kit" that engineers master in their education.

3. *No means without ends.* Many of the varieties of innovation now pushed on the public these days amount to "tools looking for uses," "means looking for ends." Those who have dealt with the introduction of computers into the schools in recent years can give many colorful examples of this phenomenon. The current promotion of high definition television and renewed efforts to push President Reagan's Star Wars project offer even more stark illustration. For HDTV and SDI bear little relationship to any significant human need. As we study the prospects offered by new technologies, it is always essential to ask: Why? Why are we doing this? What are the ends we have chosen and how well do they fit the pattern of means available? In many cases of high tech planning, suitable background music would be the theme from *The Twilight Zone.*

If you were to look for examples of places in which something similar to these three maxims are actually being put to work, I would begin by pointing to some recent experiments in the Scandinavian democracies where a positive, creative politics of technology has recently become a focus of research and development. In one such project, workers in the Swedish newspaper industry—printers, typographers, lithographers, and the like—joined with representatives from management and with university computer scientists to design a new system of computerized graphics used in newspaper layout and typesetting. The name of the project was UTOPIA, a Swedish acronym that means "training, technology and products from a skilled worker's perspective."

UTOPIA's goal was to fashion a system that would be highly advanced technically, but also one designed in ways that would take into account the skills, needs and perspectives of all those who would eventually be using it. Rather than develop a system under management directives and then impose it on workers, the project included representation of the people concerned. UTOPIA became the focus of a rigorous program of research and development at a government-sponsored laboratory: The Center for Working Life in Stockholm. Here was a case in which the purely instrumental and economic thrust of a technological innovation encountered a legitimate set of political ends and enlightened artifact/ideas. The result was democratization expressed in hardware, software and human relationships.

The technological world of the twenty-first century beckons. Will it be better than the one we now inhabit or worse? Will it realize the promise of human freedom or curtail it? And whose interests will be decisive?

If ordinary citizens are to be empowered in shaping the world to come, we must become very skillful in areas where we are now profoundly ignorant, using ideas and abilities that enable us to define and realize human freedom and social justice within the realm of technology itself: within things like new machines for the workplace, computerized systems of information management, biotechnologies in agriculture and medicine, communications devices introduced into our homes. If we cannot develop these skills or do not care to, if we fail to confront the world-shaping powers that new technologies present, then human freedom and dignity could well become obsolete remnants of a bygone era.

Part IV
USING TECHNOLOGIES AND
CONFRONTING THEIR DILEMMAS

Specific, tangible technologies—some in existence today, others on the near horizon—are the subject of the readings in the last section of *Technology and the Future*. Throughout the first three parts of the book, nearly all of the selections address technology as a concept rather than technologies as specific devices, instrumentalities, or systems. In a sense, the first three parts represent the theory of this book, while the selections in Part IV are the case studies, the practical examples. The readings here bring out many of the same issues discussed in the earlier selections—including power and control in society, the role of work in human fulfillment, equity, and the distinction between ends and means—all in relation to specific existing (or developing) technologies. As elsewhere in the book, there are more problems posed than solutions offered. The issues raised by these technologies do not have easy answers.

Genetic testing is an ideal place to start. Advances in molecular genetics, spurred by developments in biotechnology over the past two decades, have begun to raise a host of ethical and legal dilemmas—dilemmas that humankind is likely to be wrestling with over the next several decades. These are explored with great lucidity in the opening essay of Part IV, "The Dark Side of the Genome," by biologist Robert Weinberg.

Two readings on computers and society follow. In "Computer Ethics," Tom Forester and Perry Morrison survey the social problems and the range of ethical issues raised by developments in information technology—issues that face computer users as well as computer professionals. Sherry Turkle's selection, "Identity in the Age of the Internet," takes a very different approach to related issues, examining the impacts of computer-mediated communication on individual identity as well as the cultural implications of life in cyberspace.

Technology and work is the subject of the next two essays. Ruth Schwartz Cowan examines the relatively low technology of the American household and explains why "labor-saving" technology has not resulted in less household work for women. Shoshana Zuboff reports on a major study of how information technology ("the smart machine") is influencing the working lives of men and women and the nature of work itself.

Finally, from a special issue of *Wired* magazine on the future, Douglas Coupland presents a lighthearted—yet provocative and ultimately instructive—checklist of things that you might want to take on a trip to the year 2195. The future awaits. As Coupland says, "Bon voyage."

21. The Dark Side of the Genome

ROBERT A. WEINBERG

Among the most rapid and important scientific advances of the past two decades have been developments in molecular biology. The breaking of the genetic code and the development of new techniques to analyze genetic materials have given scientists the ability to understand the relationship between the biochemical building blocks of cells and the traits and characteristics of living organisms, including humans.

During the past several years, life scientists in several countries have begun a coordinated, systematic effort to create a complete biochemical description of the human genome (i.e., the DNA contained in the chromosomes in human cells) and to develop a map or atlas indicating which components of this genetic material determine which human traits, from susceptibility to particular disorders to eye color to mathematical or artistic ability. Already, geneticists have identified the location of genes associated with dozens of disorders, including cystic fibrosis, fragile-X syndrome (a form of mental retardation), and Huntington's disease.

These new capabilities offer the prospect of eliminating a great deal of human suffering, but they also present some serious ethical dilemmas and risks to society. Use of genetic information by insurance companies, by employers, and by government agencies could infringe on individual rights to privacy and could even make it difficult for some people to get health insurance, find employment, or find a marriage partner.

Robert A. Weinberg, one of the leading figures in molecular genetics, discusses some of these perplexing issues in his essay, "The Dark Side of the Genome." Weinberg is a professor of biology at MIT and a member of the Whitehead Institute for Biomedical Research. His laboratory was among the first to recognize the existence of human oncogenes, which are responsible for converting normal cells into cancer cells. Weinberg holds

Source: *Technology Review* (April 1991). Copyright © 1991. Reprinted with permission of *Technology Review.*

a Ph.D. in biology from MIT. He is a member of the National Academy of Sciences and the recipient of a long list of honors, scientific prizes, and honorary degrees.

In the past 10 years biology has undergone a revolution that has repeatedly attracted wide attention. At first, controversy swirled over whether the genetic cloning technology that powers this revolution could create new and possibly dangerous forms of life. These fears have dissipated as thousands of investigators have found that the organisms created by gene splicing pose no threat to human health or the ecosystem around us.

A much larger stream of headlines next touted the power of genetic engineering to produce great quantities of valuable medical and agricultural products cheaply. Without doubt, over the next decades these fruits of biotechnology will enormously benefit health and economic productivity.

Largely lost amidst these stories, however, are developments that will ultimately have a far larger social impact. Recently gained abilities to analyze complex genetic information, including our own, will soon allow us to predict human traits from simple DNA tests. By the end of the 1990s, routine tests will detect predispositions to dozens of diseases as well as indicate a wide range of normal human traits. We have only begun to confront the problems engendered by the power of genetic diagnosis.

Consider, for example, the societal problems that will likely develop from the recent isolation of the gene that in a defective form causes cystic fibrosis. Genetic counselors can now trace that version of the gene in families, thereby revealing those couples who could have children with cystic fibrosis. While providing extraordinarily useful information for cystic fibrosis carriers, this technique raises questions about the marriageability and reproductive decisions of gene carriers, and the terms under which their offspring will be able to obtain health and life insurance.

Individual successes like the isolation of the cystic fibrosis gene will soon be overshadowed by the avalanche of genetic information flowing out of research labs. The engine that will drive these advances in gene analysis is the biologists' moonshot, the Human Genome Project. (See box.) The ambitious goal of this international effort is to read out the sequence of the 3 billion bases of DNA that, strung end to end, carry the information of all the body's genes. Given a clear, easily read atlas

DNA and Babylonian Tablets

To find every human gene, scientists will have to determine the sequence of the 3 billion characters in our DNA that together form the genetic blueprint known as the human genome. One can convey how daunting the effort will be by comparing the genome to a Babylonian library uncovered in some nineteenth-century archeological dig.

Imagine tens of thousands of clay tablets—individual genes—scattered about, each inscribed with thousands of cuneiform characters in a language with few known cognates. The library's chaos mirrors that encountered when the precisely ordered array of DNA molecules that is present in a living cell is extracted and introduced into a test tube. Imagine, too, that the library's full meaning will be understood only when most of its tablets have been deciphered.

Geneticists today have ways of laboriously sifting through heaps of "tablets" to find certain genes of special interest. Once a gene is located and retrieved, or "cloned," the sequence of its 5,000 or more bases of DNA—our cuneiform characters—can be determined.

While biologists are proud of having sequenced more than one percent of the "tablets" so far, these achievements represent only a piecemeal solution to a very large problem. Gene cloning and sequencing techniques developed in the 1970s are so time-consuming and painstaking that systematic searches for many genes have been impossible.

A better answer, in the form of the human genome project, will begin by mapping the genome—cataloguing all the Babylonian tablets. In effect, geneticists will gather and systematically shelve the scattered tablets, reconstructing their original order.

Initially, groups of tablets (DNA fragments) thought to derive from a common section (chromosomal region) of the library will be placed together on a shelf. Then geneticists will order the tablets within a group and give each a label. They will do so without any understanding of the tablets' contents.

How is this possible? Imagine that our Babylonian scribes have used the final phrases at the end of one tablet as the opening phrases of the next one. Short redundant strings of characters would enable tablets to be shelved in the right order without any knowledge of the bulk of the text. Long, carefully ordered lists of the labels identifying individual tablets, in effect a complete library catalogue, will compose the human genome map.

Only after this work is completed can the reading of all the characters in each tablet proceed—the sequencing of the DNA bases. Great technical progress will be required before the work becomes economically viable.

(continued)

Sequencing a 1,000-base stretch of DNA now costs $5,000 to $10,000. And some genes are giants; the one involved in muscular dystrophy was recently found to encompass 2 million bases. The cost will have to drop by a factor of 10 through automation before sequencing can begin in earnest.

Think of the technology required to develop automated readers that could photograph 3,000-year-old tablets, analyze and read the characters with greater than 99 percent accuracy, flag ambiguous ones, and introduce everything into a computerized database. The details of the automated DNA-sequencing equipment under development differ, but the technical problems are no less challenging.—*Robert A. Weinberg*

of our genetic endowment, researchers will be able to accelerate the rate at which they discover important genes—now several dozen each year—by 10-fold and eventually maybe even 100-fold. Scientists will then be able to study how the normal versions of these genes work, and how their aberrant versions cause disease.

Some fear that by reading through the entire library of human gene sequences we will rapidly come to understand the ultimate secrets of life and the essence of our humanity. For my part, such fears are far astray of the mark. Our bodies function as complex networks of interacting components that are often influenced by our variable environment. By enumerating and studying individual components—genes, in this case—we will only begin to scratch the surface of our complexity.

Nonetheless, certain genes can be especially influential in determining one or another aspect of human form and function. Herein lie the seeds of the substantial problems we will begin to encounter over the next decade.

MAPPING THE GENETIC TERRAIN

Ten to fifteen years from now—barring unforeseen technical obstacles—scientists will have described every bump in our complex genetic terrain. Yet long before this project is finished, information yielded by "mapping" this landscape—breaking it into sectors of manageable size and placing them in a logical array—will make possible powerful genetic analysis techniques. These, in turn, will engender a host of ethical issues.

To understand why, it is important to know a little about the underlying biology. The human genetic landscape—our genome—consists of all the DNA information carried on the 22 pairs of chromosomes in our cells plus the X and Y chromosomes involved in determining sex. Each chromosome carries a linear molecule of DNA ranging in size from 50 million to 250 million pairs of four kinds of chemical bases. They are commonly referred to by the letters A, C, G, and T. In all, 3 billion base pairs of DNA lie on the chromosomes. Some 50,000 to 100,000 discrete segments of DNA—each several thousand or more base pairs long—constitute the genes that store our genetic information. The trick is to figure out where these genes lie, and what information each encodes.

As a first step in understanding this enormous information base, investigators have started mapping each chromosome by labeling small segments along its length. The labels used are actually built-in features of the genome. They consist of minor genetic variations called polymorphisms that occur frequently throughout human DNA sequences and distinguish one person's DNA from another's. For example, at a certain chromosomal site, one person's DNA bases may read AAGCTT while a second person's may read AAGTTT. Such polymorphisms, widely scattered throughout the genome, are readily detected using existing techniques, even without any knowledge of the genome's detailed structure.

Polymorphisms are not only important for their usefulness in marking the genome at specific places. The location of a particular gene in the human genome is usually obscure. Geneticists can track down such a gene by localizing it near one or another polymorphic marker. To do this, they ascertain the presence of markers in DNA samples collected from members of large families and even large, unrelated populations.

Researchers have already used a polymorphic marker to determine the rough location of the gene that in one variant form, or "allele," leads to Huntington's disease. This illness appears as a severe neurological deterioration in midlife. Within a large kin group studied in Venezuela, all the relatives showing the disease were found to carry a distinct polymorphic marker on a particular chromosome, while their middle-aged, disease-free relatives did not. This concordance means that the still unknown gene lies close to the polymorphic marker on that chromosome, and therefore that detection of the marker signals the presence of the gene that causes the disease. The marker will prove invaluable in helping researchers to directly identify the Huntington's gene, isolation of which offers the only real hope for understanding and treating the disease.

Genes linked to terrible diseases are not the only ones geneticists study. During the next 10 years, researchers may well make associations between polymorphic markers and normal, highly variable traits such as height, eye color, hair shape, and even foot width without knowledge of the genes that serve as blueprints for these traits. Not much further down the road, scientists may uncover links between certain markers and more complex, subtle traits, such as aspects of physical coordination, mood, and maybe even musical ability. At that point, we will confront social problems that will bedevil us for decades to come.

Imagine that investigators could predict with some accuracy certain aspects of intelligence through simple analysis of an individual's DNA. Consider the power this would give some people and the vulnerable position in which it would put others.

The magnitude of the problems of genetic diagnosis depends on one's view of how many complex human traits will be successfully associated with polymorphic markers. Some observers, such as geneticists Richard Lewontin and Jonathan Beckwith of Harvard University, believe that few such associations will be made correctly. Some people argue that traits such as perfect pitch and mathematical ability depend on the workings of dozens of genes. Yet others think that the contributions of nature and nurture can never be teased apart.

Most likely, the doubters will be correct in many cases but wrong in others. Mathematical analysis has led some geneticists to conclude that the expression of many complex traits is strongly influenced by the workings of a few genes operating amid a large number of more silent collaborators. Moreover, scientists can most easily explain rapid organismic evolution, such as humans have experienced over the last several million years, by attributing important roles to a small number of especially influential genes. According to this hypothesis, each such gene has undergone alterations over the course of evolution that have in turn resulted in profound changes in our embryological development and adult functioning.

For these reasons, I believe that a number of genetic markers will be strongly linked to certain discrete aspects of human behavior and mental functioning. Yet other traits will, as some argue, prove to be influenced by many interacting genes and the environment, and will not lend themselves to the genetic analysis soon to be at our fingertips.

What type of higher functions will be understood and predictable by genetic methods? One can only speculate. The list of possibilities—say, shyness, aggressiveness, foreign-language aptitude, chess-playing ability, heat tolerance, or sex drive—is limited only by one's imagination. Likewise, the consequences of one or another identification—and there will surely be some successes—can barely begin to be foreseen.

THE LONG REACH OF GENETIC SCREENING

From cradle to grave—even from *conception* to grave—the coming genetic diagnostic technology will have profound effects on our descendants' lives. Parents-to-be in the latter part of the 1990s will confront an ever-lengthening menu of prenatal genetic tests that will affect a variety of reproductive decisions. Terminating a pregnancy may come relatively easily to some whose offspring carry genes dooming them to crippling diseases that appear early in life, such as Tay-Sachs and cystic fibrosis. But the mutant gene leading to Huntington's disease usually permits normal life until one's 40s or 50s, typically after the trait has been passed on to half of the next generation. Will its detection in a fetus justify abortion?

As the years pass, this gray area of decision making will widen inexorably. Sooner or later, an enterprising graduate student will uncover a close association between a polymorphic marker and some benign aspect of human variability like eye color or body shape. And then genetic decision making will hinge on far more than avoiding dread disease.

Such knowledge and the tests it makes possible could lead to eugenics through elective abortion. In India, thousands of abortions are said to be performed solely on the basis of fetal sex. It would seem to be but a small step for many to use the genetic profile of a fetus to justify abortion for a myriad of other real or perceived genetic insufficiencies.

This prospect may appear remote, seemingly encumbered by complicated laboratory procedures that will limit these analyses to a privileged elite. And the revulsion built up against eugenics would seem to present a significant obstacle. But the onward march of technology will change all this. Current programs for developing new diagnostic instruments should, by the end of this decade, yield machines able to automatically detect dozens of markers in a single, small DNA sample. As genetic diagnosis becomes more automated, it will become cheap and widely available. And the responsibility for children's genetic fitness will shift from the uncontrollable hand of fate into the hands of parents. By 1999, the birth of a cystic fibrosis child will, in the minds of many, reflect more the negligence of parents than God's will or the whims of nature.

Still other specters loom as the coming generation matures. Twenty-five years hence, educators and guidance counselors intent on optimizing educational "efficiency" could find children's genetic profiles irresistible tools. Once correlations are developed between performance and the frequency of certain genetic sequences—and once computers can forecast the interactions of multiple genes—such analyses could be used in attempts to predict various aspects of cognitive function and general educability.

The dangers here are legion. Some will use tests that will at best provide only probabilistic predictors of performance as precise gauges of competence. And factors strongly affecting education, including personality and environment, will likely be overlooked, leading to gross misreadings of individual ability.

Only slightly less insidious could be the effects of genetic analysis on future marriages. Will courtships be determined by perceptions of the genetic fitness of prospective partners? Over the past decade, how many Jewish couples who have discovered that their children could be born with Tay-Sachs disease, and black couples with similar concerns about sickle-cell anemia, have opted to forgo marriage altogether? As we uncover genes affecting traits that fall well within the range of normal variability, will these too become the object of prenuptial examination?

Once again, such an Orwellian vision would seem to reach far beyond current realities. Yet nightmares have already occurred. Two decades ago, genetic screening among the population in central Greece for the blood disease sickle-cell anemia revealed a number of normal individuals carrying genes that predispose their offspring to the disease. Because the test results were inappropriately disclosed, these individuals became publicly identified and stigmatized, and formed an unmarriageable genetic underclass.

Along with facing new issues around marriage, young adults with unfavorable constellations of genes may be limited in their employment possibilities. Employers want to hire productive, intelligent people. Will they exploit genetic screening to decide how rapidly a prospective employee will adapt to a new job or contribute to a company's productivity?

Even more likely will be attempts to use genetic markers to predict

susceptibility to dangers in the workplace. People have different toler-
ances to on-the-job chemical exposures, dictated by their genetic variability. There is therefore great interest in uncovering polymorphic
markers that would allow companies to predict employees' susceptibility to certain chemicals encountered in the workplace.

Employers will also feel pressure to use the expanding powers of
genetic diagnosis to predict lifelong disease susceptibility among workers. The staggering rise in health-insurance costs has already generated
strong economic incentives for employers to improve the health of
their workers by promoting smoke-free environments, routine medical
screening, and healthy life-styles. Hiring only those people who pass
genetic profile tests might be seen as a means to reduce health-
insurance costs further.

Most employers have until now been unwilling to enter so deeply
into employees' private lives. But insurance providers have shown no
such reticence. For example, many have been interested in learning
whether their insured carry the AIDS virus. Genetic tests predicting
heart disease at an early age or susceptibility to cancer will be tempting
targets for insurance companies intent on establishing allocation of risk
and premiums as precisely as possible. Such logic might dictate that the
risks now shared within large insurance pools should be allocated instead on the basis of individual genetic profiles.

Genetic profiles could be widely available by the year 2000, when
many primary-care physicians will routinely order certain genetic tests
along with the usual blood pressure reading and urinalysis. Overlook-
ing a standard genetic test will increasingly be seen as tantamount to
malpractice. And as genetic profiles are routinely entered into health
records, limiting insurers' access to such data may prove difficult.

Surely legislation could limit the direct viewing of confidential genetic data by insurers, but they might circumvent even the best attempts at regulation. Imagine the health policy of 2001 that offers
substantially reduced premiums to nonsmokers having a desirable genetic makeup. Such incentives will drive many people to flaunt their
DNA profiles. As the genetically fit flock to the low-risk, low-premium
pool, those left behind will have to pay higher premiums or even forgo
insurance. In time, the concept of pooling genetic risk will seem a
quaint relic of a pretechnological era.

While these developments are unsettling and even frightening, they
pale beside the possibility that our ever-advancing understanding of
human genetics could stoke the fires of racism.

Imagine in the not-so-distant future a survey of the prevalence of certain polymorphic alleles among different ethnic and racial groups. Ten years from now, will our enterprising graduate student find a polymorphic marker correlated with acute visual perception that is unusually common among Tibetans, or another correlated with impaired mathematical ability that crops up frequently among coastal Albanians? Given the vagaries of human history and population genetics, it is more than likely that different versions of genes are unevenly distributed throughout the human species.

Will such ostensibly innocent measurements of distributions of polymorphic markers ultimately provide a scientific basis for the type of virulent racism that inflamed Europe a half-century ago? Nazi racial theories were based on a pseudoscience that today looks ludicrous. But surely some observed variations in gene frequencies will place solid scientific data in the hands of those with an openly racist agenda.

BEYOND LEGISLATION

Policies governing the use of genetic information need to be debated and put in place early in this decade, not after problems emerge. Bioethics is already a thriving cottage industry, but the problems many of its practitioners wrestle with—issues like surrogate motherhood and *in vitro* fertilization—will be dwarfed by those surrounding genetic analysis. The groups organizing the human genome project have already assembled experts to confront the ethical, legal, and social dimensions of this work. But these individuals have yet to plumb the depths of the problems.

Even if we as a society can anticipate and rein in most misuse of genetic data, we will also need to address a more insidious and ultimately far more corrosive problem of DNA profiling: the rise of an ethic of genetic determinism.

For the past century, the prevailing winds of ideology have largely driven the ebb and flow of the nature versus nurture debate. A widespread reaction against social Darwinism and Nazi racism buoyed the strong nurturist sentiments of the past half-century, but the tide is turning, pushed by the ever more frequent success of genetics. As this decade progresses, a growing proportion of the lay public will come to accept genes as the all-powerful determinants of the human condition. This uncritical embrace of genetics will not be deterred by scientists' reminders that the powers of genetic predictions are limited.

Even some experts who, through appropriate channels, will gain access to genetic profiles may overinterpret the data. DNA profiles will never be clear, fully reliable predictors of all traits. For many complex traits, such as those involved in behavior and cognition, genetics will at best provide only a probability of development. After all, many traits are governed by the interplay between genetics and the environment. Environmental variations can cause genetically similar individuals to develop in dramatically different ways. Interpreters of genetic information who overlook this fact will repeatedly and disastrously misjudge individual ability.

What a tragedy this would be. The world we thrive in was built by many people who were not shackled by their pedigree. They saw their origins as vestiges to be transcended. By and large, we Americans have viewed our roots as interesting historical relics, hardly as rigid molds that dictate all that we are and will be. What will come of a world view that says people live and struggle to fulfill an agenda planned in detail by their genes? Such a surrender to genetic determinism may disenfranchise generations of children who might come to believe that genes, rather than spunk, ambition, and passion, must guide their life course.

A belief that each of us is ultimately responsible for our own behavior has woven our social fabric. Yet in the coming years, we will hear more and more from those who write off bad behavior to the inexorable forces of biology and who embrace a new astrology in which alleles rather than stars determine individuals' lives. It is hard to imagine how far this growing abdication of responsibility will carry us.

As a biologist, I find this prospect a bitter pill. The biological revolution of the past decades has proven extraordinarily exciting and endlessly fascinating, and it will, without doubt, spawn enormous benefit. But as with most new technologies, we will pay a price unless we anticipate the human genome project's dark side. We need to craft an ethic that cherishes our human ability to transcend biology, that enshrines our spontaneity, unpredictability, and individual uniqueness. At the moment, I find myself and those around me ill equipped to respond to the challenge.

22. Computer Ethics

TOM FORESTER AND PERRY MORRISON

The technologies of computers, computer networks, and information processing figure in numerous readings throughout this book. As Forester and Morrison put it, "Computers are the core technology of our times." During the past forty-five years (and especially in the last two decades), they have become central to the functioning of our society. As we have become increasingly dependent on computers and networks, we have become more vulnerable to their malfunctions and their misuse. Examples abound. Forester and Morrison cite case after case, including hacker break-ins to military computers, software bugs causing aircraft accidents, software and hardware sabotage disruptions of telephone service, and the shutdown of four major U.S. air traffic control centers caused by a farmer cutting a fiber optic cable while burying a dead cow.

The pervasiveness of computer technology and its susceptibility to misuse and malfunction raise a great many ethical, social, and legal issues. How can the intellectual property rights of software developers be protected when copying software is easy and widely practiced? Are hackers criminals or just pranksters? Is electronic mail private, or do employers have the right to monitor their employees' communications? Are computer professionals legally or ethically responsible for the consequences of flaws in the systems they have created? These and other questions are the subject of Forester and Morrison's highly readable and provocative introduction to the increasingly important subject of computer ethics.

Until his death in late 1993, Tom Forester was a senior lecturer in the School of Computer and Information Technology at Griffith University in Queensland, Australia. He was the editor or author of seven books on

Source: From *Computer Ethics: Cautionary Tales and Ethical Dilemmas in Computing*, Second Edition. (Cambridge, MA: The MIT Press, 1993), pp. 1–22. Reprinted by permission of MIT Press.

social aspects of computing. Perry Morrison lectures on psychology at the National University of Singapore.

Computers are the core technology of our times. They are the new paradigm, the new "common sense." In the comparatively short space of forty years, computers have become central to the operations of industrial societies. Without computers and computer networks, much of manufacturing industry, commerce, transport and distribution, gov- ernment, the military, health services, education, and research would simply grind to a halt.

Computers are certainly the most important technology to have come along this century, and the current Information Technology Revolution may in time equal or even exceed the Industrial Revolution in terms of social significance. We are still trying to understand the full implications of the computerization that has already taken place in key areas of society such as the workplace. Computers and computer-based information and communication systems will have an even greater impact on our way of life in the next millennium—now just a few years away.

Yet as society becomes more dependent on computers and computer networks, we also become more and more vulnerable to computer mal- functions (usually caused by unreliable software) and to computer misuse—that is, to the misuse of computers and computer networks by human beings. Malfunctioning computers and the misuse of computers have created a whole new range of social problems, such as computer crime, software theft, hacking, the creation of viruses, invasions of privacy, overreliance on intelligent machines, and workplace stress. In turn, each of these problems creates ethical dilemmas for computer professionals and users. Ethical theory and professional codes of ethics can help us resolve these ethical dilemmas to some extent, while comput- ing educators have a special responsibility to try to ensure more ethical behavior among future generations of computer users.

OUR COMPUTERIZED SOCIETY

When computers hit the headlines, it usually results in bad publicity for them. When power supplies fail, phone systems go down, air traffic control systems seize up, or traffic lights go on the blink, there is nearly always a spokesperson ready to blame the problem on a luckless com-

316 Technology and the Future

puter. When public utilities, credit-checking agencies, the police, tax departments, or motor vehicle license centers make hideous mistakes, they invariably blame it on computer error. When the bank or the airline cannot process our transaction, we're told that "the computer is down" or that "we're having problems with our computer." The poor old computer gets the blame on these and many other occasions, although frequently something else is at fault. Even when the problem is computer-related, the ultimate cause of failure is human error rather than machine error, because humans design the computers and write the software that tells computers what to do.

Computers have been associated with some major blunders in recent times. For instance, the infamous hole in the ozone layer remained undetected for seven years because of a program design error. No less than twenty-two U.S. servicemen died in the early 1980s in five separate crashes of the U.S. Air Force's Blackhawk helicopter as a result of radio interference with its novel, computer-based fly-by-wire system. At least four people died in North America because of computer glitches in the Therac-25 cancer radiotherapy machine, while similar disasters have been reported recently in England and Spain. During the 1991 Gulf war, software failure in the Patriot missile defense system enabled an Iraqi Scud missile to penetrate the U.S. military barracks in Dhahran, killing twenty-eight people, while the notorious trouble with the Hubble space telescope in the same year was exacerbated by a programming error that shut down the onboard computer.[1]

In fact, computers have figured one way or another in almost every famous system failure, from Three Mile Island, Chernobyl, and the Challenger space shuttle disaster, to the Air New Zealand antarctic crash and the downing of the Korean Air Lines flight 007 over Sakhalin Island, not to mention the sinking of HMS *Sheffield* in the Falklands war and the shooting down of an Iranian Airbus by the USS *Vincennes* over the Persian Gulf. A software bug lay behind the massive New York phone failure of January 1990, which shut down AT&T's phone network and New York's airports for nine hours, while a system design error helped shut down New York's phones for another four hours in September 1991 (key AT&T engineers were away at a seminar on how to cope with emergencies). A whole series of aerospace accidents such as the French, Indian, and Nepalese A320 Airbus disasters, the Bell V-22 Osprey and Northrop YF-23 crashes, and the downing of the Lauda Air Boeing 767 in Thailand has been attributed to unreliable software in computerized fly-by-wire systems. Undeterred, engineers are now developing sail-by-wire navigation systems for ships and drive-by-wire systems for our cars.[2]

Computers and computer networks are vulnerable to physical breaches such as fires, floods, earthquakes, and power cuts—including very short power spikes or voltage sags ("dirty power") that can be enough to knock out a sensitive system. A good example was the fire in the Setagaya telephone office in Tokyo in 1984 that instantly cut 3,000 data and 89,000 telephone lines and resulted in huge losses for Japanese businesses. Communication networks are also vulnerable to inadvertent human or animal intervention. For instance, increasingly popular fiber optic cables, containing thousands of phone circuits, have been devoured by hungry beavers in Missouri, foxes in outback Australia, and sharks and beam-trawling fishermen in the Pacific Ocean. In January 1991, a clumsy New Jersey repair crew sliced through a major optical fiber artery, shutting down New York's phones for a further six hours, while similar breaks have been reported from Chicago, Los Angeles, and Washington, D.C. The Federal Aviation Administration recently recorded the shutdown of four major U.S. air traffic control centers. The cause? "Fiber cable cut by farmer burying dead cow," said the official report.[3]

Computers and communication systems are also vulnerable to physical attacks by humans and to software sabotage by outside hackers and inside employees. For example, a saboteur entered telecommunications tunnels in Sydney, Australia, one day in 1987 and carefully severed twenty-four cables, knocking out 35,000 telephone lines in forty Sydney suburbs and bringing down hundreds of computers, automated teller machines (ATMs), and point of sale (POS), telex, and fax terminals with it. Some businesses were put out of action for forty-eight hours as engineers battled to restore services. Had the saboteur not been working with an out-of-date plan, the whole of Australia's telecommunications system might have been blacked out. In Chicago in 1986, a disgruntled employee at Encyclopaedia Brittanica, angry at having been laid off, merely tapped into the encyclopedia's database and made a few alterations to the text being prepared for a new edition of the renowned work—like changing references to Jesus Christ to Allah and inserting the names of company executives in odd positions. As one executive commented, "In the computer age, this is exactly what we have nightmares about."[4]

Our growing dependency on computers has been highlighted further in recent years by such incidents as the theft in the former Soviet

Union in 1990 of computer disks containing medical information on some 670,000 people exposed to radiation in the Chernobyl nuclear disaster. The disks were simply wiped and then resold by the teenaged thieves. In 1989, vital information about the infamous Alaskan oil spill was "inadvertently" destroyed at a stroke by an Exxon computer operator. In the same year, U.S. retailer Montgomery Ward allegedly discovered one of its warehouses in California that had been lost for three years because of an error in its master inventory program. Apparently, one day the trucks stopped arriving at the warehouse: nothing came in or went out. But the paychecks were issued on a different system, so for three whole years (so the story goes) the employees went to work every day, moved boxes around, and submitted timecards—without ever telling company headquarters. "It was a bit like a job with the government," said one worker after the blunder had been discovered.[5]

In Amsterdam, Holland, in 1991, the body of an old man who had died six months earlier was found in an apartment by a caretaker who had been concerned about a large pile of mail for him. The man had been something of a recluse, but because his rent, gas, and electricity bills were paid automatically by computer, he wasn't missed. His pension also had been transferred into his bank account every month, so all the relevant authorities assumed that he was still alive. Another particularly disturbing example of computer dependency came from London during the Gulf war, when computer disks containing the Allies' plans for Desert Storm disappeared, along with a laptop computer, from a parked car belonging to Wing Commander David Farquhar of the Royal Air Force Strike Command. Luckily for the Allies, the thieves did not recognize the value of the unencrypted data, which did not fall into Iraqi hands. But a court-martial for negligence and breach of security awaited Farquhar.[6]

Computers are changing our way of life in all sorts of ways. At work, we may have our performance monitored by computer and our electronic mail read by the boss. It's no good trying to delete embarrassing e-mail statements because someone probably will have a backup copy of what you wrote. This is what happened to White House adviser Colonel Oliver North and to John Poindexter, the former national security adviser to President Ronald Reagan, when they tried to cover up evidence of the Iran-Contra scandal. Poindexter allegedly sat up all night deleting 5,012 e-mail messages, while North destroyed a further 736, but unknown to Poindexter and North the messages were all preserved on backup tapes that were subsequently read by congressional investigators. And if you use a spell-checker or language-corrector in your word processing program, be sure that it doesn't land you in trouble. For example, the *Fresno Bee* newspaper in California recently had to run a correction that read: "An item in Thursday's Nation Digest about the Massachusetts budget crisis made reference to new taxes that will help 'put Massachusetts back in the African-American.' This item should have read 'put Massachusetts back in the black.'"[7]

Recent government reports have confirmed that our growing dependence on computers leaves society increasingly vulnerable to software bugs, physical accidents, and attacks on critical systems. In 1989, a report to the U.S. Congress from one of its subcommittees, written by James H. Paul and Gregory C. Simon, found that the U.S. government was wasting millions of dollars a year on software that was overdue, inadequate, unsafe, and riddled with bugs. In 1990, the Canadian auditor-general, Ken Dye, warned that most of the Canadian government's computer systems were vulnerable to physical or logical attack: "That's like running a railroad without signals or a busy airport without traffic controls," he said. In 1991, a major report by the System Security Study Committee of the U.S. National Academy of Sciences, published as *Computers at Risk*, called for improved security, safety, and reliability in computer systems. The report declared that society was becoming more vulnerable to "poor system design, accidents that disable systems, and attacks on computer systems."[8]

SOME NEW SOCIAL PROBLEMS CREATED BY COMPUTERS

Although society as a whole derives benefit from the use of computers and computer networks, computerization has created some serious problems for society that were largely unforeseen.

We classify the new social problems created by computers into seven main categories: computer crime and the problem of computer security; software theft and the question of intellectual property rights; the new phenomena of hacking and the creation of viruses; computer unreliability and the key question of software quality; data storage and the invasion of privacy; the social implications of artificial intelligence and expert systems; and the many problems associated with workplace computerization.

These new problems have proved to be costly: computer crime costs companies millions of dollars a year, while software producers lose staggering sums as a result of widespread software theft. In recent years, huge amounts of time and money have had to be devoted to repairing the damage to systems caused by the activities of malicious hackers and virus creators. Unreliable hardware and software costs society untold billions every year in terms of downtime, cost overruns, and abandoned systems, while invasions of privacy and database mix-ups have resulted in expensive lawsuits and much individual stress. Sophisticated expert systems lie unused for fear of attracting lawsuits, and workplace stress caused by inappropriate computerization costs society millions in absenteeism, sickness benefits, and reduced productivity.

Computer crime is a growing problem for companies, according to recent reports. Every new technology introduced into society creates new opportunities for crime, and information technology is no exception. A new generation of high-tech criminals is busy stealing data, doctoring data, and threatening to destroy data for monetary gain. New types of fraud made possible by computers include ATM fraud, EFT (electronic funds transfer) fraud, EDI (electronic data interchange) fraud, mobile phone fraud, cable TV fraud, and telemarketing fraud. Desktop printing (DTP) has even made desktop forgery possible. Perhaps the biggest new crime is phone fraud, which may be costing American companies as much as $2 billion a year. Most analysts think that reported computer crime is just the tip of an iceberg of underground digital deviance that sees criminals and the crime authorities competing to stay one jump ahead of each other.

Software theft or the illegal copying of software is a major problem that is costing software producers an estimated $12 billion a year. Recent cases of software piracy highlight the prevalence of software copying and the worldwide threat posed by organized software pirates. Computer users and software developers tend to have very different ethical positions on the question of copying software, while the law in most countries is confusing and out of date. There is an ongoing debate about whether copyright law or patent law provides the most appropriate protection for software. Meanwhile the legal position in the United States, for example, has been confused further by the widely varying judgments handed down by U.S. courts in recent years. The recent rash of look and feel suits launched by companies such as Lotus and Apple have muddied the waters still further. The central question facing the information technology (IT) industry is how to reward innovation without stifling creativity, but there is no obvious answer to this conundrum and no consensus as to what constitutes ethical practice.

Attacks by hackers and virus creators on computer systems have proved enormously costly to computer operators. In recent cases, hackers have broken into university computers in order to alter exam results, downloaded software worth millions, disrupted the 911 emergency phone system in the United States, stolen credit card numbers, hacked into U.S. military computers and sold the stolen data to the KGB, and blackmailed London banks into employing them as security advisers. Hackers also have planted viruses that have caused computer users untold misery in recent years. Viruses have erased files, damaged disks, and completely shut down systems. For example, the famous Internet worm, let loose by Cornell student Robert Morris in 1988, badly damaged 6,000 systems across the United States. There is ongoing debate about whether hackers can sometimes function as guardians of our civil liberties, but in most countries the response to the hacking craze has been new security measures, new laws such as Britain's Computer Misuse Act (1990), and new calls for improved network ethics. Peter J. Denning, editor-in-chief of communications of the ACM, says that we must expect increasing attacks on computers by hackers and virus creators in the years ahead; Professor Lance J. Hoffman has called for all new computers to be fitted with antiviral protection as standard equipment, rather like seat belts on cars.[9]

Unreliable computers are proving to be a major headache for modern society. Computer crashes or downtime—usually caused by buggy software—are estimated to cost the United States as much as $4 billion a year, according to a recent report. When bug-ridden software has been used to control fly-by-wire aircraft, railroad signals, and ambulance dispatch systems, the cost of unreliable computers sometimes has had to be measured in terms of human lives. Computers tend to be unreliable because they are digital devices prone to total failure and because their complexity ensures that they cannot be tested thoroughly before use. Massive complexity can make computer systems completely unmanageable and can result in huge cost overruns or budget runaways. For example, in 1988 the Bank of America had to abandon an $80 million computer system that failed to work, while in 1992 American Airlines announced a loss of over $100 million on a runaway computer project. The Wessex Regional Health Authority in England scrapped a system in 1990 that had cost $60 million, and Blue Cross &

Blue Shield of Massachusetts pulled the plug in 1992 on a project that had cost a staggering $120 million. U.S. Department of Defense runaways are rumored to have easily exceeded these sums. Computer scientists are exploring a variety of ways to improve software quality, but progress with this key problem is slow.[10]

The problem of safeguarding privacy in a society where computers can store, manipulate, and transmit at a stroke vast quantities of information about individuals is proving to be intractable. In recent years, a whole series of database disasters involving mistaken identities, data mix-ups, and doctored data have indicated that we probably place too much faith in information stored on computers. People have had their driver's license and credit records altered or stolen and their lives generally made a misery by inaccurate computer records. There is growing concern about the volume and the quality of the data stored by the FBI's National Crime Information Center (NCIC), the United Kingdom's Police National Computer (PNC), and other national security agencies. (Such concerns even led to a public riot in Switzerland in 1990.) Moreover, new controversies have erupted over the privacy aspects of such practices as calling number identification (CNID) on phone networks, the monitoring of e-mail (by employers such as Nissan and Epson and, it seems, the mayor of Colorado Springs), and the phenomenon of database marketing, which involves the sale of mailing lists and other personal information to junk mailers (in 1990, Lotus and Equifax were forced to drop their Lotus Marketplace: Households, which put on disk personal information about 120 million Americans). Governments around the world are now being pushed into tightening privacy laws.[11]

The arrival of expert systems and primitive forms of artificial intelligence (AI) have generated a number of technical, legal, and ethical problems that have yet to be resolved. Technical problems have seriously slowed progress toward the holy grail of AI, while many are now asking whether computers could ever be trusted to make medical, legal, judicial, political, and administrative judgments. Given what we know about bugs in software, some are saying that it will never be safe to let computers run, for instance, air traffic control systems and nuclear power stations without human expert backup. Legal difficulties associated with product liability laws have meant that nobody dares use many of the expert systems that have been developed. In addition, AI critics are asking serious ethical questions, such as, Is AI a proper goal for humanity? Do we really need to replace humans in so many tasks when there is so much unemployment?[12]

Because paid employment is still central to the lives of most people and, according to the U.S. Bureau of Labor Statistics, about 46 million Americans now work with computers, workplace computerization is clearly an important issue. Indeed, it has proved to be fertile ground for controversies, debates, and choices about the quantity of work available and the quality of working life. While the 1980s did not see massive technological unemployment precisely because of the slow and messy nature of IT implementation, there is now renewed concern that computers are steadily eroding job opportunities in manufacturing and services. Moreover, concern about the impact of computers on the quality of working life has increased with the realization that managers can go in very different directions with the design and implementation of new work systems. Computers have the ability to enhance or degrade the quality of working life, depending upon the route chosen. Computer monitoring of employees has become a controversial issue, as have the alleged health hazards of computer keyboard usage, which has resulted recently in some celebrated RSI (repetitive strain injury) legal suits against employers and computer vendors.[13]

ETHICAL DILEMMAS FOR COMPUTER USERS

Each of the new social problems just outlined generates all sorts of ethical dilemmas for computer users. Some of these dilemmas—such as whether or not to copy software—are entirely new, while others are new versions of old moral issues such as right and wrong, honesty, loyalty, responsibility, confidentiality, trust, accountability, and fairness. Some of these ethical dilemmas are faced by all computer users; others are faced only by computer professionals. But many of these dilemmas constitute new gray areas for which there are few accepted rules or social conventions, let alone established legal case law.

Another way of saying that computers create new versions of old moral issues is to say that information technology transforms the context in which old ethical issues arise and adds interesting new twists to old problems.[14] These issues arise from the fact that computers are machines that control other machines and from the specific, revolutionary characteristics of IT. Thus new storage devices allow us to store massive amounts of information, but they also generate new ethical choices about access to that information and about the use or misuse of that information. Ethical issues concerning privacy, confidentiality, and security thus come to the fore. The arrival of new media such as e-mail, bulletin boards, faxes, mobile phones, and EDI [Electronic Data Interchange] has generated new ethical and legal issues concerning user identity, authenticity, the legal status of such communications, and whether or not free speech protection and/or defamation law applies to them.

IT provides powerful new capabilities such as monitoring, surveillance, data linking, and database searching. These capabilities can be utilized wisely and ethically, or they can be used to create mischief, to spy on people, and to profit from new scams. IT transforms relationships between people, depersonalizing human contact and replacing it with instant, paperless communication. This phenomenon can sometimes lead people into temptation by creating a false sense of reality and by disguising the true nature of their actions, such as breaking into a computer system. IT transforms relationships between individuals and organizations, raising new versions of issues such as accountability and responsibility. Finally, IT unreliability creates new uncertainties and a whole series of ethical choices for those who operate complex systems and those who design and build them. Computer producers and vendors too often neglect to adequately consider the eventual users of their systems, yet they should not escape responsibility for the consequences of their system design.[15]

Under the heading of computer crime and security, a number of ethical issues have been raised—despite the fact that the choice of whether or not to commit a crime should not present a moral dilemma for most people. For example, some have sought to make a distinction between crimes against other persons and so-called victimless crimes against, for example, banks, phone companies, and computer companies. While not wishing to excuse victimless crimes, some have suggested that they somehow be placed in a less serious category, especially when it comes to sentencing. Yet it is hard to accept that a company is any less a victim than an individual when it is deprived of its wealth. Because so many computer criminals appear to be first-time offenders who have fallen victim to temptation, do employers bear any responsibility for misdeeds that have occurred on their premises? And how far should employers or security agencies be allowed to go in their attempts to prevent or detect crime? (Should they be allowed to monitor e-mail or spy on people in toilets, for example?)

The ease with which computer software can be copied presents ethical dilemmas to computer users and professionals almost every day of the year. Some justify the widespread copying of software because everybody else does it or because the cost of well-known software packages is seen as too high. But copying software is a form of stealing and a blatant infringement of the developer's intellectual property rights. In the past, intellectual property such as literary works and mechanical inventions was protected by copyright and patents, but software is a new and unique hybrid. How do we protect the rights of software developers so as to ensure that innovation in the industry continues? What does the responsible computer professional do? Is all copying of software wrong, or are some kinds worse than others? How should the individual user behave when the law is unclear and when people in the industry disagree as to what constitutes ethical practice?

The new phenomena of computer hacking and the creation of computer viruses have raised many unresolved ethical questions. Is hacking merely harmless fun or is it the computer equivalent of burglary, fraud, and theft? When do high-tech high jinks become seriously criminal behavior? Because hacking almost always involves unauthorized access to other people's systems, should all hacking activity be considered unethical? What are we to make of hackers themselves? Are they well-intentioned guardians of our civil liberties and useful amateur security advisers, or are they mixed-up adolescents whose stock in trade is malicious damage and theft? Can the creation of viruses ever be justified in any circumstances? If not, what punishment should be meted out to virus creators? Finally, what should responsible individuals do if they hear of people who are hacking?

The reality of computer unreliability creates many ethical dilemmas, mainly for the computer professionals who are charged with creating and installing systems. Who is responsible when things go wrong? When a system malfunctions or completely crashes because of an error in a computer program, who is to blame—the original programmers, the system designer, the software supplier, or someone else? More to the point, should system suppliers warn users that computer systems are prone to failure, are often too complex to be fully understood, have not been thoroughly tested before sale, and are likely to contain buggy software? Should software producers be made to provide a warranty on software? And to whom should individual computer professionals ultimately be responsible—the companies they work for, their colleagues, the customers, or the wider society?

The recurring issue of privacy confronts computer professionals and users in all sorts of contexts. First, there are general questions, such as what is privacy and how much of it are individuals entitled to, even in today's society. Does individual information stored on databases pose a threat to privacy? What right do governments and commercial organizations have to store personal information on individuals? What steps should be taken to ensure the accuracy of such information? Then there are the dilemmas faced by computer professionals and users over whether or not to use information collected for another purpose, whether to purchase personal information illicitly obtained, whether to link information in disparate databases, and so on. Practically every attempt to improve security (and sometimes even productivity) in organizations involves choices about the degree of privacy to which employees are entitled, while new controversies have arisen over the privacy aspects of e-mail and caller ID.

In a sense, the quest for artificial intelligence is one big ethical problem for the computing world because we have yet to determine whether AI is a proper goal—let alone a realistic goal—for humanity. Should computer professionals work on systems and devices that they know will make yet more humans redundant? Should we really be aiming to replace humans in more tasks? Isn't it somehow demeaning to human intelligence to put so much emphasis on making a machine version of it? Perhaps even more to the point, given what we know about computer unreliability, can we afford to trust our lives to artificially intelligent expert systems? What should be the attitude of responsible computer professionals: should they warn users of the risks involved or refuse to work on life-critical applications? Should they refuse to work on the many AI projects funded by the military? Moreover, should institutional users trust computers to make judicial, administrative, and medical judgments when human judgment has often proved to be superior?

Some might think that the workplace does not provide an obviously rich source of ethical dilemmas for computer professionals and users. Yet workplace computerization involves numerous choices for management about the type of system to be implemented, and different systems have radically different impacts on both the quantity and the quality of work. Generally speaking, computers in factories can be used to enhance the quality of working life, to improve job satisfaction, to provide more responsibility, and to upgrade or reskill the workforce; or, they can be used to get rid of as many people as possible and to turn those remaining into deskilled, degraded machine-minders, pressing buttons in a soulless, depersonalized environment. Office computerization can increase stress levels and thus health hazards if the new work process is badly designed or even if the new office furniture and equipment are badly designed. Employee monitoring often makes matters worse, while speedups and the creation of excessively repetitive tasks like keying-in data for hours can result in cases of repetitive strain injury. Computer professionals and managers have a responsibility to ensure that these outcomes are avoided.

HOW ETHICAL THEORY CAN HELP

"Ethics" has been defined as the code or set of principles by which people live. Ethics is about what is considered to be right and what is considered to be wrong. When people make ethical judgments, they are making prescriptive or normative statements about what ought to be done, not descriptive statements about what is being done.

But when people face ethical dilemmas in their everyday lives, they tend to make very different judgments about what is the right and what is the wrong thing to do. Ensuing discussions between the parties often remain unresolved because individuals find it hard to explain the reasoning behind their subjective, moral judgments. It is virtually impossible to conclude what ought to be the most appropriate behavior. Ethical theory—sometimes referred to as moral philosophy—is the study of the rules or principles that lie behind moral decisions. This theory helps provide us with a rational basis for moral judgments, enables us to classify and compare different ethical positions, and enables people to defend a particular position on a given issue. Thus the use of ethical theory can help us throw some light on the moral dilemmas faced by computer professionals and users and may even go some way toward determining how people ought to behave when using computers.[16]

Classical ethical theories are worth knowing because they provide useful background in some of the terminology, but they have limited relevance to everyday behavior in the IT industry. For example, Plato (429–347 B.C.) talked about the "good life," and much of his life was spent searching for the one good life. He also believed that an action was right or wrong in itself—a so-called objectivist (later, deontological) position. Aristotle (384–322 B.C.), on the other hand, adopted a more relativist and empiricist approach, arguing that there were many good and bad lives and that good lives were happy lives created by practicing "moderation in all things." Epicurus (341–270 B.C.) was the exact opposite, promoting hedonism, or the pursuit of pleasure, as the sole goal of life (although modern hedonists tend to forget that he also warned that too much pleasure was harmful and that the highest form of pleasure was practicing virtue and improving one's mind).

Diogenes (413–323 B.C.) was leader of the cynics, who believed that the world was fundamentally evil. The cynics were antisocial; they shunned public life and led an ascetic, privatized life—rising early, eating frugally, working hard, sleeping rough, and so on. Individual cynics found salvation in themselves and their honest life-style, not in worldly possessions. Modern cynics don't necessarily do this, but they are very distrustful of what they see as a thoroughly corrupt world. Finally, the stoics, such as Zeno (ca. 335–263 B.C.) and Epictetus (ca. 55–135 A.D.), were the essential fatalists, arguing that people should learn to accept whatever happened to them and that everything in the world occurs according to a plan that we do not understand. A true stoic believes that there is no such thing as good or evil and seeks to rise above the circumstances of everyday life, rejecting temptations, controlling emotions, and eschewing ambitions.

But probably the three most influential ethical theories of recent times—and the three of most likely relevance for our purposes—are ethical relativism (associated with Spinoza, 1632–1677), utilitarianism (J. S. Mill, 1806–1873) or consequentialism, and Kantianism (Kant, 1724–1804) or deontologism. Ethical relativism, which says that there are no universal moral norms, need not detain us for long, for it offers no guidance as to what is correct behavior. Ethical relativists merely point to the variety of behaviors in different cultures and conclude that the issue of right and wrong is all relative. Ethical relativism is a descriptive account of what is being done rather than a normative theory of what should be done. While it is true that people in different societies have different moralities, this does not prove that one morality might not be the correct one or that one might not constitute the universal moral code. Ethical relativism is not much use when trying to decide what is the right thing to do in today's world of computing.

Consequentialism and deontologism are much more relevant for our purposes. Consequentialism says simply that an action is right or wrong depending upon its consequences, such as its effects on society. Utilitarianism, as outlined by J. S. Mill and Jeremy Bentham, is one form of consequentialism. Its basic principle is that everyone should behave in such a way as to bring about the greatest happiness of the greatest number of people. Utilitarians arrive at this cardinal principle by arguing that happiness is the ultimate good because everything else in life is desired as a means of achieving happiness. Happiness is the ultimate goal of humans, and thus all actions must be evaluated on the basis of whether they increase or decrease human happiness. An action is therefore right or wrong depending upon whether it contributes to the sum total of human happiness.

By contrast, deontologism says that an action is right or wrong in itself. Deontologists stress the intrinsic character of an act and disregard motives or consequences. Thus a deontologist might say that the act of copying software is always wrong, regardless of other considerations, while a utilitarian might say that it was justified if it had a beneficial effect on society as a whole. Deontologists appear to be on particularly strong ground when they state, for example, that killing is wrong no matter what the circumstances, but they are on weaker ground when they say that lying is always wrong. Utilitarians would say that lying can be justified in certain circumstances, as in the case of white lies. On the other hand, utilitarians can find themselves in the position of defending actions that are morally wrong (like lying) or condoning actions that penalize the few in order to benefit the many (such as exploiting labor in a third world manufacturing plant). Consequentialists tend to look at the overall impact on society, whereas deontologists tend to focus on individuals and their rights. Kantians, in particular, argue strongly that people should always be treated as ends and never merely as means.

The distinction between consequentialists and deontologists is quite useful when we consider the ethical issues confronting computer professionals and users.

ETHICS AND THE COMPUTER PROFESSIONAL

Because computing is a relatively new field, the emerging computer profession has had neither the time nor the organizational capability to establish a binding set of moral rules or ethics. Older professions, like medicine and the law, have had centuries to formulate their codes of ethics and professional conduct. And there is another problem, too: the practice of computing, unlike the practice of medicine or the law, goes on outside the profession—this is an open field, with unfenced boundaries.

Computing, with its subdisciplines like software engineering, has not yet emerged as a full-fledged profession. Classic professions involve mental work, a high level of skill, and a lengthy period of training, and they perform some vital service to society—just like computing. But more than that, the classic profession is highly organized, with a central body that admits members only when they have achieved a certain level of skill. Although members have a considerable degree of autonomy, they are expected to exercise their professional judgment within the framework of a set of ethical principles laid down by the profession's central organization. Transgressors can be disciplined or even thrown out of the profession altogether. Some see the development of professions as a sign of a well-ordered, mature society, whereas critics have seen them as little more than self-serving protection rackets (the British author and playwright George Bernard Shaw once described all professions as "a conspiracy against the people").

So what sort of profession is computing? Members of the fledgling computer profession do not yet have the social status of doctors or lawyers. Instead, their status has been likened to that of engineers, who work mostly as employees rather than in their own right, who have esoteric knowledge but quite limited autonomy, and who often work in teams or on small segments of large projects rather than alone. Worryingly, they are often distant from the effects of their work. Yet despite the lower social status of computer professionals, the widespread use of information technology for storing all sorts of vital information puts considerable power into their hands, from the humble operator to the top systems developer. This power has not been sought specifically but arises from the nature of the technology. Computer professionals often find themselves in positions of power over employers, clients, coprofessionals, and the wider public, and this power can be abused easily by those without scruples or those who easily fall victim to temptation.[17]

Computer professionals face all sorts of ethical dilemmas in their everyday work life. First, although they have obligations to their employers, to the customers, to their coprofessionals, and to the general public, these obligations often come into conflict and need to be resolved one way or another. For example, what should be the response of a systems analyst whose employer insists on selling an overengineered, expensive system to gullible customers? Go along with the scam, or tell the customers that they are being duped? Second, almost every day the computer professional is confronted with issues of responsibility, intellectual property, and privacy. Who should take the blame when a system malfunctions or crashes? What attitude should professionals take when someone's intellectual property rights are clearly being infringed? How should they balance the need for greater system security with the right of individuals to privacy?

In an effort to help computer professionals cope with these kinds of conflicts, professional organizations such as the ACM (Association for Computing Machinery), the IEEE (Institute of Electrical and Electronics Engineers), the British Computer Society (BCS), and IFIP (International Federation for Information Processing) have been formulating and revising codes of ethics and professional conduct applicable to the IT industry. One problem with these codes is that they often have consisted mainly of motherhood statements like "I will avoid harm to others" and "I will always be honest and trustworthy." These proclamations could just as easily apply to any profession or walk of life and say nothing of specific relevance to computing. However, the new ACM code is much improved in this respect in that it talks about specific IT industry responsibilities. A more serious criticism is that these codes contain little in the way of sanctions by which their laudable aims could be enforced. A number of critics have pointed out that these codes have never been used and their language never interpreted. Furthermore, the codes usually have talked purely in terms of individuals being at fault and not whole organizations (although this, too, is addressed to some extent in the new IFIP and ACM codes).[18]

An even more fundamental difficulty with all such codes of ethics is that they don't necessarily do much to make people behave more ethically. The pressure, financial or otherwise, to conform with unethical industry practices is often too great. Thus, in a classic critique of professional ethics, John Ladd argued that attempts to develop professional codes of ethics are not only marked by intellectual and moral confusion (such as describing a code of conduct as ethics), they are also likely to fail. Codes of conduct, he says, are widely disregarded by members of professions. Worthy and inspirational though such codes may be, he says that their existence leads to complacency and to self-congratulation—and maybe even to the cover-up of unethical conduct. "Look, we have a code of ethics," professionals might say, "so everything we do must be ethical." The real objectives of such codes, Ladd says, are to enhance the image of the profession in the outside world and to protect the monopoly of the profession. In other words, they are a bit of window dressing designed to improve the status and the income of members.[19]

Another debate has arisen over suggestions that computer professionals be licensed or certified. Under this proposal, a programmer would have to obtain a certificate of competence before being allowed to work on major projects—especially those involving life-critical systems—and perhaps every computer user would have to obtain a kind of driver's license before being allowed onto the computer networks. Certification of software developers might help reduce the number of software project runaways. But there would be endless difficulties involved in measuring programming competence, and these problems could perhaps lead to religious wars in the profession![20] Moreover, there is a danger that certification could create a closed shop or craft guild that might exclude talented and innovative newcomers. On the other hand, it seems likely that some sort of certification safeguards will have to be introduced in the future to cover high-risk systems.

THE RESPONSIBILITY OF COMPUTING EDUCATORS

Recent well-publicized incidents of hacking, virus creation, computer-based fraud, and invasions of privacy have increased the pressure on computing educators to help instill a greater sense of responsibility in today's students. The world of computing has been portrayed in the media as a kind of electronic frontier society where a "shoot from the hip" mentality prevails. It is widely believed that there is far too much computerized anarchy and mayhem.

We believe that computing educators need to do three things. They must encourage tomorrow's computer professionals to behave in a more ethical, responsible manner for the long-term good of the IT industry. They also need to help make students aware of the social problems caused by computers and the social context in which computerization occurs. And they need to sensitize students to the kinds of moral dilemmas they will face in their everyday lives as computer professionals. Many of today's computer science undergraduates will go on to create systems that will have major impacts on people, organizations, and society in general. If those systems are to be successful economically and socially, graduates will need to know the lessons from the computerization story so far, the ethical and social issues involved, and the range of choices available to computer professionals.

NOTES

1. Sources for the ozone hole: *The New York Times*, Science section, 29 July 1986, page C1; the Blackhawk crashes: B. Cooper and D. Newkirk, *Risks to the Public in Computers and Related Systems*, on Internet, compiled by Peter G. Neumann, November 1987; Therac-25 and other radiation therapy cases: Jonathan Jacky, "Programmed for Disaster—Software Errors Imperil Lives," in *The Sciences*, September–October, 1989; Jonathan Jacky, "Risks in Medical Electronics," *Communications of the ACM*, vol. 33, no. 12, December 1990, page 138; "Patients Die After Radiation Mix-Up," *The Guardian*, London, 23 February 1991; John Arlidge, "Hospital Admits Error in Treating Cancer Patients," *The Independent*, London, 7 February 1992; Patriot missile: *New York Times*, 21 May 1991, and *Patriot Missile Defense: Software Problem Led to System Failure at Dhahran, Saudi Arabia*, U.S. General Accounting Office, February 1992; Hubble trouble: *Software Engineering Notes*, vol. 17, no. 1, January 1992, page 3.

2. AT&T phone outages, January 1990 and September 1991: *Software Engineering Notes*, vol. 15, no. 2, April 1990, and vol. 16, no. 4, October 1991, pages 6–7, *Time*, 30 September 1991; *Fortune*, 13 January 1992; A320 crashes: "Airbus Safety Claim 'Cannot Be Proved,' " *New Scientist*, 7 September 1991, page 16, and successive reports in *Software Engineering Notes*, esp. vol. 13, 14 and 15; Osprey crashes: *Flight International*, 18–24 September 1991; *New Scientist*, 15 August 1992; YF-23 and other fly-by-wire glitches: *Software Engineering Notes*, vol. 16, no. 3, July 1991, pages 21–22; Lauda crash: various reports in *Software Engineering Notes*, vol. 16 and 17.

3. "The 'Dirty Power' Clogging Industry's Pipeline," *Business Week*, 8 April 1991; Naruko Taknashi et al., "The Achilles Heel of the Information Society: Socioeconomic Impacts of the Telecommunications Cable Fire in the Setagaya Telephone Office, Tokyo," *Technological Forecasting and Social Change*, vol. 34, no. 1, 1988, pages 27–52; Beavers and dead cows: AP report in *Software Engineering Notes*, vol. 17, no. 1, January 1992; Foxes: *The Riverine Grazier*, Hay, New South Wales, Australia, 10 April 1991; Trawlers: *The Australian*, 16 April 1991; New Jersey and others: *Software Engineering Notes*, vol. 16, no. 2, April 1991, page 4, and vol. 16, no. 3, July 1991, pages 16–17.

4. "Saboteur Tries to Blank Out Oz," *The Australian*, 23 November 1987, page 1; "Laid-Off Worker Sabotages Encyclopaedia," *San Jose Mercury News*, 5 September 1986.

5. "Thieves Destroy Data on Chernobyl Victims," *New Scientist*, 22 September 1990; "Exxon Man Destroys Oil Spill Documents," UPI report in *The Australian*, 4 July 1989; " 'Losing' a Warehouse," *Software Engineering Notes*, vol. 16, no. 3, July 1991, page 7.

6. "Inhabitant of Amsterdam Lies Dead in Apartment for Half a Year," *Software Engineering Notes*, vol. 16, no. 2, April 1991, page 11; "Defence of the Data," *New Scientist*, 19 January 1991, and "Theft of Computer Puts Allies' Plan at Risk," report from *The Times* (London) in *The Australian*, 14 March 1991.

7. "Poindexter Deleted 5,000 Computer Notes," Reuters and AP reports in *The Weekend Australian*, 17–18 March 1990; "Terminally Dumb Substitutions," *Software Engineering Notes*, vol. 15, no. 5, October 1990, page 4.
8. James H. Paul and Gregory C. Simon, *Bugs in the Program: Problems in Federal Government Computer Software Development and Regulation* (Subcommittee on Investigations and Oversight of the House Committee on Science, Space and Technology, U.S. Government Printing Office, Washington, DC, September 1989); Shawn McCarthy, "Dye Fears Computer Sabotage," *Toronto Star*, 31 October 1990; *Computers at Risk: Safe Computing in the Information Age* (National Academy Press, Washington, DC, 1991).
9. Peter J. Denning (ed.), *Computers Under Attack: Intruders, Worms and Viruses* (ACM Press/Addison-Wesley, Reading, MA, 1990), page iii; Lance J. Hoffman (ed.), *Rogue Programs: Viruses, Worms, and Trojan Horses* (Van Nostrand Reinhold, New York, 1990), page 1.
10. Reports in *Business Week*, 7 November 1988, 3 April 1989, 15 June 1992, and 27 July 1992; *The Australian*, 4 August 1992.
11. Peter G. Neumann, "What's in a Name?" *Communications of the ACM*, vol. 35, no. 1, January 1992, page 186; *Software Engineering Notes*, vol. 14, no. 5, July 1989, page 11; vol. 16, no. 3, July 1991, pages 3–4; and vol. 17, no. 1, January 1992, pages 12–13; *Business Week*, 18 June 1990, 18 May 1992, and 8 June 1992; Marc Rotenberg, "Protecting Privacy," *Communications of the ACM*, vol. 35, no. 4, page 164; *The New York Times*, 4 May 1990, page A12; *The Los Angeles Times*, 8 January 1991; *Computing Australia*, 20 August 1990; Langdon Winner, "A Victory for Computer Populism," *Technology Review*, May–June 1991, page 66.
12. "Expert Systems Fail to Flourish," *The Australian*, 22 May 1990; Harvey P. Newquist III, "Experts at Retail," *Datamation*, 1 April 1990, pages 53–56; Dianne Berry and Anna Hart (eds.), *Expert Systems: Human Issues* (MIT Press, Cambridge, MA, 1990); Roger Penrose, *The Emperor's New Mind* (Oxford University Press, New York, 1989).
13. Reports in *Business Week*, 19 August 1991, 15 June 1992, 13 July 1992; *Fortune*, 4 November 1991, 24 February 1992, 24 August 1992; Barbara Goldoftas, "Hands That Hurt: Repetitive Motion Injuries on the Job," *Technology Review*, January 1991, pages 43–50.
14. Deborah C. Johnson, *Computer Ethics* (Prentice-Hall, Englewood Cliffs, NJ, 1985), page 3; John Ladd, "Computers and Moral Responsibility: A Framework for an Ethical Analysis," in Carol C. Gould (ed.), *The Information Web: Ethical and Social Implications of Computer Networking* (Westview Press, Boulder, CO, 1989), pages 218–220.
15. Peter G. Neumann, "Computers, Ethics and Values," *Communications of the ACM*, vol. 34, no. 7, July 1991, page 106; Leslie S. Chalmers, "A Question of Ethics," *Journal of Accounting and EDP*, vol. 5, no. 2, Summer 1989, pages 50–53.
16. See Deborah C. Johnson, op. cit., 1985, chapter 1; and M. David Ermann, Mary B. Williams, and Claudio Gutierrez (eds.), *Computers, Ethics and Society* (Oxford University Press, New York, 1990), part 1.
17. Deborah C. Johnson, op. cit., 1985, chapter 2; Deborah C. Johnson and

John W. Snapper (eds.), *Ethical Issues in the Use of Computers* (Wadsworth, Belmont, CA, 1985), part 1; Donn B. Parker, Susan Swope, and Bruce N. Baker (eds.), *Ethical Conflicts in Information and Computer Science, Technology, and Business* (QED, Wellesley, MA, 1990), parts 2, 4, 5, and 6.

18. Donn B. Parker et al., op. cit., 1990, page 5; Charles Dunlop and Rob Kling (eds.), *Computerization and Controversy: Value Conflicts and Social Choices* (Academic Press, San Diego, CA, 1991), pages 656–657; D. Dianne Martin and David H. Martin, "Professional Codes of Conduct and Computer Ethics Education," *Social Science Computer Review*, vol. 8, no. 1, Spring 1990, pages 96–108.

19. John Ladd, "The Quest for a Code of Professional Ethics: An Intellectual and Moral Confusion," in Deborah C. Johnson and John W. Snapper (eds.), op. cit., 1985, page 813.

20. Peter G. Neumann, "Certifying Professionals," *Communications of the ACM*, vol. 34, no. 2, February 1991, page 130.

23.   Identity in the Age of the Internet

SHERRY TURKLE

The Internet is a technological icon—literally as well as figuratively—of the mid-1990s. As it has grown, the personal computer has evolved from a calculator, word processor, and high-priced game machine into a new communications medium—a medium radically different from any we have known until now. Our exploration of the virtual world that has opened through this medium, known widely as "cyberspace," is not only changing our relations with computers, it is, as Sherry Turkle points out, changing our relations with our "selves" as we conceive of them. In the following essay, taken from her 1995 book, Life on the Screen, *Turkle explores the nature and implications of some of these changes.*

Turkle seems most interested in the interactions among participants in "MUDs," Multi-User Domains (originally Multi-User Dungeons). These are elaborate computer games in which many individuals partici-pate simultaneously either through the Internet or through dial-up tele-phone connections. By joining, participants enter virtual spaces in which they are able to navigate, construct "rooms," and converse and interact with other participants. Turkle observes that as people play characters in these environments, they construct new selves through their social inter-actions. What is "real" and what is "virtual" in such experiences? What do they say about us and how are they changing us? How are these experiences likely to influence our identities and our culture? Turkle's is a novel and provocative approach to the subject of computers and soci-ety, one that will engage those readers who are already residents of cyberspace and may tempt those who are not to begin exploring it.

Sherry Turkle is a professor of the sociology of science at the Massachu-

setts Institute of Technology. She is also a licensed clinical psychologist. Her 1984 book The Second Self: Computers and the Human Spirit *is widely regarded as a seminal work in the area of computers, culture, and society.*

There was a child went forth every day,
And the first object he look'd upon, that object he became.
— Walt Whitman

We come to see ourselves differently as we catch sight of our images in the mirror of the machine. A decade ago, when I first called the computer a second self, these identity-transforming relationships were almost always one-on-one, a person alone with a machine. This is no longer the case. A rapidly expanding system of networks, collectively known as the Internet, links millions of people in new spaces that are changing the way we think, the nature of our sexuality, the form of our communities, our very identities.

At one level, the computer is a tool. It helps us write, keep track of our accounts, and communicate with others. Beyond this, the computer offers us both new models of mind and a new medium on which to project our ideas and fantasies. Most recently, the computer has become even more than tool and mirror: We are able to step through the looking glass. We are learning to live in virtual worlds. We may find ourselves alone as we navigate virtual oceans, unravel virtual mysteries, and engineer virtual skyscrapers. But increasingly, when we step through the looking glass, other people are there as well.

The use of the term "cyberspace" to describe virtual worlds grew out of science fiction,[1] but for many of us, cyberspace is now part of the routines of everyday life. When we read our electronic mail or send postings to an electronic bulletin board or make an airline reservation over a computer network, we are in cyberspace. In cyberspace, we can talk, exchange ideas, and assume personae of our own creation. We have the opportunity to build new kinds of communities, virtual communities, in which we participate with people from all over the world, people with whom we converse daily, people with whom we may have fairly intimate relationships but whom we may never physically meet.

[*Life on the Screen,* from which this chapter is drawn] describes how a nascent culture of simulation is affecting our ideas about mind, body, self, and machine. We shall encounter virtual sex and cyberspace marriage, computer psychotherapists, robot insects, and researchers who are trying to build artificial two-year-olds. Biological children, too, are in the story as their play with computer toys leads them to speculate about whether computers are smart and what it is to be alive. Indeed, in much of this, it is our children who are leading the way, and adults who are anxiously trailing behind.

In the story of constructing identity in the culture of simulation, experiences on the Internet figure prominently, but these experiences can only be understood as part of a larger cultural context. That context is the story of the eroding boundaries between the real and the virtual, the animate and the inanimate, the unitary and the multiple self, which is occurring both in advanced scientific fields of research and in the patterns of everyday life. From scientists trying to create artificial life to children "morphing" through a series of virtual personae, we shall see evidence of fundamental shifts in the way we create and experience human identity. But it is on the Internet that our confrontations with technology as it collides with our sense of human identity are fresh, even raw. In the real-time communities of cyberspace, we are dwellers on the threshold between the real and virtual, unsure of our footing, inventing ourselves as we go along.

In an interactive, text-based computer game designed to represent a world inspired by the television series *Star Trek: The Next Generation*, thousands of players spend up to eighty hours a week participating in intergalactic exploration and wars. Through typed descriptions and typed commands, they create characters who have casual and romantic sexual encounters, hold jobs and collect paychecks, attend rituals and celebrations, fall in love and get married. To the participants, such goings-on can be gripping; "This is more real than my real life," says a character who turns out to be a man playing a woman who is pretending to be a man. In this game the self is constructed and the rules of social interaction are built, not received.[2]

In another text-based game, each of nearly ten thousand players creates a character or several characters, specifying their genders and other physical and psychological attributes. The characters need not be human and there are more than two genders. Players are invited to help build the computer world itself. Using a relatively simple programming language, they can create a room in the game space where they are able to set the stage and define the rules. They can fill the room with objects and specify how they work; they can, for instance, create a virtual dog that barks if one types the command "bark Rover." An eleven-year-old player built a room she calls the condo. It is beautifully furnished. She has created magical jewelry and makeup for her dressing table. When she visits the condo, she invites her cyberfriends to join her there, she chats, orders a virtual pizza, and flirts.

LIVING IN THE MUD

The *Star Trek* game, TrekMUSE, and the other, LambdaMOO, are both computer programs that can be accessed through the Internet. The Internet was once available only to military personnel and technical researchers. It is now available to anyone who can buy or borrow an account on a commercial on-line service. TrekMUSE and LambdaMOO are known as MUDs, Multi-User Domains or, with greater historical accuracy, Multi-User Dungeons, because of their genealogy from Dungeons and Dragons, the fantasy role-playing game that swept high schools and colleges in the late 1970s and early 1980s.

The multiuser computer games are based on different kinds of software (this is what the MUSE or MOO or MUSH part of their names stands for). For simplicity, here I use the term MUD to refer to all of them.

MUDs put you in virtual spaces in which you are able to navigate, converse, and build. You join a MUD through a command that links your computer to the computer on which the MUD program resides. Making the connection is not difficult; it requires no particular technical sophistication. The basic commands may seem awkward at first but soon become familiar. For example, if I am playing a character named ST on LambdaMOO, any words I type after the command "say" will appear on all players' screens as "ST says." Any actions I type after the command "emote" will appear after my name just as I type them, as in "ST waves hi" or "ST laughs uncontrollably." I can "whisper" to a designated character and only that character will be able to see my words. As of this writing there are over five hundred MUDs in which hundreds of thousands of people participate.[3] In some MUDs, players are represented by graphical icons; most MUDs are purely text-based. Most players are middle class. A large majority are male. Some players are over thirty, but most are in their early twenties and late teens. However, it is no longer unusual to find MUDs where eight- and nine-year-olds "play" such grade-school icons as Barbie or the Mighty Morphin Power Rangers.

MUDs are a new kind of virtual parlor game and a new form of community. In addition, text-based MUDs are a new form of collaboratively written literature. MUD players are MUD authors, the creators as well as consumers of media content. In this, participating in a MUD has much in common with script writing, performance art, street theater, improvisational theater—or even commedia dell'arte. But MUDs are something else as well.

As players participate, they become authors not only of text but of themselves, constructing new selves through social interaction. One player says, "You are the character and you are not the character, both at the same time." Another says, "You are who you pretend to be." MUDs provide worlds for anonymous social interaction in which one can play a role as close to or as far away from one's "real self" as one chooses. Since one participates in MUDs by sending text to a computer that houses the MUD's program and database, MUD selves are constituted in interaction with the machine. Take it away and the MUD selves cease to exist: "Part of me, a very important part of me, only exists inside PernMUD," says one player. Several players joke that they are like "the electrodes in the computer," trying to express the degree to which they feel part of its space.

On MUDs, one's body is represented by one's own textual description, so the obese can be slender, the beautiful plain, the "nerdy" sophisticated. A *New Yorker* cartoon captures the potential for MUDs as laboratories for experimenting with one's identity. In it, one dog, paw on a computer keyboard, explains to another, "On the Internet, nobody knows you're a dog." The anonymity of MUDs—one is known on the MUD only by the name of one's character or characters—gives people the chance to express multiple and often unexplored aspects of the self, to play with their identity and to try out new ones. MUDs make possible the creation of an identity so fluid and multiple that it strains the limits of the notion. Identity, after all, refers to the sameness between two qualities, in this case between a person and his or her persona. But in MUDs, one can be many.

Dedicated MUD players are often people who work all day with computers at their regular jobs—as architects, programmers, secretaries, students, and stockbrokers. From time to time when playing on MUDs, they can put their characters "to sleep" and pursue "real life" (MUD players call this RL) activities on the computer—all the while remaining connected, logged on to the game's virtual world. Some leave special programs running that send them signals when a particular character logs on or when they are "paged" by a MUD acquaintance. Some leave behind small artificial intelligence programs called bots (derived from the word "robot") running in the MUD that may serve as their alter egos, able to make small talk or answer simple questions. In the course of a day, players move in and out of the active game space. As they do so, some experience their lives as a "cycling through" between the real world, RL, and a series of virtual worlds. I say a series because people are frequently connected to several MUDs at a time. In an MIT computer cluster at 2 A.M., an eighteen-year-old freshman sits at a networked machine and points to the four boxed-off areas on his vibrantly colored computer screen. "On this MUD I'm relaxing, shooting the breeze. On this other MUD I'm in a flame war.[4] On this last one I'm into heavy sexual things. I'm traveling between the MUDs and a physics homework assignment due at 10 tomorrow morning."

This kind of cycling through MUDs and RL is made possible by the existence of those boxed-off areas on the screen, commonly called windows. Windows provide a way for a computer to place you in several contexts at the same time. As a user, you are attentive to only one of the windows on your screen at any given moment, but in a sense you are a presence in all of them at all times. For example, you might be using your computer to help you write a paper about bacteriology. In that case, you would be present to a word-processing program you are using to take notes, to communications software with which you are collecting reference materials from a distant computer, and to a simulation program, which is charting the growth of virtual bacterial colonies. Each of these activities takes place in a window; your identity on the computer is the sum of your distributed presence.

Doug is a midwestern college junior. He plays four characters distributed across three different MUDs. One is a seductive woman. One is a macho, cowboy type whose self-description stresses that he is a "Marlboros rolled in the T-shirt sleeve kind of guy." The third is a rabbit of unspecified gender who wanders its MUD introducing people to each other, a character he calls Carrot. Doug says, "Carrot is so low key that people let it be around while they are having private conversations. So I think of Carrot as my passive, voyeuristic character." Doug's fourth character is one that he plays only on a MUD in which all the characters are furry animals. "I'd rather not even talk about the character because my anonymity there is very important to me," Doug says. "Let's just say that on FurryMUDs I feel like a sexual tourist."[5] Doug talks about playing his characters in windows and says that using windows has made it possible for him to "turn pieces of my mind on and off."

> I split my mind. I'm getting better at it. I can see myself as being two or three or more. And I just turn on one part of my mind and then another when I go from window to window. I'm in some kind of argument in one window and trying to come on to a girl in a MUD in another, and another window might be running a spreadsheet program or some other technical thing for school. . . . And then I'll get a real-time message [that flashes on the screen as soon as it is sent from another system user], and I guess that's RL. It's just one more window.

"RL is just one more window," he repeats, "and it's not usually my best one."

The development of windows for computer interfaces was a technical innovation motivated by the desire to get people working more efficiently by cycling through different applications. But in the daily practice of many computer users, windows have become a powerful metaphor for thinking about the self as a multiple, distributed system. The self is no longer simply playing different roles in different settings at different times, something that a person experiences when, for example, she wakes up as a lover, makes breakfast as a mother, and drives to work as a lawyer. The life practice of windows is that of a decentered self that exists in many worlds and plays many roles at the same time. In traditional theater and in role-playing games that take place in physical space, one steps in and out of character; MUDs, in contrast, offer parallel identities, parallel lives. The experience of this parallelism encourages treating on-screen and off-screen lives with a surprising degree of equality. Experiences on the Internet extend the metaphor of windows—now RL itself, as Doug said, can be "just one more window."

MUDs are dramatic examples of how computer-mediated communication can serve as a place for the construction and reconstruction of identity. There are many others. On the Internet, Internet Relay Chat (commonly known as IRC) is another widely used conversational forum in which any user can open a channel and attract guests to it, all of whom speak to each other as if in the same room. Commercial services such as America Online and CompuServe provide online chat rooms that have much of the appeal of MUDs—a combination of real time interaction with other people, anonymity (or, in some cases, the illusion of anonymity), and the ability to assume a role as close to or as far from one's "real self" as one chooses.

As more people spend more time in these virtual spaces, some go so far as to challenge the idea of giving any priority to RL at all. "After all," says one dedicated MUD player and IRC user, "why grant such superior status to the self that has the body when the selves that don't have bodies are able to have different kinds of experiences?" When people can play at having different genders and different lives, it isn't surprising that for some this play has become as real as what we conventionally think of as their lives, although for them this is no longer a valid distinction.

FRENCH LESSONS

In the late 1960s and early 1970s, I lived in a culture that taught that the self is constituted by and through language, that sexual congress is the exchange of signifiers, and that each of us is a multiplicity of parts, fragments, and desiring connections. This was the hothouse of Paris intellectual culture whose gurus included Jacques Lacan, Michel Foucault, Gilles Deleuze, and Félix Guattari.[6] But despite such ideal conditions for learning, my "French lessons" remained merely abstract exercises. These theorists of poststructuralism and what would come to be called postmodernism spoke words that addressed the relationship between mind and body but, from my point of view, had little or nothing to do with my own.

In my lack of connection with these ideas, I was not alone. To take one example, for many people it is hard to accept any challenge to the idea of an autonomous ego. While in recent years, many psychologists, social theorists, psychoanalysts, and philosophers have argued that the self should be thought of as essentially decentered, the normal requirements of everyday life exert strong pressure on people to take responsibility for their actions and to see themselves as intentional and unitary actors. This disjuncture between theory (the unitary self is an illusion) and lived experience (the unitary self is the most basic reality) is one of the main reasons why multiple and decentered theories have been slow to catch on—or when they do, why we tend to settle back quickly into older, centralized ways of looking at things.

Today I use the personal computer and modem on my desk to access MUDs. Anonymously, I travel their rooms and public spaces (a bar, a lounge, a hot tub). I create several characters, some not of my biological gender, who are able to have social and sexual encounters with other characters. On different MUDs, I have different routines, different friends, different names. One day I learned of a virtual rape. One MUD player had used his skill with the system to seize control of another player's character. In this way the aggressor was able to direct the seized character to submit to a violent sexual encounter. He did all this against the will and over the distraught objections of the player usually "behind" this character, the player to whom this character "belonged." Although some made light of the offender's actions by saying that the episode was just words, in text-based virtual realities such as MUDs, words *are* deeds.

Thus, more than twenty years after meeting the ideas of Lacan, Foucault, Deleuze, and Guattari, I am meeting them again in my new life on the screen. But this time, the Gallic abstractions are more concrete. In my computer-mediated worlds, the self is multiple, fluid, and constituted in interaction with machine connections; it is made and transformed by language; sexual congress is an exchange of signifiers; and understanding follows from navigation and tinkering rather than analysis. And in the machine-generated world of MUDs, I meet characters who put me in a new relationship with my own identity.

One day on a MUD, I came across a reference to a character named Dr. Sherry, a cyberpsychologist with an office in the rambling house that constituted this MUD's virtual geography. There, I was informed, Dr. Sherry was administering questionnaires and conducting interviews about the psychology of MUDs. I suspected that the name Dr. Sherry referred to my long career as a student of the psychological impact of technology. But I didn't create this character. I was not playing her on the MUD. Dr. Sherry was (she is no longer on the MUD) a derivative of me, but she was not mine. The character I played on this MUD had another name—and did not give out questionnaires or conduct interviews. My formal studies were conducted offline in a traditional clinical setting where I spoke face-to-face with people who participate in virtual communities. Dr. Sherry may have been a character someone else created as an efficient way of communicating an interest in questions about technology and the self, but I was experiencing her as a little piece of my history spinning out of control. I tried to quiet my mind. I told myself that surely one's books, one's intellectual identity, one's public persona, are pieces of oneself that others may use as they please. I tried to convince myself that this virtual appropriation was a form of flattery. But my disquiet continued. Dr. Sherry, after all, was not an inanimate book but a person, or at least a person behind a character who was meeting with others in the MUD world.

I talked my disquiet over with a friend who posed the conversation-stopping question, "Well, would you prefer it if Dr. Sherry were a bot trained to interview people about life on the MUD?" (Recall that bots are computer programs that are able to roam cyberspace and interact with characters there.) The idea that Dr. Sherry might be a bot had not occurred to me, but in a flash I realized that this too was possible, even likely. Many bots roam MUDs. They log onto the games as though they were characters. Players create these programs for many reasons: bots help with navigation, pass messages, and create a background atmosphere of animation in the MUD. When you enter a virtual café, you are usually not alone. A waiter bot approaches who asks if you want a drink and delivers it with a smile.

Characters played by people are sometimes mistaken for these little artificial intelligences. This was the case for Doug's character Carrot, because its passive, facilitating persona struck many as one a robot could play. I myself have made this kind of mistake several times, assuming that a person was a program when a character's responses seemed too automatic, too machinelike. And sometimes bots are mistaken for people. I have made this mistake too, fooled by a bot that flattered me by remembering my name or our last interaction. Dr. Sherry could indeed have been one of these. I found myself confronted with a double that could be a person or a program. As things turned out, Dr. Sherry was neither; it was a composite character created by two college students who wished to write a paper on the psychology of MUDs and who were using my name as a kind of trademark or generic descriptor for the idea of a cybershrink.[7] On MUDs, the one can be many and the many can be one.

So not only are MUDs places where the self is multiple and constructed by language, they are places where people and machines are in a new relation to each other, indeed can be mistaken for each other. In such ways, MUDs are evocative objects for thinking about human identity and, more generally, about a set of ideas that have come to be known as "postmodernism."

These ideas are difficult to define simply, but they are characterized by such terms as "decentered," "fluid," "nonlinear," and "opaque." They contrast with modernism, the classical worldview that has dominated Western thinking since the Enlightenment. The modernist view of reality is characterized by such terms as "linear," "logical," "hierarchical," and by having "depths" that can be plumbed and understood. MUDs offer an experience of the abstract postmodern ideas that had intrigued yet confused me during my intellectual coming of age. In this, MUDs exemplify a phenomenon we shall meet often in these pages, that of computer-mediated experiences bringing philosophy down to earth.

In a surprising and counterintuitive twist, in the past decade, the mechanical engines of computers have been grounding the radically nonmechanical philosophy of postmodernism. The online world of the Internet is not the only instance of evocative computer objects and experiences bringing postmodernism down to earth. One of my students at MIT dropped out of a course I teach on social theory, complaining that the writings of the literary theorist Jacques Derrida were simply beyond him. He found that Derrida's dense prose and far-flung philosophical allusions were incomprehensible. The following semester I ran into the student in an MIT cafeteria. "Maybe I wouldn't have to drop out now," he told me. In the past month, with his roommate's acquisition of new software for his Macintosh computer, my student had found his own key to Derrida. That software was a type of hypertext, which allows a computer user to create links between related texts, songs, photographs, and video, as well as to travel along the links made by others. Derrida emphasized that writing is constructed by the audience as well as by the author and that what is absent from the text is as significant as what is present. The student made the following connection:

> Derrida was saying that the messages of the great books are no more written in stone than are the links of a hypertext. I look at my roommate's hypertext stacks and I am able to trace the connections he made and the peculiarities of how he links things together. . . . And the things he might have linked but didn't. The traditional texts are like [elements in] the stack. Meanings are arbitrary, as arbitrary as the links in a stack.

"The cards in a hypertext stack," he concluded, "get their meaning in relation to each other. It's like Derrida. The links have a reason but there is no final truth behind them."[8]

Like experiences on MUDs, the student's story shows how technology is bringing a set of ideas associated with postmodernism—in this case, ideas about the instability of meanings and the lack of universal and knowable truths—into everyday life. In recent years, it has become fashionable to poke fun at postmodern philosophy and lampoon its allusiveness and density. Indeed, I have done some of this myself. But in [*Life on the Screen*] we shall see that through experiences with computers, people come to a certain understanding of postmodernism and to recognize its ability to usefully capture certain aspects of their own experience, both online and off.

In *The Electronic Word*, the classicist Richard A. Lanham argues that open-ended screen text subverts traditional fantasies of a master narrative, or definitive reading, by presenting the reader with possibilities for changing fonts, zooming in and out, and rearranging and replacing text. The result is "a body of work active not passive, a canon not frozen in perfection but volatile with contending human motive."[9] Lanham puts technology and postmodernism together and concludes that the computer is a "fulfillment of social thought." But I believe the relationship is better thought of as a two-way process. Computer technology not only "fulfills the postmodern aesthetic" as Lanham would have it, heightening and concretizing the postmodern experience, but helps that aesthetic hit the street as well as the seminar room. Computers embody postmodern theory and bring it down to earth.

As recently as ten to fifteen years ago, it was almost unthinkable to speak of the computer's involvement with ideas about unstable meanings and unknowable truths.[10] The computer had a clear intellectual identity as a calculating machine. Indeed, when I took an introductory programming course at Harvard in 1978, the professor introduced the computer to the class by calling it a giant calculator. Programming, he reassured us, was a cut and dried technical activity whose rules were crystal clear.

These reassurances captured the essence of what I shall be calling the modernist computational aesthetic. The image of the computer as calculator suggested that no matter how complicated a computer might seem, what happened inside it could be mechanically unpacked. Programming was a technical skill that could be done a right way or a wrong way. The right way was dictated by the computer's calculator essence. The right way was linear and logical. My professor made it clear that this linear, logical calculating machine combined with a structured, rule-based method of writing software offered guidance for thinking not only about technology and programming, but about economics, psychology, and social life. In other words, computational ideas were presented as one of the great modern metanarratives, stories of how the world worked that provided unifying pictures and analyzed complicated things by breaking them down into simpler parts. The modernist computational aesthetic promised to explain and unpack, to reduce and clarify. Although the computer culture was never monolithic, always including dissenters and deviant subcultures, for many years its professional mainstream (including computer scientists, engineers, economists, and cognitive scientists) shared this clear intellectual direction. Computers, it was assumed, would become more powerful, both as tools and as metaphors, by becoming better and faster calculating machines, better and faster analytical engines.

FROM A CULTURE OF CALCULATION
TOWARD A CULTURE OF SIMULATION

Most people over thirty years old (and even many younger ones) have had an introduction to computers similar to the one I received in that programming course. But from today's perspective, the fundamental lessons of computing that I was taught are wrong. First of all, programming is no longer cut and dried. Indeed, even its dimensions have become elusive. Are you programming when you customize your word-processing software? When you design "organisms" to populate a simulation of Darwinian evolution in a computer game called SimLife? Or when you build a room in a MUD so that opening a door to it will cause "Happy Un-Birthday" to ring out on all but one day of the year? In a sense, these activities are forms of programming, but that sense is radically different from the one presented in my 1978 computer course.

The lessons of computing today have little to do with calculation and rules; instead they concern simulation, navigation, and interaction. The very image of the computer as a giant calculator has become quaint and dated. Of course, there is still "calculation" going on within the computer, but it is no longer the important or interesting level to think about or interact with. Fifteen years ago, most computer users were limited to typing commands. Today they use off-the-shelf products to manipulate simulated desktops, draw with simulated paints and brushes, and fly in simulated airplane cockpits. The computer culture's center of gravity has shifted decisively to people who do not think of themselves as programmers. The computer science research community as well as industry pundits maintain that in the near future we can expect to interact with computers by communicating with simulated people on our screens, agents who will help organize our personal and professional lives.

On my daughter's third birthday she received a computer game called The Playroom, among the most popular pieces of software for the preschool set. If you ask for help, The Playroom offers an instruction that is one sentence long: "Just move the cursor to any object, click on it, explore and have fun." During the same week that my daughter learned to click in The Playroom, a colleague gave me my first lesson on how to use the World Wide Web, a cyberconstruct that links text, graphics, video, and audio on computers all over the world. Her instructions were almost identical to those I had just read to my daughter: "Just move the cursor to any underlined word or phrase, click on it, explore, and have fun." When I wrote this text in January 1995, the Microsoft Corporation had just introduced Bob, a "social" interface for its Windows operating system, the most widely used operating system for personal computers in the world.[11] Bob, a computer agent with a human face and "personality," operates within a screen environment designed to look like a living room that is in almost every sense a playroom for adults. In my daughter's screen playroom, she is presented with such objects as alphabet blocks and a clock for learning to tell time. Bob offers adults a word processor, a fax machine, a telephone. Children and adults are united in the actions they take in virtual worlds. Both move the cursor and click.

The meaning of the computer presence in people's lives is very different from what most expected in the late 1970s. One way to describe what has happened is to say that we are moving from a modernist culture of calculation toward a postmodernist culture of simulation.

The culture of simulation is emerging in many domains. It is affecting our understanding of our minds and our bodies. For example, fifteen years ago, the computational models of mind that dominated academic psychology were modernist in spirit: Nearly all tried to describe the mind in terms of centralized structures and programmed rules. In contrast, today's models often embrace a postmodern aesthetic of complexity and decentering. Mainstream computer researchers no longer aspire to program intelligence into computers but expect intelligence to emerge from the interactions of small subprograms. If these emergent simulations are "opaque," that is, too complex to be completely analyzed, this is not necessarily a problem. After all, these theorists say, our brains are opaque to us, but this has never prevented them from functioning perfectly well as minds.

Fifteen years ago in popular culture, people were just getting used to the idea that computers could project and extend a person's intellect.

Today people are embracing the notion that computers may extend an individual's physical presence. Some people use computers to extend their physical presence via real-time video links and shared virtual conference rooms. Some use computer-mediated screen communication for sexual encounters. An Internet list of "Frequently Asked Questions" describes the latter activity—known as netsex, cybersex, and (in MUDs) TinySex—as people typing messages with erotic content to each other, "sometimes with one hand on the keyset, sometimes with two."

Many people who engage in netsex say that they are constantly surprised by how emotionally and physically powerful it can be. They insist that it demonstrates the truth of the adage that ninety percent of sex takes place in the mind. This is certainly not a new idea, but netsex has made it commonplace among teenage boys, a social group not usually known for its sophistication about such matters. A seventeen-year-old high school student tells me that he tries to make his erotic communications on the net "exciting and thrilling and sort of imaginative." In contrast, he admits that before he used computer communication for erotic purposes he thought about his sexual life in terms of "trying [almost always unsuccessfully] to get laid." A sixteen-year-old has a similar report on his cyberpassage to greater sensitivity: "Before I was on the net, I used to masturbate with *Playboy*; now I do netsex on DinoMUD[12] with a woman in another state." When I ask how the two experiences differ, he replies:

> With netsex, it is fantasies. My MUD lover doesn't want to meet me in RL. With *Playboy*, it was fantasies too, but in the MUD there is also the other person. So I don't think of what I do on the MUD as masturbation. Although, you might say that I'm the only one who's touching me. But in netsex, I have to think of fantasies she will like too. So now, I see fantasies as something that's part of sex with two people, not just me in my room.

Sexual encounters in cyberspace are only one (albeit well-publicized) element of our new lives on the screen. Virtual communities ranging from MUDs to computer bulletin boards allow people to generate experiences, relationships, identities, and living spaces that arise only through interaction with technology. In the many thousands of hours that Mike, a college freshman in Kansas, has been logged on to his favorite MUD, he has created an apartment with rooms, furniture, books, desk, and even a small computer. Its interior is exquisitely detailed, even though it exists only in textual cyberspace. "It's where I live," Mike says. "More than I do in my dingy dorm room. There's no place like home."

As human beings become increasingly intertwined with the technology and with each other via the technology, old distinctions between what is specifically human and specifically technological become more complex. Are we living life *on* the screen or life *in* the screen? Our new technologically enmeshed relationships oblige us to ask to what extent we ourselves have become cyborgs, transgressive mixtures of biology, technology, and code.[13] The traditional distance between people and machines has become harder to maintain.

Writing in his diary in 1832, Ralph Waldo Emerson reflected that "Dreams and beasts are two keys by which we are to find out the secrets of our nature . . . they are our test objects."[14] Emerson was prescient. Freud and his heirs would measure human rationality against the dream. Darwin and his heirs would insist that we measure human nature against nature itself—the world of the beasts seen as our forbears and kin. If Emerson had lived at the end of the twentieth century, he would surely have seen the computer as a new test object. Like dreams and beasts, the computer stands on the margins. It is a mind that is not yet a mind. It is inanimate yet interactive. It does not think, yet neither is it external to thought. It is an object, ultimately a mechanism, but it behaves, interacts, and seems in a certain sense to know. It confronts us with an uneasy sense of kinship. After all, we too behave, interact, and seem to know, and yet are ultimately made of matter and programmed DNA. We think we can think. But can *it* think? Could it have the capacity to feel? Could it ever be said to be alive?

Dreams and beasts were the test objects for Freud and Darwin, the test objects for modernism. In the past decade, the computer has become the test object for postmodernism. The computer takes us beyond a world of dreams and beasts because it enables us to contemplate mental life that exists apart from bodies. It enables us to contemplate dreams that do not need beasts. The computer is an evocative object that causes old boundaries to be renegotiated.

[*Life on the Screen*] traces a set of such boundary negotiations. It is a reflection on the role that technology is playing in the creation of a new social and cultural sensibility. I have observed and participated in settings, physical and virtual, where people and computers come together. Over the past decade, I have talked to more than a thousand people, nearly three hundred of them children, about their experience of using computers or computational objects to program, to navigate, to write, to build, to experiment, or to communicate. In a sense, I have interrogated the computers as well. What messages, both explicit and implicit, have they carried for their human users about what is possible and what is impossible, about what is valuable and what is unimportant?

In the spirit of Whitman's reflections on the child, I want to know what we are becoming if the first objects we look upon each day are simulations into which we deploy our virtual selves. In other words, [I am not writing] about computers. Rather, [I am writing] about the intense relationships people have with computers and how these relationships are changing the way we think and feel. Along with the movement from a culture of calculation toward a culture of simulation have come changes in what computers do *for* us and in what they do *to* us—to our relationships and our ways of thinking about ourselves.

We have become accustomed to opaque technology. As the processing power of computers increased exponentially, it became possible to use that power to build graphical user interfaces, commonly known by the acronym GUI, that hid the bare machine from its user. The new opaque interfaces—most specifically, the Macintosh iconic style of interface, which simulates the space of a desktop as well as communication through dialogue—represented more than a technical change. These new interfaces modeled a way of understanding that depended on getting to know a computer through interacting with it, as one might get to know a person or explore a town.

The early personal computers of the 1970s and the IBM PC of the early 1980s presented themselves as open, "transparent," potentially reducible to their underlying mechanisms. These were systems that invited users to imagine that they could understand its "gears" as they turned, even if very few people ever tried to reach that level of understanding. When people say that they used to be able to "see" what was "inside" their first personal computers, it is important to keep in mind that for most of them there still remained many intermediate levels of software between them and the bare machine. But their computer systems encouraged them to represent their understanding of the technology as knowledge of what lay beneath the screen surface. They were encouraged to think of understanding as looking beyond the magic to the mechanism.

In contrast, the 1984 introduction of the Macintosh's iconic style presented the public with simulations (the icons of file folders, a trash can, a desktop) that did nothing to suggest how their underlying structure could be known. It seemed unavailable, visible only through its effects. As one user said, "The Mac looked perfect, finished. To install a program on my DOS machine, I had to fiddle with things. It clearly wasn't perfect. With the Mac, the system told me to stay on the surface." This is the kind of involvement with computers that has come to dominate the field; no longer associated only with the Macintosh, it is nearly universal in personal computing.

We have learned to take things at interface value. We are moving toward a culture of simulation in which people are increasingly comfortable with substituting representations of reality for the real. We use a Macintosh-style "desktop" as well as one on four legs. We join virtual communities that exist only among people communicating on computer networks as well as communities in which we are physically present. We come to question simple distinctions between real and artificial. In what sense should one consider a screen desktop less real than any other? The screen desktop I am currently using has a folder on it labeled "Professional Life." It contains my business correspondence, date book, and telephone directory. Another folder, labeled "Courses," contains syllabuses, reading assignments, class lists, and lecture notes. A third, "Current Work," contains my research notes and this book's drafts. I feel no sense of unreality in my relationship to any of these objects. The culture of simulation encourages me to take what I see on the screen "at (inter)face value." In the culture of simulation, if it works for you, it has all the reality it needs.

The habit of taking things at interface value is new, but it has gone quite far. For example, a decade ago, the idea of a conversation with a computer about emotional matters, the image of a computer psychotherapist, struck most people as inappropriate or even obscene. Today, several such programs are on the market, and they tend to provoke a very different and quite pragmatic response. People are most likely to say, "Might as well try it. It might help. What's the harm?"

We have used our relationships with technology to reflect on the human. A decade ago, people were often made nervous by the idea of thinking about computers in human terms. Behind their anxiety was distress at the idea that their own minds might be similar to a computer's "mind." This reaction against the formalism and rationality of the machine was romantic.

I use this term to analogize our cultural response to computing to nineteenth-century Romanticism. I do not mean to suggest that it was merely an emotional response. [Rather,] it expressed serious philosophical resistance to any view of people that denied their complexity and continuing mystery. This response emphasized not only the richness of human emotion but the flexibility of human thought and the degree to which knowledge arises in subtle interaction with the environment. Humans, it insists, have to be something very different from mere calculating machines.

In the mid-1980s, this romantic reaction was met by a movement in computer science toward the research and design of increasingly "romantic machines." These machines were touted not as logical but as biological, not as programmed but as able to learn from experience. The researchers who worked on them said they sought a species of machine that would prove as unpredictable and undetermined as the human mind itself. The cultural presence of these romantic machines encouraged a new discourse; both persons and objects were reconfigured, machines as psychological objects, people as living machines.

But even as people have come to greater acceptance of a kinship between computers and human minds, they have also begun to pursue a new set of boundary questions about things and people. After several decades of asking, "What does it mean to think?" the question at the end of the twentieth century is, "What does it mean to be alive?" We are positioned for yet another romantic reaction, this time emphasizing biology, physical embodiment, the question of whether an artifact can be a life.[15]

These psychological and philosophical effects of the computer presence are by no means confined to adults. Like their parents, and often before their parents, the children of the early 1980s began to think of computers and computer toys as psychological objects because these machines combined mind activities (talking, singing, spelling, game playing, and doing math), an interactive style, and an opaque surface. But the children, too, had a romantic reaction, and came to define people as those emotional and unprogrammable things that computers were not. Nevertheless, from the moment children gave up on mechanistic understandings and saw the computer as a psychological entity, they began to draw computers closer to themselves. Today children may refer to the computers in their homes and classrooms as "just machines," but qualities that used to be ascribed only to people are now ascribed to computers as well. Among children, the past decade has seen a movement from defining people as what machines are not to believing that the computational objects of everyday life think and know while remaining "just machines."

In the past decade, the changes in the intellectual identity and cultural impact of the computer have taken place in a culture still deeply attached to the quest for a modernist understanding of the mechanisms of life. Larger scientific and cultural trends, among them advances in psychopharmacology and the development of genetics as a computational biology, reflect the extent to which we assume ourselves to be like machines whose inner workings we can understand. "Do we have our emotions," asks a college sophomore whose mother has been transformed by taking antidepressant medication, "or do our emotions have us?" To whom is one listening when one is "listening to Prozac"?[16] The aim of the Human Genome Project is to specify the location and role of all the genes in human DNA. The Project is often justified on the grounds that it promises to find the pieces of our genetic code responsible for many human diseases so that these may be better treated, perhaps by genetic reengineering. But talk about the Project also addresses the possibility of finding the genetic markers that determine human personality, temperament, and sexual orientation. As we contemplate reengineering the genome, we are also reengineering our view of ourselves as programmed beings.[17] Any romantic reaction that relies on biology as the bottom line is fragile, because it is building on shifting ground. Biology is appropriating computer technology's older, modernist models of computation while at the same time computer scientists are aspiring to develop a new opaque, emergent biology that is closer to the postmodern culture of simulation.[18]

Today, more lifelike machines sit on our desktops, computer science uses biological concepts, and human biology is recast in terms of deciphering a code. With descriptions of the brain that explicitly invoke computers and images of computers that explicitly invoke the brain, we have reached a cultural watershed. The rethinking of human and machine identity is not taking place just among philosophers but "on the ground," through a philosophy in everyday life that is in some measure both provoked and carried by the computer presence.

We have sought out the subjective computer. Computers don't just do things for us, they do things to us, including to our ways of thinking about ourselves and other people. A decade ago, such subjective effects of the computer presence were secondary in the sense that they were not the ones being sought.[19] Today, things are often the other way around. People explicitly turn to computers for experiences that they hope will change their ways of thinking or will affect their social and emotional lives. When people explore simulation games and fantasy worlds or log on to a community where they have virtual friends and lovers, they are not thinking of the computer as what Charles Babbage, the nineteenth-century mathematician who invented the first programmable machine, called an analytical engine. They are seeking out the computer as an intimate machine.

You might think from its title that [*Life on the Screen*] was a book about filmgoers and the ways that a fan—the heroine of Woody Allen's *The Purple Rose of Cairo*, for example—might project himself or herself into favorite movies. But here I argue that it is computer screens where we project ourselves into our own dramas, dramas in which we are producer, director, and star. Some of these dramas are private, but increasingly we are able to draw in other people. Computer screens are the new location for our fantasies, both erotic and intellectual. We are using life on computer screens to become comfortable with new ways of thinking about evolution, relationships, sexuality, politics, and identity. . . .

NOTES

1. William Gibson, *Neuromancer* (New York: Ace, 1984).
2. For a general introduction to LambdaMOO and MUDding, see Pavel Curtis, "Mudding: Social Phenomena in Text-Based Virtual Realities," available via anonymous ftp://parcftp.xerox.com/pub/MOO/papers/DIAC92.*; Amy Bruckman, "Identity Workshop: Emergent Social and Psychological Phenomena in Text-Based Virtual Reality," unpub. ms., March 1992, via anonymous ftp://media.mit.edu/pub/asb/papers/identity-workshop.*; and the chapter on MUDs in Howard Rheingold's *Virtual Community: Homesteading on the Electronic Frontier* (New York: Addison-Wesley, 1993). On virtual community in general, see Allucquere Rosanne Stone, "Will the Real Body Please Stand Up?: Boundary Stories about Virtual Cultures," in *Cyberspace: First Steps*, ed. Michael Benedikt (Cambridge, Mass.: MIT Press, 1992), pp. 81–118. The asterisk in a net address indicates that the document is available in several formats.
3. The number of MUDs is changing rapidly. Most estimates place it at over five hundred, but an increasing number are private and so without any official "listing." The software on which they are based (and which gives them their names as MOOs, MUSHes, MUSEs, etc.) determines several things about the game; among these is the general layout of the game space. For example, in the class of MUDs known as AberMUDs, the center of town is similar from one game to another, but the mountains, castles, and forests that surround the town are different in different games, because these have been built specifically for that game by its resident "wizards." MUDs also differ in their governance. In MUD parlance, wizards are administra-

tors; they usually achieve this status through virtuosity in the game. In AberMUDs only wizards have the right to build onto the game. In other kinds of MUDs, all players are invited to build. Who has the right to build and how building is monitored (for example, whether the MUD government should allow a player to build a machine that would destroy other players' property or characters) is an important feature that distinguishes types of MUDs. Although it may be technically correct to refer to being in a MUD (as in a dungeon), it is also common to speak of being on a MUD (as in logging on to a program). To me, the dual usage reflects the ambiguity of cyberspace as both space and program. I (and my informants) use both in this chapter.

4. A flame war is computer culture jargon for an incendiary expression of differences of opinion. In flame wars, participants give themselves permission to state their positions in strong, even outrageous terms with little room for compromise.

5. I promised Doug anonymity, a promise I made to all the people I interviewed in researching [Life on the Screen]. Doug has been told that his name will be changed, his identity disguised, and the names and distinguishing features of his MUD characters altered. It is striking that even given these reassurances, which enable him to have an open conversation with me about his social and sexual activities on MUDs, he wants to protect his FurryMUD character.

6. I immersed myself in these "French lessons," first in the aftermath of the May 1968 student revolt, a revolt in which Lacan and Foucault became intellectual heroes. Later, in 1973–74, the immersion continued while I studied the intellectual fallout of that failed revolution. That fallout included a love affair with things Freudian and an attack on unitary models of self. While followers of Lacan relied on interpretations of Freud that challenged models of centralized ego, Deleuze and Guattari proposed more radical views that described the self as a multiplicity of desiring machines. See Gilles Deleuze and Félix Guattari, Anti-Oedipus: Capitalism and Schizophrenia, trans. Robert Hurley, Mark Seem, and Helen R. Lane (Minneapolis: University of Minnesota Press, 1983).

7. Jill Serpentelli, "Conversational Structure and Personality Correlates of Electronic Communication," unpub. ms., 1992.

8. The student's association of Derrida and hypertext may be unsophisticated, but it is far from outlandish. See, for example, George P. Landow, Hypertext: The Convergence of Critical Theory and Technology (Baltimore: Johns Hopkins, 1992), pp. 1–34; and in George P. Landow and Paul Delany, eds., Hypermedia and Literary Studies (Cambridge, Mass.: MIT Press, 1991).

9. Richard A. Lanham, The Electronic Word: Democracy, Technology, and the Arts (Chicago: The University of Chicago Press, 1993), p. 51. George Landow sees critical theory and technology in the midst of a "convergence." See Landow, Hypertext.

10. I say almost unthinkable because a small number of postmodern writers had begun to associate their work with the possibilities of computer technology. See, in particular, Jean-François Lyotard, The Postmodern Condition: A Re-

port on Knowledge, trans. Geoff Bennington and Brian Massumi (Minneapolis: University of Minnesota Press, 1984).

11. *The Wall Street Journal,* 3 January 1995: A3, A4, and *The Wall Street Journal,* 10 January 1995: B1, B3.
12. Here I have changed the name of the MUD (there is to my knowledge no DinoMUD) to protect the confidentiality I promise all informants. I use the real name of a MUD when it is important to my account and will not compromise confidentiality.
13. See, for example, Donna Haraway, "A Manifesto for Cyborgs: Science, Technology, and Socialist Feminism in the 1980s," *Socialist Review* 80 (March–April 1985): 65–107.
14. The quotation is from a journal entry by Emerson in January 1832. The passage reads in full, "Dreams and beasts are two keys by which we are to find out the secrets of our nature. All mystics use them. They are like comparative anatomy. They are our test objects." See Joel Porte, ed., *Emerson in His Journals* (Cambridge, Mass.: Belknap Press, 1982), p. 81.
15. In a recent review of the place of genetics in contemporary popular culture, Dorothy Nelkin and Susan Lindee have said: "DNA has taken on the social and cultural functions of the soul." See their *The DNA Mystique: The Gene as a Cultural Icon* (San Francisco and New York: W. H. Freeman, 1995), p. 42.
16. Peter Kramer, *Listening to Prozac: A Psychiatrist Explores Mood-Altering Drugs and the New Meaning of the Self* (New York: Viking, 1993).
17. Nelkin and Lindee's *The DNA Mystique* documents the degree to which genetic essentialism dominates American popular culture today.
18. Evelyn Fox Keller, "The Body of a New Machine: Situating the Organism Between Telegraphs and Computers," *Perspectives on Science* 2, no. 3 (1994): 302–23.
19. For a view of this matter from the perspective of the 1980s, see Sherry Turkle, *The Second Self: Computers and the Human Spirit* (New York: Simon & Schuster, 1984).

24.   Less Work for Mother?

RUTH SCHWARTZ COWAN

In many ways household technology represents a polar opposite to the kinds of high technology involved in genetic research or computer networks. Vacuum cleaners, refrigerators, electric ranges, and washing machines—to most Americans these are the mundane features of everyday life. Yet, within the memory of many people living today, they were the leading edge of a revolutionary new laborsaving technology that promised to liberate American women from a life of drudgery.

Although few would argue that these and other household technologies (such as microwave ovens) have not made life easier for millions of women (and men), the average woman who is not employed outside of the home devotes just about the same amount of time to taking care of a household as did her mother and grandmother earlier in this century. Add to that fact that today a great many women not only take care of a family and household but also work full-time outside the home, and one can easily be led to question whether these improved technologies mean progress.

In "Less Work for Mother?" Ruth Schwartz Cowan, a professor of history at the State University of New York at Stony Brook, explores the social context and meaning of changing household technology in the United States. Although the selection views the topic from a historical perspective rather than speculating about future developments, it is highly illuminating with regard to how technological change interacts with social change to shape human experience. Cowan is author of More Work for Mother: The Ironies of Household Technology from the Open Hearth to the Microwave (1983) and The Social History of American Technology (1997).

Source: From *American Heritage of Invention and Technology*, vol. 2, no. 3 (1987). With the permission of the author.

Things are never what they seem. Skimmed milk masquerades as cream. And laborsaving household appliances often do not save labor. This is the surprising conclusion reached by a small army of historians, sociologists, and home economists who have undertaken, in recent years, to study the one form of work that has turned out to be most resistant to inquiry and analysis—namely, housework.

During the first half of the twentieth century, the average American household was transformed by the introduction of a group of machines that profoundly altered the daily lives of housewives; the forty years between 1920 and 1960 witnessed what might be aptly called the "industrial revolution in the home." Where once there had been a wood- or coal-burning stove there now was a gas or electric range. Clothes that had once been scrubbed on a metal washboard were now tossed into a tub and cleansed by an electrically driven agitator. The dryer replaced the clothesline; the vacuum cleaner replaced the broom; the refrigerator replaced the icebox and the root cellar; an automatic pump, some piping, and a tap replaced the hand pump, the bucket, and the well. No one had to chop and haul wood any more. No one had to shovel out ashes or beat rugs or carry water; no one even had to toss egg whites with a fork for an hour to make an angel food cake.

And yet American housewives in 1960, 1970, and even 1980 continued to log about the same number of hours at their work as their grandmothers and mothers had in 1910, 1920, and 1930. The earliest time studies of housewives date from the very same period in which time studies of other workers were becoming popular—the first three decades of the twentieth century. The sample sizes of these studies were usually quite small, and they did not always define housework in precisely the same way (some counted an hour spent taking children to the playground as "work," while others called it "leisure"), but their results were more or less consistent: whether rural or urban, the average American housewife performed fifty to sixty hours of unpaid work in her home every week, and the only variable that significantly altered this was the number of small children.

A half century later not much had changed. Survey research had become much more sophisticated, and sample sizes had grown considerably, but the results of the time studies remained surprisingly consistent. The average American housewife, now armed with dozens of motors and thousands of electronic chips, still spends fifty to sixty hours a week doing housework. The only variable that significantly altered the size of that number was full-time employment in the labor force; "working" housewives cut down the average number of hours that they spend cooking and cleaning, shopping and chauffeuring, to a not insignificant thirty-five—virtually the equivalent of another full-time job.

How can this be true? Surely even the most sophisticated advertising copywriter of all times could not fool almost the entire American population over the course of at least three generations. Laborsaving devices must be saving something or Americans would not continue, year after year, to plunk down their hard-earned dollars for them.

And if laborsaving devices have not saved labor in the home, then what is it that has suddenly made it possible for more than 70 percent of the wives and mothers in the American population to enter the work force and stay there? A brief glance at the histories of some of the technologies that have transformed housework in the twentieth century will help us answer some of these questions.

The portable vacuum cleaner was one of the earliest electric appliances to make its appearance in American homes, and reasonably priced models appeared on the retail market as early as 1910. For decades prior to the turn of the century, inventors had been trying to create a carpet-cleaning system that would improve on the carpet sweeper with adjustable rotary brushes (patented by Melville Bissell in 1876), or the semiannual ritual of hauling rugs outside and beating them, or the practice of regularly sweeping the dirt out of a rug that had been covered with dampened, torn newspapers. Early efforts to solve the problem had focused on the use of large steam, gasoline, or electric motors attached to piston-type pumps and lots of hoses. Many of these "stationary" vacuum-cleaning systems were installed in apartment houses or hotels, but some were hauled around the streets in horse-drawn carriages by entrepreneurs hoping to establish themselves as "professional housecleaners."

In the first decade of the twentieth century, when fractional-horsepower electric motors became widely—and inexpensively—available, the portable vacuum cleaner intended for use in an individual household was born. One early model—invented by a woman, Corrine Dufour—consisted of a rotary brush, an electrically driven fan, and a wet sponge for absorbing the dust and dirt. Another, patented by David E. Kenney in 1907, had a twelve-inch nozzle, attached to a metal tube, attached to a flexible hose that led to a vacuum pump and separating devices. The Hoover, which was based on a brush, a fan, and a collecting bag, was on the market by 1908.

The Electrolux, the first of the canister types of cleaner, which could vacuum something above the level of the floor, was brought over from Sweden in 1924 and met with immediate success.

These early vacuum cleaners were hardly a breeze to operate. All were heavy, and most were extremely cumbersome to boot. One early home economist mounted a basal metabolism machine on the back of one of her hapless students and proceeded to determine that more energy was expended in the effort to clean a sample carpet with a vacuum cleaner than when the same carpet was attacked with a hard broom. The difference, of course, was that the vacuum cleaner did a better job, at least on carpets, because a good deal of what the broom stirred up simply resettled a foot or two away from where it had first been lodged. Whatever the liabilities of the early vacuum cleaners may have been, Americans nonetheless appreciated their virtues; according to a market survey done in Zanesville, Ohio, in 1926, slightly more than half the households owned one. Eventually improvements in the design made these devices easier to operate. By 1960 vacuum cleaners could be found in 70 percent of the nation's homes.

When the vacuum cleaner is viewed in a historical context, however, it is easy to see why it did not save housewifely labor. Its introduction coincided almost precisely with the disappearance of the domestic servant. The number of persons engaged in household service dropped from 1,851,000 in 1910 to 1,411,000 in 1920, while the number of households enumerated in the census rose from 20.3 million to 24.4 million. Moreover, between 1900 and 1920 the number of household servants per thousand persons dropped from 98.9 to 58.0, while during the 1920s the decline was even more precipitous as the restrictive immigration acts dried up what had once been the single most abundant source of domestic labor.

For the most economically comfortable segment of the population, this meant just one thing: the adult female head of the household was doing more housework than she had ever done before. What Maggie had once done with a broom, Mrs. Smith was now doing with a vacuum cleaner. Knowing that this was happening, several early copywriters for vacuum cleaner advertisements focused on its implications. The vacuum cleaner, General Electric announced in 1918, is better than a maid: it doesn't quit, get drunk, or demand higher wages. The switch from Maggie to Mrs. Smith shows up, in time-study statistics, as an increase in the time that Mrs. Smith is spending at her work.

For those—and they were the vast majority of the population—who were not economically comfortable, the vacuum cleaner implied something else again: not an increase in the time spent in housework but an increase in the standard of living. In many households across the country, acquisition of a vacuum cleaner was connected to an expansion of living space, the move from a small apartment to a small house, the purchase of wall-to-wall carpeting. If this did not happen during the difficult 1930s, it became more possible during the expansive 1950s. As living quarters grew larger, standards for their upkeep increased; rugs had to be vacuumed every week, in some households every day, rather than semiannually, as had been customary. The net result, of course, was that when armed with a vacuum cleaner, housewives whose parents had been poor could keep more space cleaner than their mothers and grandmothers would have ever believed possible. We might put this everyday phenomenon in language that economists can understand: The introduction of the vacuum cleaner led to improvements in productivity but not to any significant decrease in the amount of time expended by each worker.

The history of the washing machine illustrates a similar phenomenon. "Blue Monday" had traditionally been, as its name implies, the bane of a housewife's existence—especially when Monday turned out to be "Monday . . . and Tuesday to do the ironing." Thousands of patents for "new and improved" washers were issued during the nineteenth century in an effort to cash in on the housewife's despair. Most of these early washing machines were wooden or metal tubs combined with some kind of hand-cranked mechanism that would rub or push or twirl laundry when the tub was filled with water and soap. At the end of the century, the Sears catalog offered four such washing machines, ranging in price from $2.50 to $4.25, all sold in combination with hand-cranked wringers.

These early machines may have saved time in the laundering process (four shirts could be washed at once instead of each having to be rubbed separately against a washboard), but they probably didn't save much energy. Lacking taps and drains, the tubs still had to be filled and emptied by hand, and each piece still had to be run through a wringer and hung up to dry.

Not long after the appearance of fractional-horsepower motors, several enterprising manufacturers had the idea of hooking them up to the crank mechanisms of washers and wringers—and the electric washer was born. By the 1920s, when mass production of such machines began, both the general structure of the machine (a central-shaft agitator rotating within a cylindrical tub, hooked up to the household water supply) and the general structure of the industry (oligopolistic—with a very few firms holding most of the patents and controlling most of the market) had achieved their final form. By 1926 just over a quarter of the families in Zanesville had an electric washer, but by 1941 fully 52 percent of all American households either owned or had interior access (which means that they could use coin-operated models installed in the basements of apartment houses) to such a machine. The automatic washer, which consisted of a vertically rotating washer cylinder that could also act as a centrifugal extractor, was introduced by the Bendix Home Appliance Corporation in 1938, but it remained expensive, and therefore inaccessible, until after World War II. This machine contained timing devices that allowed it to proceed through its various cycles automatically; by spinning the clothes around in the extractor phase of its cycle, it also eliminated the wringer. Although the Bendix subsequently disappeared from the retail market (versions of this sturdy machine may still be found in Laundromats), its design principles are replicated in the agitator washers that currently chug away in millions of American homes.

Both the early wringer washers and their more recent automatic cousins have released American women from the burden of drudgery. No one who has ever tried to launder a sheet by hand, and without the benefits of hot running water, would want to return to the days of the scrubboard and tub. But "labor" is composed of both "energy expenditure" and "time expenditure," and the history of laundry work demonstrates that the one may be conserved while the other is not.

The reason for this is, as with the vacuum cleaner, twofold. In the early decades of the century, many households employed laundresses to do their wash; this was true, surprisingly enough, even for some very poor households when wives and mothers were disabled or employed full-time in field or factory. Other households—rich and poor—used commercial laundry services. Large, mechanized "steam" laundries were first constructed in this country in the 1860s, and by the 1920s they could be found in virtually every urban neighborhood and many rural ones as well.

But the advent of the electric home washer spelled doom both for the laundress and for the commercial laundry; since the housewife's labor was unpaid, and since the washer took so much of the drudgery out of washday, the one-time expenditure for a machine seemed, in many families, a more sensible arrangement than continuous expenditure for domestic services. In the process, of course, the time spent on laundry work by the individual housewife, who had previously employed either a laundress or a service, was bound to increase.

For those who had not previously enjoyed the benefits of relief from washday drudgery, the electric washer meant something quite different but equally significant: an upgrading of household cleanliness. Men stopped wearing removable collars and cuffs, which meant that the whole of their shirts had to be washed and then ironed. Housewives began changing two sheets every week, instead of moving the top sheet to the bottom and adding only one that was fresh. Teenagers began changing their underwear every day instead of every weekend. In the early 1960s, when synthetic no-iron fabrics were introduced, the size of the household laundry load increased again; shirts and skirts, sheets and blouses that had once been sent out to the dry cleaner or the corner laundry were now being tossed into the household wash basket. By the 1980s the average American housewife, armed now with an automatic washing machine and an automatic dryer, was processing roughly ten times (by weight) the amount of laundry that her mother had been accustomed to. Drudgery had disappeared, but the laundry hadn't. The average time spent on this chore in 1925 had been 5.8 hours per week; in 1964 it was 6.2.

And then there is the automobile. We do not usually think of our cars as household appliances, but that is precisely what they are since housework, as currently understood, could not possibly be performed without them. The average American housewife is today more likely to be found behind a steering wheel than in front of a stove. While writing this article I interrupted myself five times: once to take a child to field-hockey practice, then a second time, to bring her back when practice was finished; once to pick up some groceries at the supermarket; once to retrieve my husband, who was stranded at the train station; once for a trip to a doctor's office. Each time I was doing housework, and each time I had to use my car.

Like the washing machine and the vacuum cleaner, the automobile started to transform the nature of housework in the 1920s. Until the introduction of the Model T in 1908, automobiles had been playthings for the idle rich, and although many wealthy women learned to drive early in the century (and several participated in well-publicized auto races), they were hardly the women who were likely to be using their cars to haul groceries.

But by 1920, and certainly by 1930, all this had changed. Helen and

Robert Lynd, who conducted an intensive study of Muncie, Indiana, between 1923 and 1925 (reported in their famous book *Middletown*), estimated that in Muncie in the 1890s only 125 families, all members of the "elite," owned a horse and buggy, but by 1923 there were 6,222 passenger cars in the city, "roughly one for every 7.1 persons, or two for every three families." By 1930, according to national statistics, there were roughly 30 million households in the United States—and 26 million registered automobiles.

What did the automobile mean for the housewife? Unlike public transportation systems, it was convenient. Located right at her doorstep, it could deposit her at the doorstep that she wanted or needed to visit. And unlike the bicycle or her own two feet, the automobile could carry bulky packages as well as several additional people. Acquisition of an automobile therefore meant that a housewife, once she had learned how to drive, could become her own door-to-door delivery service. And as more housewives acquired automobiles, more business-men discovered the joys of dispensing with delivery services—par-ticularly during the Depression.

To make a long story short, the iceman does not cometh anymore. Neither does the milkman, the bakery truck, the butcher, the grocer, the knife sharpener, the seamstress, or the doctor. Like many other businessmen, doctors discovered that their earnings increased when they stayed in their offices and transferred the responsibility for trans-portation to their ambulatory patients.

Thus a new category was added to the housewife's traditional job description: chauffeur. The suburban station wagon is now "Mom's Taxi." Children who once walked to school now have to be transported by their mothers; husbands who once walked home from work now have to be picked up by their wives; groceries that once were dispensed from pushcarts or horse-drawn wagons now have to be packed into paper bags and hauled home in family cars. "Contemporary women," one time-study expert reported in 1974, "spend about one full working day per week on the road and in stores compared with less than two hours per week for women in the 1920s." If everything we needed to maintain our homes and sustain our families were delivered right to our doorsteps—and every member of the family had independent means for getting where she or he wanted to go—the hours spent in house-work by American housewives would decrease dramatically.

The histories of the vacuum cleaner, the washing machine, and the automobile illustrate the varied reasons why the time spent in housework has not markedly decreased in the United States during the last half century despite the introduction of so many ostensibly laborsaving appliances. But these histories do not help us understand what has made it possible for so many American wives and mothers to enter the labor force full-time during those same years. Until recently, one of the explanations most often offered for the startling increase in the participation of married women in the work force (up from 24.8 percent in 1950 to 50.1 percent in 1980) was household technology. What with microwave ovens and frozen foods, washer and dryer combinations and paper diapers, the reasoning goes, housework can now be done in no time at all, and women have so much time on their hands that they find they must go out and look for a job for fear of going stark, raving mad.

As every "working" housewife knows, this pattern of reasoning is itself stark, raving mad. Most adult women are in the work force today quite simply because they need the money. Indeed, most "working" housewives today hold down not one but two jobs; they put in what has come to be called a "double day." Secretaries, lab technicians, janitors, sewing machine operators, teachers, nurses, or physicians for eight (or nine or ten) hours, they race home to become chief cook and bottle washer for another five, leaving the cleaning and the marketing for Saturday and Sunday. Housework, as we have seen, still takes a lot of time, modern technology notwithstanding.

Yet household technologies have played a major role in facilitating (as opposed to causing) what some observers believe to be the most significant social revolution of our time. They do it in two ways, the first of which we have already noted. By relieving housework of the drudgery that it once entailed, washing machines, vacuum cleaners, dishwashers, and water pumps have made it feasible for a woman to put in a double day without destroying her health, to work full-time and still sustain herself and her family at a reasonably comfortable level.

The second relationship between household technology and the participation of married women in the work force is considerably more subtle. It involves the history of some technologies that we rarely think of as technologies at all—and certainly not as household appliances. Instead of being sheathed in stainless steel or porcelain, these devices appear in our kitchens in little brown bottles and bags of flour; instead of using switches and buttons to turn them on, we use hypodermic needles and sugar cubes. They are various forms of medication, the products not only of modern medicine but also of modern industrial chemistry: polio vaccines and vitamin pills; tetanus toxins and ampicillin; enriched breads and tuberculin tests.

Before any of these technologies had made their appearance, nursing may well have been the most time-consuming and most essential aspect of housework. During the eighteenth and nineteenth centuries and even during the first five decades of the twentieth century, it was the woman of the house who was expected (and who had been trained, usually by *her* mother) to sit up all night cooling and calming a feverish child, to change bandages on suppurating wounds, to clean bed linens stained with excrement, to prepare easily digestible broths, to cradle colicky infants on her lap for hours on end, to prepare bodies for burial. An attack of the measles might mean the care of a bedridden child for a month. Pneumonia might require six months of bed rest. A small knife cut could become infected and produce a fever that would rage for days. Every summer brought the fear of polio epidemics, and every polio epidemic left some group of mothers with the perpetual problem of tending to the needs of a handicapped child. Cholera, diphtheria, typhoid fever—if they weren't fatal—could mean weeks of sleepless nights and hardpressed days. "Just as soon as the person is attacked," one experienced mother wrote to her worried daughter during a cholera epidemic in Oklahoma in 1885, "be it ever so slightly, he or she ought to go to bed immediately and stay there; put a mustard [plaster] over the bowels and if vomiting over the stomach. See that the feet are kept warm, either by warm iron or brick, or bottles of hot water. If the disease progresses the limbs will begin to cramp, which must be prevented by applying cloths wrung out of hot water and wrapping round them. When one is vomiting so terribly, of course, it is next to impossible to keep medicine down, but in cholera it must be done."

These were the routines to which American women were once accustomed, routines regarded as matters of life and death. To gain some sense of the way in which modern medicines have altered not only the routines of housework but also the emotional commitment that often accompanies such work, we need only read out a list of the diseases to which most American children are unlikely to succumb today, remembering how many of them once were fatal or terribly disabling: diphtheria, whooping cough, tetanus, pellagra, rickets, measles, mumps, tuberculosis, smallpox, cholera, malaria, and polio.

And many of today's ordinary childhood complaints, curable within a few days of the ingestion of antibiotics, once might have entailed weeks, or even months, of full-time attention: bronchitis; strep throat;

scarlet fever; bacterial pneumonia; infections of the skin, or the eyes, or the ears, or the airways. In the days before the introduction of modern vaccines, antibiotics, and vitamin supplements, a mother who was employed full-time was a serious, sometimes life-endangering threat to the health of her family. This is part of the reason why life expectancy was always low and infant mortality high among the poorest segment of the population—those most likely to be dependent upon a mother's wages.

Thus modern technology, especially modern medical technology, has made it possible for married women to enter the work force by releasing housewives not just from drudgery but also from the dreaded emotional equation of female employment with poverty and disease. She may be exhausted at the end of her double day, but the modern "working" housewife can at least fall into bed knowing that her efforts have made it possible to sustain her family at a level of health and comfort that not so long ago was reserved only for those who were very rich.

25. In the Age of the Smart Machine

SHOSHANA ZUBOFF

As discussed in several previous readings, developments in computers and information technology are a driving force in virtually all areas of technological advance. Nowhere are the impacts likely to be more profound than in the nature of work, its organization, and its management. In her widely discussed and influential book In the Age of the Smart Machine: The Future of Work and Power, *Shoshana Zuboff argues that the computerized workplace is qualitatively different from its predecessors. Traditional approaches to organizing and managing work will not work in the new "informated" environment. An alternative vision is needed for the organization of the twenty-first century.*

Changes in the workplace will affect the fundamental nature of society, as authority relationships, class structure, individuals' control over their own fates, and opportunities for self-fulfillment all take new shapes. The new information technology "offers a historical opportunity to more fully develop the economic and human potential of our work organizations," according to Zuboff. If we don't seize this opportunity, we run the risk of seeing the future become nothing but a "stale reproduction of the past." Like many of the other works represented in this anthology, In the Age of the Smart Machine *is a rich and complex work, and the brief excerpt that follows can do little but give the reader an idea of its flavor and introduce a few of its basic ideas.*

Shoshana Zuboff is professor of organizational behavior and human resource management in Harvard University's Graduate School of Business Administration. She holds a Ph.D. in social psychology from Har-

vard and an undergraduate degree from the University of Chicago. She has published numerous articles and cases on the subject of information technology in the workplace and lectures and consults widely on this subject.

Piney Wood, one of the nation's largest pulp mills, was in the throes of a mass modernization effort that would place every aspect of the production process under computer control. Six workers were crowded around a table in the snack area outside what they called the Star Trek Suite, one of the first control rooms to have been completely converted to microprocessor-based instrumentation. It looked enough like a NASA control room to have earned its name.

It was almost midnight, but despite the late hour and the approach of the shift change, each of the six workers was at once animated and thoughtful. "Knowledge and technology are changing so fast," they said, "what will happen to us?" Their visions of the future foresaw wrenching change. They feared that today's working assumptions could not be relied upon to carry them through, that the future would not resemble the past or the present. More frightening still was the sense of a future moving out of reach so rapidly that there was little opportunity to plan or make choices. The speed of dissolution and renovation seemed to leave no time for assurances that we were not heading toward calamity—and it would be all the more regrettable for having been something of an accident.

The discussion around the table betrayed a grudging admiration for the new technology—its power, its intelligence, and the aura of progress surrounding it. That admiration, however, bore a sense of grief. Each expression of gee-whiz-Buck-Rogers breathless wonder brought with it an aching dread conveyed in images of a future that rendered their authors obsolete. In what ways would computer technology transform their work lives? Did it promise the Big Rock Candy Mountain or a silent graveyard?

In fifteen years there will be nothing for the worker to do. The technology will be so good it will operate itself. You will just sit there behind a desk running two or three areas of the mill yourself and get bored.

The group concluded that the worker of the future would need "an extremely flexible personality" so that he or she would not be "mentally affected" by the velocity of change. They anticipated that workers would need a great deal of education and training in order to "breed flexibility." "We find it all to be a great stress," they said, "but it won't be that way for the new flexible people." Nor did they perceive any real choice, for most agreed that without an investment in the new technology, the company could not remain competitive. They also knew that without their additional flexibility, the technology would not fly right. "We are in a bind," one man groaned, "and there is no way out." The most they could do, it was agreed, was to avoid thinking too hard about the loss of overtime pay, the diminished probability of jobs for their sons and daughters, the fears of seeming incompetent in a strange new milieu, or the possibility that the company might welsh on its promise not to lay off workers.

During the conversation, a woman in stained overalls had remained silent with her head bowed, apparently lost in thought. Suddenly, she raised her face to us. It was lined with decades of hard work, her brow drawn together. Her hands lay quietly on the table. They were calloused and swollen, but her deep brown eyes were luminous, youthful, and kind. She seemed frozen, chilled by her own insight, as she solemnly delivered her conclusion:

> I think the country has a problem. The managers want everything to be run by computers. But if no one has a job, no one will know how to do anything anymore. Who will pay the taxes? What kind of society will it be when people have lost their knowledge and depend on computers for everything?

Her voice trailed off as the men stared at her in dazzled silence. They slowly turned their heads to look at one another and nodded in agreement. The forecast seemed true enough. Yes, there was a problem. They looked as though they had just run a hard race, only to stop short at the edge of a cliff. As their heels skidded in the dirt, they could see nothing ahead but a steep drop downward.

Must it be so? Should the advent of the smart machine be taken as an invitation to relax the demands upon human comprehension and critical judgment? Does the massive diffusion of computer technology throughout our workplaces necessarily entail an equally dramatic loss of meaningful employment opportunities? Must the new electronic milieu engender a world in which individuals have lost control over their daily work lives? Do these visions of the future represent the price of economic success or might they signal an industrial legacy that must be overcome if intelligent technology is to yield its full value? Will the new information technology represent an opportunity for the rejuvenation of competitiveness, productive vitality, and organizational ingenuity? Which aspects of the future of working life can we predict, and which will depend upon the choices we make today?

The workers outside the Star Trek Suite knew that the so-called technological choices we face are really much more than that. Their consternation puts us on alert. There is a world to be lost and a world to be gained. Choices that appear to be merely technical will redefine our lives together at work. This means more than simply contemplating the implications or consequences of a new technology. It means that a powerful new technology, such as that represented by the computer, fundamentally reorganizes the infrastructure of our material world. It eliminates former alternatives. It creates new possibilities. It necessitates fresh choices.

The choices that we face concern the conception and distribution of knowledge in the workplace. Imagine the following scenario: Intelligence is lodged in the smart machine at the expense of the human capacity for critical judgment. Organizational members become ever more dependent, docile, and secretly cynical. As more tasks must be accomplished through the medium of information technology (I call this "computer-mediated work"), the sentient body loses its salience as a source of knowledge, resulting in profound disorientation and loss of meaning. People intensify their search for avenues of escape through drugs, apathy, or adversarial conflict, as the majority of jobs in our offices and factories become increasingly isolated, remote, routine, and perfunctory. Alternatively, imagine this scenario: Organizational leaders recognize the new forms of skill and knowledge needed to truly exploit the potential of an intelligent technology. They direct their resources toward creating a work force that can exercise critical judgment as it manages the surrounding machine systems. Work becomes more abstract as it depends upon understanding and manipulating information. This marks the beginning of new forms of mastery and provides an opportunity to imbue jobs with more comprehensive meaning. A new array of work tasks offer unprecedented opportunities for a wide range of employees to add value to products and services.

The choices that we make will shape relations of authority in the workplace. Once more, imagine: Managers struggle to retain their traditional sources of authority, which have depended in an important way upon their exclusive control of the organization's knowledge base. They use the new technology to structure organizational experience in ways that help reproduce the legitimacy of their traditional roles. Managers insist on the prerogatives of command and seek methods that protect the hierarchical distance that distinguishes them from their subordinates. Employees barred from the new forms of mastery relinquish their sense of responsibility for the organization's work and use obedience to authority as a means of expressing their resentment. Imagine an alternative: This technological transformation engenders a new approach to organizational behavior, one in which relationships are more intricate, collaborative, and bound by the mutual responsibilities of colleagues. As the new technology integrates information across time and space, managers and workers each overcome their narrow functional perspectives and create new roles that are better suited to enhancing value-adding activities in a data-rich environment. As the quality of skills at each organizational level becomes similar, hierarchical distinctions begin to blur. Authority comes to depend more upon an appropriate fit between knowledge and responsibility than upon the ranking rules of the traditional organizational pyramid.

The choices that we make will determine the techniques of administration that color the psychological ambiences and shape communicative behavior in the emerging workplace. Imagine this scenario: The new technology becomes the source of surveillance techniques that are used to ensnare organizational members or to subtly bully them into conformity. Managers employ the technology to circumvent the demanding work of face-to-face engagement, substituting instead techniques of remote management and automated administration. The new technological infrastructure becomes a battlefield of techniques, with managers inventing novel ways to enhance certainty and control while employees discover new methods of self-protection and even sabotage. Imagine the alternative: The new technological milieu becomes a resource from which are fashioned innovative methods of information sharing and social exchange. These methods in turn produce a deepened sense of collective responsibility and joint ownership, as access to ever-broader domains of information lend new objectivity to data and preempt the dictates of hierarchical authority.

[The book from which this selection is drawn] is about these alternative futures. Computer-based technologies are not neutral; they embody essential characteristics that are bound to alter the nature of work within our factories and offices, and among workers, professionals, and managers. New choices are laid open by these technologies, and these choices are being confronted in the daily lives of men and women across the landscape of modern organizations. [The] book is an effort to understand the deep structure of these choices—the historical, psychological, and organizational forces that imbue our conduct and sensibility. It is also a vision of a fruitful future, a call for action that can lead us beyond the stale reproduction of the past into an era that offers a historical opportunity to more fully develop the economic and human potential of our work organizations.

THE TWO FACES OF INTELLIGENT TECHNOLOGY

The past twenty years have seen their share of soothsayers ready to predict with conviction one extreme or another of the alternative futures I have presented. From the unmanned factory to the automated cockpit, visions of the future hail information technology as the final answer to "the labor question," the ultimate opportunity to rid ourselves of the thorny problems associated with training and managing a competent and committed work force. These very same technologies have been applauded as the hallmark of a second industrial revolution, in which the classic conflicts of knowledge and power associated with an earlier age will be synthesized in an array of organizational innovations and new procedures for the production of goods and services, all characterized by an unprecedented degree of labor harmony and widespread participation in management process.[1] Why the paradox? How can the very same technologies be interpreted in these different ways? Is this evidence that the technology is indeed neutral, a blank screen upon which managers project their biases and encounter only their own limitations? Alternatively, might it tell us something else about the interior structure of information technology?

Throughout history, humans have designed mechanisms to reproduce and extend the capacity of the human body as an instrument of work. The industrial age has carried this principle to a dramatic new level of sophistication with machines that can substitute for and amplify the abilities of the human body. Because machines are mute, and because they are precise and repetitive, they can be controlled according to a set of rational principles in a way that human bodies cannot.

There is no doubt that information technology can provide substitutes for the human body that reach an even greater degree of certainty and precision. When a task is automated by a computer, it must first be broken down to its smallest components. Whether the activity involves spraying paint on an automobile or performing a clerical transaction, it is the information contained in this analysis that translates human agency into a computer program. The resulting software can be used to automatically guide equipment, as in the case of a robot, or to execute an information transaction, as in the case of an automated teller machine.

A computer program makes it possible to rationalize activities more comprehensively than if they had been undertaken by a human being. Programmability means, for example, that a robot will respond with unwavering precision because the instructions that guide it are themselves unvarying, or that office transactions will be uniform because the instructions that guide them have been standardized. Events and processes can be rationalized to the extent that human agency can be analyzed and translated into a computer program.

What is it, then, that distinguishes information technology from earlier generations of machine technology? As information technology is used to reproduce, extend, and improve upon the process of substituting machines for human agency, it simultaneously accomplishes something quite different. The devices that automate by translating information into action also register data about those automated activities, thus generating new streams of information. For example, computer-based, numerically controlled machine tools or microprocessor-based sensing devices not only apply programmed instructions to equipment but also convert the current state of equipment, product, or process into data. Scanner devices in supermarkets automate the checkout process and simultaneously generate data that can be used for inventory control, warehousing, scheduling of deliveries, and market analysis. The same systems that make it possible to automate office transactions also create a vast overview of an organization's operations, with many levels of data coordinated and accessible for a variety of analytical efforts.

Thus, information technology, even when it is applied to automatically reproduce a finite activity, is not mute. It not only imposes information (in the form of programmed instructions) but also produces information. It both accomplishes tasks and translates them into information. The action of a machine is entirely invested in its object, the product. Information technology, on the other hand, introduces an additional dimension of reflexivity: it makes its contribution to the product, but it also reflects back on its activities and on the system of activities to which it is related. Information technology not only produces action but also produces a voice that symbolically renders events, objects, and processes so that they become visible, knowable, and shareable in a new way.

Viewed from this interior perspective, information technology is characterized by a fundamental duality that has not yet been fully appreciated. On the one hand, the technology can be applied to automating operations according to a logic that hardly differs from that of the nineteenth-century machine system—replace the human body with a technology that enables the same processes to be performed with more continuity and control. On the other, the same technology simultaneously generates information about the underlying productive and administrative processes through which an organization accomplishes its work. It provides a deeper level of transparency to activities that had been either partially or completely opaque. In this way information technology supersedes the traditional logic of automation. The word that I have coined to describe this unique capacity is *informate*. Activities, events, and objects are translated into and made visible by information when a technology *informates* as well as *automates*.

The information power of intelligent technology can be seen in the manufacturing environment when microprocessor-based devices such as robots, programmable logic controllers, or sensors are used to translate the three-dimensional production process into digitized data. These data are then made available within a two-dimensional space, typically on the screen of a video display terminal or on a computer printout, in the form of electronic symbols, numbers, letters, and graphics. These data constitute a quality of information that did not exist before. The programmable controller not only tells the machine what to do—imposing information that guides operating equipment— but also tells what the machine has done—translating the production process and making it visible.

In the office environment, the combination of on-line transaction systems, information systems, and communications systems creates a vast information presence that now includes data formerly stored in people's heads, in face-to-face conversations, in metal file drawers, and on widely dispersed pieces of paper. The same technology that processes documents more rapidly, and with less intervention, than a mechanical typewriter or pen and ink can be used to display those documents in a communications network. As more of the underlying transactional and communicative processes of an organization become automated, they too become available as items in a growing organizational data base.

In its capacity as an automating technology, information technology has a vast potential to displace the human presence. Its implications as an informating technology, on the other hand, are not well understood. The distinction between *automate* and *informate* provides one way to understand how this technology represents both continuities and discontinuities with the traditions of industrial history. As long as the technology is treated narrowly in its automating function, it perpetuates the logic of the industrial machine that, over the course of this century, has made it possible to rationalize work while decreasing the dependence on human skills. However, when the technology also informates the processes to which it is applied, it increases the explicit information content of tasks and sets into motion a series of dynamics that will ultimately reconfigure the nature of work and the social relationships that organize productive activity.

Because this duality of intelligent technology has not been clearly recognized, the consequences of the technology's informating capacity are often regarded as unintended. Its effects are not planned, and the potential that it lays open remains relatively unexploited. Because the informating process is poorly defined, it often evades the conventional categories of description that are used to gauge the effects of industrial technology.

These dual capacities of information technology are not opposites; they are hierarchically integrated. Informating derives from and builds upon automation. Automation is a necessary but not sufficient condition for informating. It is quite possible to proceed with automation without reference to how it will contribute to the technology's informating potential. When this occurs, informating is experienced as an unintended consequence of automation. This is one point at which choices are laid open. Managers can choose to exploit the emergent informating capacity and explore the organizational innovations required to sustain and develop it. Alternatively, they can choose to ignore or suppress the informating process. In contrast, it is possible to consider informating objectives at the start of an automation process. When this occurs, the choices that are made with respect to how and what to automate are guided by criteria that reflect developmental goals associated with using the technology's unique informating power.

Information technology is frequently hailed as "revolutionary." What are the implications of this term? *Revolution* means a pervasive, marked, radical change, but *revolution* also refers to a movement around a fixed course that returns to the starting point. Each sense of the word has relevance for the central problem of this [study]. The informating capacity of the new computer-based technologies brings about radical change as it alters the intrinsic character of work—the way millions of people experience daily life on the job. It also poses fundamentally new choices for our organizational futures, and the ways in which labor and management respond to these new choices will finally determine whether our era becomes a time for radical change or a return to the familiar patterns and pitfalls of the traditional workplace. An emphasis on the informating capacity of intelligent technology can provide a point of origin for new conceptions of work and power. A more restricted emphasis on its automating capacity can provide the occasion for that second kind of revolution—a return to the familiar grounds of industrial society with divergent interests battling for control, augmented by an array of new material resources with which to attack and defend.

The questions that we face today are finally about leadership. Will there be leaders who are able to recognize the historical moment and the choices it presents? Will they find ways to create the organizational conditions in which new visions, new concepts, and a new language of workplace relations can emerge? Will they be able to create organizational innovations that can exploit the unique capacities of the new technology and thus mobilize their organization's productive potential to meet the heightened rigors of global competition? Will there be leaders who understand the crucial role that human beings from each organizational stratum can play in adding value to the production of goods and services? If not, we will be stranded in a new world with old solutions. We will suffer through the unintended consequences of change, because we have failed to understand this technology and how it differs from what came before. By neglecting the unique informating capacity of advanced computer-based technology and ignoring the need for a new vision of work and organization, we will have forfeited the dramatic business benefits it can provide. Instead, we will find ways to absorb the dysfunctions, putting out brush fires and patching wounds in a slow-burning bewilderment.

NOTE

1. See, for example, Michael Piore and Charles F. Sabel, *The Second Industrial Divide: Possibilities for Prosperity* (New York: Basic Books, 1984).

26. Packing Tips for Your Trip (to the Year 2195)

DOUGLAS COUPLAND

What would you take with you on a trip to the future? Food? Money? A gun? How would you prove to the people you encountered there that you were a visitor from the late twentieth century and not simply a lunatic? What would you bring back to show your friends (and the rest of the world)?

Douglas Coupland has some original and amusing answers to these questions. His list may give you food for thought—and it may make you think twice before stepping into any time machines you happen to come across.

Coupland is the author of three novels—Generation X, Shampoo Planet, and Microserfs—as well as two collections of short stories. He has also written for Wired *magazine, where this piece appeared (under the pseudonym Sandra Noguchi) in a special issue on the future in October 1995.*

Congratulations! You've just won a two-week, round-trip ticket to exactly where you are right now—but in the year 2195. So pack quickly! Your only packing restrictions:

- You can't bring anything of a size that United Airlines wouldn't allow you to take along.
- You have to bring one object that *proves* you're from the year 1995.

Source: From *Wired Scenarios,* special issue, October 1995. Reprinted by permission of the author.

ISSUES TO CONSIDER BEFORE FETCHING THE SAMSONITE

Life Support

In accepting your ticket, you also accept that the world might not be biologically inhabitable in 2195. Do you bring a radiation suit? A Geiger detector? Bottled water? Freeze-dried camping meals? (*Yuck!*) Topographical maps? What does the U.S. Army pack when it wants to enter zones of possible profound uninhabitability? Is Tang called for?

- camping meals
- geological survey maps of your neighborhood
- Evian
- Tang

The Future May Well Be an Ugly Dump

Whenever people get misty-eyed about traveling backward in time, they usually assume their journey will be "vaccinated." That is to say, they want a return ticket, pox and polio vaccinations, contact lenses, appendix removal, membership in the ruling elite of the time they wish to visit, vitamins, etc. To reverse this situation, how might this sensibility work if a person were to travel *forward* in time? Is the future a Third World country? What hygiene essentials will a traveler require?

- antibiotics and tetanus booster
- toilet paper
- Halazone tablets to purify water

Arms, Armor & Ammo

Imagine that you live in George Washington's time—and you visit 1995. You might bring a blunderbuss and look awfully silly walking around your neighborhood. Would you even *need* a blunderbuss or any other weapon in 2195? It's your decision.

- handgun and bullets
- Swiss Army knife
- Mace

Prove That You're from 200 Years Ago

As far as I know, you could *already* be from 200 years ago. Assuming people 200 years from now still speak your language, they'll probably look at you and say, *"Get out of my face, loser,"* before you get a chance to elaborate. Therefore, you want to bring along something that has no conceivable chance of existing 200 years from now:

- a Komodo dragon or some other animal on the cusp of extinction (but what if factories make pandas and spotted owls in 2195?)
- a box of Pop-Tarts with 1995 packaging, plus expiration dates
- CDs, this week's favorites

(Question: How will these people know that you're not really from the *future* and are tricking them into thinking you're from the past? It can get dodgy here.)

Money? What's Money?

Gold, platinum, or thorium may well be valueless in 2195. What artifacts from now might you bring if you had to trade in order to get, say, *food*? Remember, chances are that the things of value in the future are the things you could have had no way of knowing would be valuable: an autograph from Button Gwinnett; IBM stock; fertilized passenger pigeon eggs. Future citizens might also end up ransacking you for semi-random items—like Monty Hall of *Let's Make A Deal* offering you U.S. $500 for every paper clip you've got. Some packing ideas:

- comic books
- fruit juices
- celebrity autographs
- Franklin Mint Star Trek™ dinner plates
- drugs or mood-altering pharmaceuticals
- baseball cards or stamps
- samples of your own blood or reproductive cells
- California chardonnay
- Freon

(Tip: Do you know any *information* right now that might be useful to people in 2195? What could somebody from 1795 tell you right now that might even be interesting, let alone valuable?)

We Come in Peace

Future humans may well be prepared to be nice to you . . . *assuming you bear a friendly enough gift as a gesture*. Would a letter from the president do? Partridge Family cassettes with a battery-loaded boom box? Perhaps simply being unarmed would be gesture enough.

- Partridge Family cassettes and child's boom box

Your Friends Will Want to See Pictures

What sort of camera will you bring? What sort of devices for trapping liquids, solids, and artifacts from the year 2195?

- Olympus Stylus camera and lots of film

The Future Might Well Suck beyond All Belief

If this is the case, perhaps you'll want to crawl into a hidden nook somewhere (surely you can find *one* safe location) and sedate yourself for 14 days until you return to 1995. In this case you might need these:

- syringe and large supply of barbiturates
- sleeping bag
- animal repellent
- Deep Woods Off!
- whatever minimal food and water might stretch over 14 days (Jenny Craig meals?)

Happy packing, future traveler. Hope to see you two weeks from now, and hope you have many tales to tell. Will Ronald Reagan be president again? Will fresh water be a form of currency? And how many United Airlines points will it take for a first-class return trip to the moon? Only *you* can tell. Bon voyage.

About the Author

ALBERT H. TEICH has been Director of Science and Policy Programs at the American Association for the Advancement of Science (AAAS) since 1990. The Association, founded in 1848, is the world's largest federation of scientific and engineering societies; it is also a professional organization with over 140,000 members and the publisher of *Science* magazine. Dr. Teich is responsible for the association's activities in science and technology policy (including the R&D Budget and Policy Program and the Congressional Science and Engineering Fellows Program), as well as for programs in science and ethics, law, and human rights, and AAAS's new Center for Science, Technology, and Congress.

Prior to joining AAAS in 1980, Dr. Teich served on the faculty and in research and administrative positions at George Washington University, the State University of New York, and Syracuse University. He is the author of numerous articles and editor of several books. He is senior author (with Jill H. Pace and several others) of *Science and Technology in the U.S.A.*, volume 5 of Longman's "World Guides to Science and Technology," published in 1986.

Dr. Teich is a Fellow of AAAS, and a member of the editorial advisory boards to the journals *Science Communication* and *Science, Technology, and Human Values.* He has served as a consultant to government agencies, national laboratories, industrial firms, and international organizations. He chaired the Visiting Committee for the Department of Science and Technology Studies at Rensselaer Polytechnic Institute and is a member of the Board of Advisors to Georgia Tech's School of Public Policy. He holds a B.S. in physics and a Ph.D. in political science, both from MIT.